Trilobites of the Middle Ordovician Elnes Formation of the Oslo Region, Norway

by

Thomas Hansen

Acknowledgement

Financial support for the publication of this number of Fossils and Strata was provided by Naturhistorisk Museum, Universitetet i Oslo, Postboks 1172, Blindern, NO-0318 Oslo, Norway.

Contents

Trilobites of the Middle Ordovician Elnes Formation of the Oslo Region, Norway

THOMAS HANSEN

Hansen, T. 2009: Trilobites of the Middle Ordovician Elnes Formation of the Oslo Region, Norway. *Fossils and Strata*, No. 56, pp. 1–215. ISSN 0024-1164.

The Elnes Formation, which succeeds the Baltoscandian 'Orthoceratite limestone', can be traced all over the Oslo Region. The largely mudstone-dominated formation, which in places has been heavily faulted and folded during the Caledonian Orogeny, reaches thicknesses of nearly 100 m in the central Oslo Region and possibly significantly more in the northern part of the region, where it extends higher in the stratigraphical record. Sediments of the Elnes Formation were laid down in what may possibly have been a foreland basin on the north-western margin of the isolated continent Baltica and where conditions were generally quiet and periods of slightly dysoxic bottom conditions prevailed around or somewhat below normal storm wave base. The fairly rich trilobite fauna mainly occurs in the lower and upper part of the formation, deposited in slightly more oxic waters at or above storm wave base. The lower part of the Elnes Formation in the northern Mjøsa district is considered equivalent to the Heggen Member further south, while the middle part is regarded as a possible distal variety of the Engervik Member from the central Oslo Region. Based on the biostratigraphic results from a number of sections of the Elnes Formation throughout the Oslo Region, the boundaries between various members have been defined more accurately, resulting in some adjustment to previous work. Of the close to 100 trilobite taxa known from the Elnes Formation only four seem to have immigrated from elsewhere, two of which have a clear Laurentian origin. All four species are restricted to the northern Mjøsa district. Inferences on the autecology of the most common trilobite groups are summarised. Systematic descriptions are presented of *Ampyx, Asaphus, Atractopyge, Botrioides, Bronteopsis, Cnemidopyge, Cybellela, Cybelurus, Cyrtometopus, Geragnostus, Gravicalymene, Icelorobergia* n. gen. (type species *Robergiella brevilingua* Fortey 1980), *Illaenus, Lonchodomas, Megistaspis, Metopolichas, Nileus, Niobe, Ogmasaphus, Ogygiocaris, Pliomera, Porterfieldia, Primaspis, Proetus, Pseudasaphus, Pseudobasilicus, Pseudomegalaspis, Pterygometopus, Robergia, Scotoharpes, Sculptaspis, Sculptella, Sphaerocoryphe, Sthenarocalymene, Telephina, Raymondaspis* (*Cyrtocybe*) and *Volchovites*. Nine new taxa are described: *Asaphus narinosus* n. sp., *A. raaenensis* n. sp., *Lonchodomas cuspicaudus* n. sp., *Megistaspis* (*M.*) *giganteus runcinatus* n. ssp., *Ogmasaphus furnensis* n. sp., *Ogygiocaris isodilatata* n. sp., *Ogygiocaris henningsmoeni* n. sp., *Pseudasaphus limatus longistriatus* n. subsp. and *P.* (*P.*) *truncatus* n. sp. The asaphid subgenus *Megistaspis* (*Heraspis*) Wandås is considered a junior subjective synonym of *M.* (*Megistaspidella*) Jaanusson. The genus *Ogygiocaris* is totally revised and all former subspecies of *Ogygiocaris sarsi* Angelin are assigned species rank. *Ogygiocaris regina* Henningsmoen is regarded a junior subjective synonym *of O. lata* Hadding, reducing the number of taxa to seven species and two subspecies. The trinucleid *Botrioides efflorescens* Hadding occurs before *B. foveolatus* Angelin, and is considered its ancestor. *Atractopyge dentata* is redescribed and its range reduced, and Upper Ordovician specimens from Norway and the United Kingdom are reassigned to *Cybellela grewingki* (Schmidt). *Lonchodomas striolatus* Månsson and *Ampyx clavifrons* Hadding are considered subjective synonyms of *L. rostratus* Sars and *A. mammilatus* Sars, respectively. The two genera *Valdaites* Balashova, 1976 and *Mischynogorites* Balashova, 1976 are both regarded as junior subjective synonyms of *Pseudasaphus* Schmidt, 1904. □ *Biogeography, biostratigraphy, Elnes Formation, lithostratigraphy, Oslo Region, palaeoecology, trilobite taxonomy, upper Darriwilian.*

Thomas Hansen, [th@geo.ku.dk], Natural History Museum, Department of Geology, University of Oslo, PO Box 1172, Blindern, NO-0318 Oslo, Norway; Present address: Institute for Geography and Geology, University of Copenhagen, DK-1350 København K, Denmark; submitted 2 October 2007; accepted in revision form 26 September 2008.

Introduction

The Middle Ordovician succession of the Oslo Region and its biostratigraphy are relatively poorly known in relation to the contemporaneous autochthoneous strata of the Baltoscandian platform to the east. This is partly a result of the Caledonian folding and faulting, making it difficult to reconstruct the former thicknesses and spatial variation of the depositional units. However, during the last 50 years, several works by Henningsmoen (1960), Maletz (1997), Nielsen (1995), Nikolaisen (1963), Owen *et al.* (1990),

Pålsson *et al.* (2002), Rasmussen (2001), Rasmussen & Bruton (1994) and Wandås (1982) have focused on the subject, leading to a better understanding of the biostratigraphy and facies relations throughout the region. The upper Darriwilian Elnes Formation is one of the less well-known deposits, but even here detailed palaeontological and sedimentological investigations have been made by Bjørlykke (1974), Hansen *et al.* (2005), Henningsmoen (1960), Maletz (1997), Maletz & Egenhoff (2005), Nikolaisen (1961, 1963, 1965, 1983), Nilssen (1985), Owen (1987), Owen *et al.* (1990), Rasmussen & Bruton (1994), Seilacher & Meischner (1965), Siveter (1977), Størmer (1930, 1953) and Wandås (1982, 1984). With the exception of graptolite studies (Berry 1964; Maletz 1997) the biostratigraphic knowledge is generally very coarse, and even though the trilobites in the collections of the Natural History Museum (Geological Section), University of Oslo (abbreviated as NHM or PMO for the collections) have received considerable attention, it has not been possibly to accomplish a complete taxonomic revision due to the poor knowledge on both geographical and stratigraphical belongings of individual specimens. The general lack of detailed biostratigraphic knowledge is also a problem when relating the facies boundaries of the various parts of the Oslo Region to each other and to lithological units outside Norway, making it harder to establish a good model for the depositional environment of southern Scandinavia.

The trilobite fauna of the Elnes Formation and its stratigraphical ranges and distribution has been investigated by detailed sampling from three main areas covering the southern, central and northern part of the Oslo Region. In addition to a thorough taxonomic revision of the groups, attention has also been directed to their palaeoecology. Faunal lists are presented for 13 sections, enabling a better understanding of correlation throughout the Oslo Region and between this area and others.

Methods and material

Rocks of the Elnes Formation were investigated and their contained fossils sampled systematically at three main localities at Fiskum, Slemmestad and Nydal, representing the southern, central and northern Oslo Region. The stratigraphical level of each fossil sampled was generally known to within ± 5 mm. A detailed lithological log of each profile was carried out simultaneously with the sampling, making it easier to relate the fossils to observed beds and lamine.

The normal procedure of sampling was to collect all possible identifiable fossils and/or note their presence on the lithological log. This was done to allow for a general estimation of the distribution between the various trilobite species and the rest of the fauna. Fossils were collected only from strata *in situ*. The material collected or noted from these three localities amounted to more than 23,000 specimens, and the identifiable fragments of which slightly over 3,400 were trilobites. The rest was mainly dominated by pelagic graptolites and phosphatic shelled brachiopods, but includes calcareous brachiopods, cephalopods, gastropods, hyolithids, bivalves, ostracodes, echinoderms, machaeridians and conulariids. The graptolite material was used to pinpoint the boundaries between the Scandinavian graptolite biozones of Maletz (1997) and Maletz *et al.* (2007).

In addition to the three main localities mentioned above, four localities from which earlier collected material from the lower part of the formation in the western Modum, north-western Hadeland and northern Mjøsa districts were investigated. The fossils collected by B. Wandås (Wandås 1984) have a general stratigraphical accuracy of ± 5 cm.

Selected well-preserved specimens from the PMO collections [= Palaeontological Museum of Oslo] housed at NHM, University of Oslo, have been included in the taxonomic descriptions. The exact provenance of this material is often unclear, but the stratigraphical level may to some degree be determined on the lithology of the sample and the associated fossils.

The trilobites are often somewhat compressed and slightly distorted. This is especially the case for the ones preserved in mudstones, whereas those in the calcareous concretions have been more protected. The original calcite shell has in most cases been dissolved or recrystalized, removing the finer details. Many genera are rare, being represented by less than 10 specimens in the PMO collections, while others, such as *Nileus*, *Pseudomegalaspis* and some species of *Ogygiocaris*, *Asaphus* and *Botrioides*, are represented by several hundred specimens. Some of the latter may be found in huge numbers in coquinas, most probably representing shell-lags formed by storm currents. These beds may in some cases be used as marker beds as they can be traced from localities at Slemmestad and down to the Eiker-Sandsvær area to the south-west, a distance of some 40 km.

The examined and figured material was prepared from marly mudstone using a vibro tool. With the exception of illustrations Plate 1, figures 1–6; Plate 2, figures 7–8; Plate 4, figures 1–2, 13, 17; Plate 5, figures 6; Plate 14, figures 6, 13; Plate 15, figures 15, 18, which were taken by P. Aas; Plate 17, figure 23 taken by Professor D. L. Bruton and Plate 27, figure 20 taken by Dr Y. Candela, all material was photographed by myself. The specimens were coated with ammonium chloride to heighten the contrast before being photographed using ring-light illumination. In most cases, an additional highlight from the NW quadrant was used.

Photographs were taken using a Nikon D50 digital camera with a 90 mm objective. Apart from digital sharpening, deepening of contrast and blackening of the background, photographs have generally not been retouched. Exceptions are found in Plate 4, figure 2; Plate16, figures 16–17; Plate 18, figures 10, 17; Plate 20, figure 8; Plate 21, figures 3, 10; Plate 27, figures 1–2 and Plate 28, figure 2, where matrix has been darkened or painted black in order to identify the specimen.

Statistical and multivariable analyses have been carried out with the statistical program PAST (Hammer *et al.* 2001).

Geological setting

The Oslo Region of south-eastern Norway is a narrow, N–S trending approximately rectilinear belt of about 220 × 45 km characterized by disjunct areas or districts of folded and faulted Palaeozoic rocks (Fig. 1). In the northern end of the region, the whole succession has been thrust southwards at least 150 km and forms part of the Osen-Røa Nappe Complex (Bockelie & Nystuen 1985; Nystuen 1981; Ribecai *et al.* 2000). The dominantly marine deposits have been preserved from later erosion by extensive downfaulting in connection with the formation of the Permian Oslo Graben and capping of associated basaltic and monzonitic lavas (Dons & Larsen 1978; Oftedahl 1966, pp. 37–39; Sundvoll & Larsen 1994).

The Oslo Region is situated on the western margin of the Baltoscandian platform (Fig. 2), which in the Middle Ordovician constituted a small and strongly peneplaned plate – the Baltica Plate. Northward drift of Baltica towards about 40°S led to a warm temperate climate for the period (Torsvik *et al.* 1996; Fortey & Cocks 2003). The continent was presumably isolated, resulting in an extremely endemic fauna until the mid to late Sandbian when faunal interchange between Baltica and the approaching Avalonia became significant (Cocks & Fortey 1982, 1990; Cocks 2000; Fortey & Cocks 2003). The first faunal immigrants, though, appeared as early as the middle to late Darriwilian when an island arc collided with the Caledonian margin of Baltica (Bruton & Harper 1981; Sturesson *et al.* 2005).

The Oslo Region formed the north-western flank of a large epicontinental bay or foreland basin bordered by the main Baltoscandian platform to the east, the Tornquist Sea to the south-west and a land

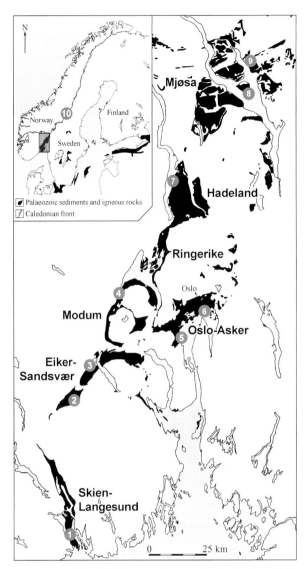

Fig. 1. Map of the Oslo Region showing the distribution of the Cambro-Silurian rocks throughout the eight districts defined by Størmer (1953). The inset of Baltoscandia presents the outcrop areas of Palaeozoic autochtonous rocks (black). Important localities are marked with numbers. 1, Stathelle; 2, Rønningfossen and Muggerudkleiva; 3, Råen, Stavlum, Vego and Krekling; 4, Vikersund skijump (Vikersundbakken); 5, Slemmestad; 6, Huk on Bygdøy; 7, Hovodden; 8, Helskjer at Hovindsholm on Helgøya; 9, Furnes Church and Nydal; 10, Andersön in central Sweden. Main map modified from Henningsmoen (1960a), while inset is based on map presented by Torsvik *et al.* (1992).

area, the Telemark Land of Skjeseth (1952, p. 152), to the north and west, shielding it from the Iapetus Ocean (Figs 3 and 4). This is reflected in the distribution of sediments with a general change from mud-dominated sediments in the east around Oslo-Asker to more coarse-grained, marly deposits to the west, with the most nearshore deposition occurring in the Skien-Langesund area in the south (Skjeseth 1952; Henningsmoen 1960; Størmer 1967; Skaar

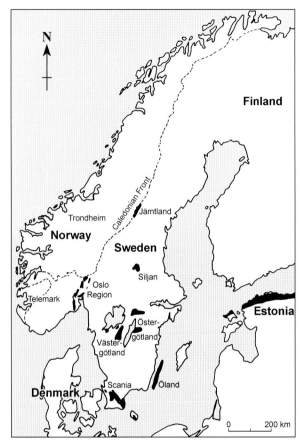

Fig. 2. Map of Baltoscandia showing major areas of Ordovician outcrops in relation to the Caledonian Front to the west. Modified from Bruton *et al.* (1985) and Nielsen (1995).

Telemark Land represented the major source of siliciclastic material for the area as suggested by extensive fine-grained turbidite deposits, largely formed by northeast running currents (present-day geography). The turbidites represents an important part of the Elnes Formation from the southernmost to central Oslo Region, becoming rare in the northern Oslo Region, where the only observations are from the western side of the lake Mjøsa (Nilssen 1985; Rasmussen 2001; Sven Egenhoff, personal communication 2005; this study). Further north they occur as thick deposits just west and north of the lake Andersö in the Jämtland area of central Sweden (Thorslund 1940, 1960; Karis & Strömberg 1998; Heuwinkel & Lindström 2007).

The deposition of the Elnes Formation was initiated by a relative sea-level rise during the mid-Darriwilian. Prior to this, most of Baltoscandia was characterized by widespread calcarenites formed on a large carbonate ramp, which supported a fairly rich shelly fauna. The transgressive phase resulted in a successive change from the initial calcarenites of a storm-dominated midramp deposit over marls and concretionary dark argillaceous mudstones to organic-rich fissile mudstones deposited during maximum flooding in an exaerobic environment. The latter facies has only been observed in the central Oslo Region, but may simply be missing due to later faulting. The highstand was followed by a long regressional phase during which the northern Oslo Region became the deepest part. In general, though, there was a gradual return to more aerobic bottom conditions and an increasingly higher marl content of the sediments, although an occasional interruption of the normal mud-dominated sedimentation occurred as a result of turbidity currents. In the southern to central Oslo Region, the depositional cycle of the Elnes Formation ended with

1972; Bockelie 1978; Nilssen 1985). Similarly from the Oslo Region and eastwards to the main Baltoscandian platform there is evidence of lower average sedimentation rates; higher carbonate/shale ratios and more breaks (Bjørlykke 1974, pp. 31–32).

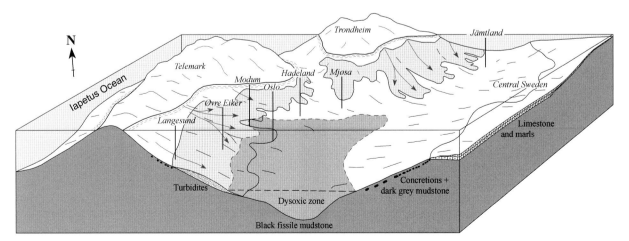

Fig. 3. Depositional model for south-western Scandinavia at the time of maximum flooding (mid-Darriwilian). Approximate palaeo-north is indicated.

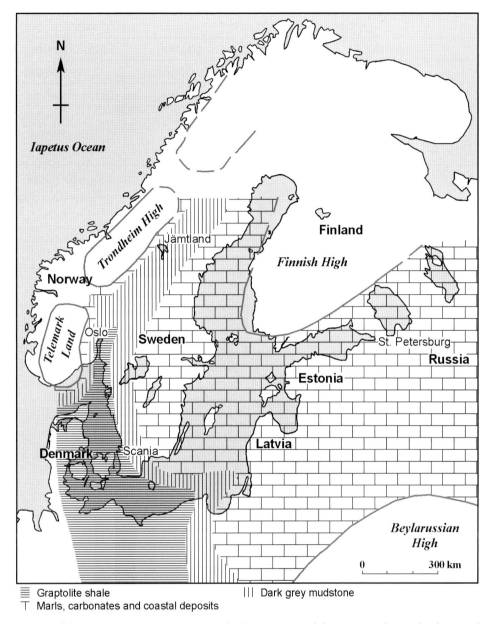

Fig. 4. Lithofacies map of the Aserian–Lasnamägian (Darriwilian) epicontinental deposits in Baltoscandia showing the approximate distribution of the different sedimentary environments. The figure is based on data from Jaanusson (1973), Männil (1966), Nilssen (1985) and Pålsson *et al.* (2002).

the return of nodular and marly mudstone deposits, but further north the calcareous deposits become very thin and are succeeded by blackish siliciclastic sediments reflecting a return to dysoxic bottom conditions. The same trend is seen in central to southern Sweden on the main Baltoscandian platform (Pålsson *et al.* 2002), suggesting that the southern to central Oslo Region experienced a tectonically induced uplift during the latest Middle Ordovician to earliest Late Ordovician times. The uplift, which most likely is connected with the island arc collision and a resulting compression and uplift of the western Telemark

Land may also be responsible for part of the turbidite deposition in the Oslo Region as some of the deposition appears unrelated to sea-level changes.

The general stratigraphic relation between the Elnes Formation and the contemporaneous formations in southern to central Sweden is summarized in Figure 5.

For practical reasons the Cambro–Silurian of the Oslo Region has been divided into districts (Størmer 1953) (Fig. 1). The districts surrounding the inland lake Mjøsa, i.e. Feiring, Hamar-Nes, Ringsaker and Toten are referred to as the Mjøsa districts. These and

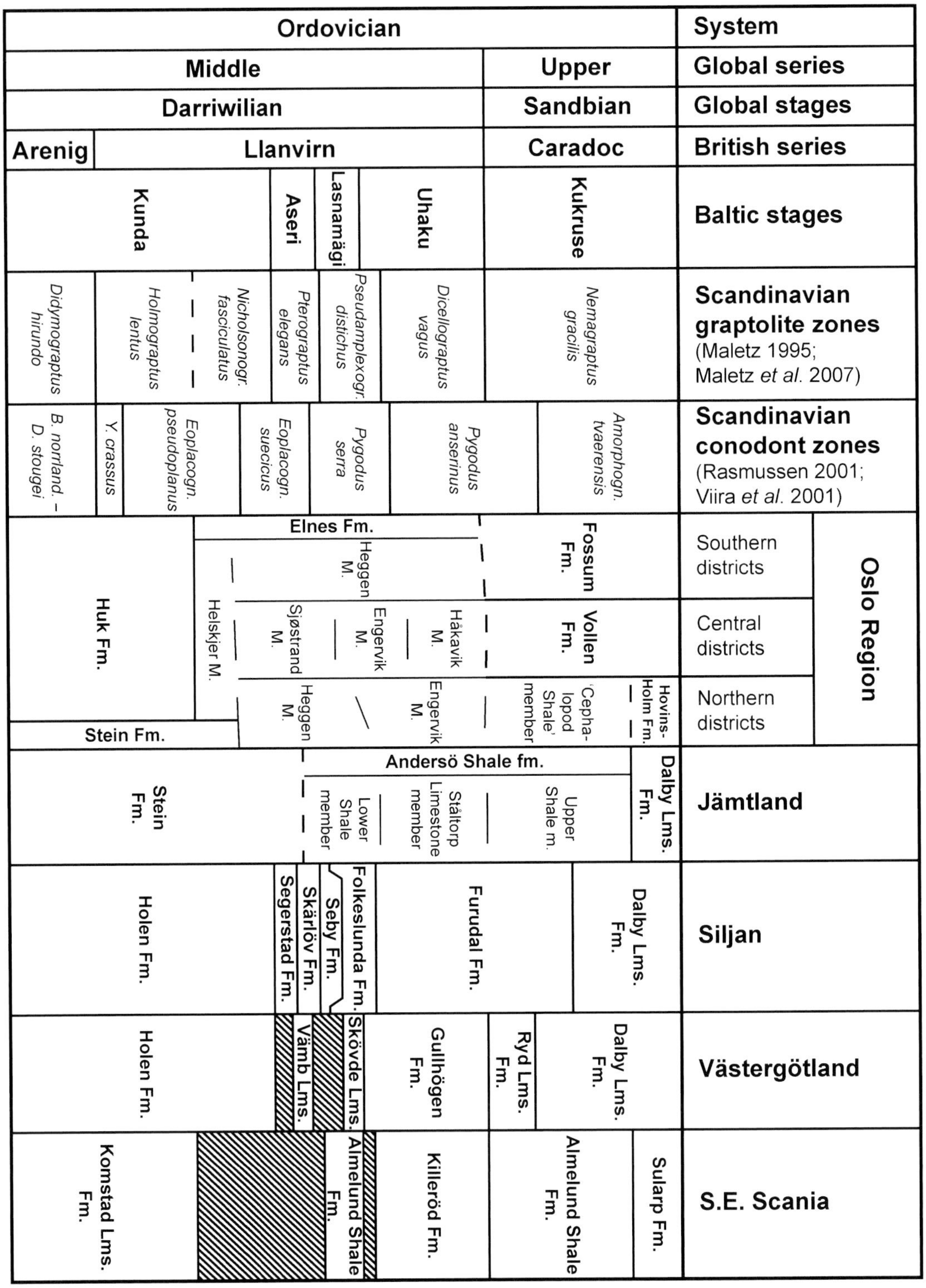

Ordovician							System
Middle						Upper	Global series
Darriwilian						Sandbian	Global stages
Arenig	Llanvirn					Caradoc	British series
Kunda			Aseri	Lasnamägi	Uhaku	Kukruse	Baltic stages
Didymograptus hirundo	Holmograptus lentus	Nicholsonogr. fasciculatus	Pterograptus elegans	Pseudamplexogr. distichus	Dicellograptus vagus	Nemagraptus gracilis	Scandinavian graptolite zones (Maletz 1995; Maletz et al. 2007)
B. norrland. – D. stougei	Y. crassus	Eoplacogn. pseudoplanus	Eoplacogn. suecicus	Pygodus serra	Pygodus anserinus	Amorphogn. tvaerensis	Scandinavian conodont zones (Rasmussen 2001; Viira et al. 2001)
Huk Fm.	Huk Fm.	Huk Fm.	Elnes Fm. — Heggen M.			Fossum Fm.	Oslo Region — Southern districts
Huk Fm.	Huk Fm.	Helskjer M.	Sjøstrand M.	Engervik M.	Håkavik M.	Vollen Fm.	Oslo Region — Central districts
Stein Fm.		Heggen M.	Engervik M.		'Cephalopod Shale' member	Hovins-Holm Fm.	Oslo Region — Northern districts
Stein Fm.	Stein Fm.	Stein Fm.	Andersö Shale fm. — Lower Shale member	Ståltorp Limestone member	Upper Shale m.	Dalby Lms. Fm.	Jämtland
Holen Fm.	Holen Fm.	Holen Fm.	Segerstad Fm. / Skärlöv Fm. / Seby Fm. / Folkeslunda Fm.	Furudal Fm.	Furudal Fm.	Dalby Lms. Fm.	Siljan
Holen Fm.	Holen Fm.	Holen Fm.	Segerstad Fm. / Vämb Lms. / Skövde Lms. / Almelund Shale Fm.	Gullhögen Fm.	Ryd Lms. Fm.	Dalby Lms. Fm.	Västergötland
Komstad Lms. Fm.	Komstad Lms. Fm.	Komstad Lms. Fm.	Almelund Shale Fm.	Killeröd Fm.	Almelund Shale Fm.	Sularp Fm.	S.E. Scania

Fig. 5. Stratigraphic classification of the Darriwilian to basal Sandbian sequence in the Oslo-Asker district of Norway and contemporaneous deposits in southern to central Sweden (based on Bergström et al. 2002; Gradstein et al. 2004; Karis 1982; Maletz 1995, 1997; Owen et al. 1990; Pålsson et al. 2002; Rasmussen 2001; Viira et al. 2001 and this study).

Hadeland constitute the northern districts. Oslo-Asker and Ringerike constitute the central districts and Eiker-Sandsvær, Modum and Skien-Langesund represent districts in the south. The Modum district was earlier grouped together with Oslo-Asker by Henningsmoen (1960a, p. 131), but has here been assigned to the southern districts with which it has a greater lithological affinity.

General lithology

The lithological development of the Elnes Formation varies somewhat within the Oslo Region, but is in general initiated by a change from the underlying limestone deposits of the Huk and Stein formations into the trilobite-rich and dominantly marly deposits at the base of the formation. This interval varies in thickness between < 1.5 m in the southern Oslo Region to > 10 m in the northern Hadeland district. In the northernmost Mjøsa district most of the dark marls have been included in the underlaying Stein Formation, and a similar situation occurs in the central Swedish Jämtland area. The lower marls and concretionary mudstones are succeeded by dark grey or black mudstones containing unfossilifereous septarian concretions together with smaller (10–20 cm) fossiliferous, limestone concretions. The septarian concretions are poorer in calcium carbonate and richer in iron than the smaller carbonate nodules. The latter must have formed fairly early since the contained fossils and burrows are largely uncompressed compared to the flattened fossils in the surrounding mudstones. The mudstones are typically structureless, indicating strong bioturbation. Intervals characterized by light grey, sandy or silty beds of up to tens of centimetres in thickness may occur. The beds are generally completely free of body fossils, although the trace-fossil *Planolites* is fairly common. Sedimentary structures include a general fining upward of the beds, planar lamination, flame structures and apparently homogeneous intervals suggesting fast deposition and current formed structures like ripples, cross-lamination, drag-structures and sole-marks indicating a general current-direction from the south-west (Seilacher & Meischner 1965; Nilssen 1985; Maletz & Egenhoff 2005; this study) (Fig. 6). No basal shell-lags are found in connection with these coarse-grained beds from the middle Elnes Formation, although they do occur independently in the intervening mudstones. The beds have been interpreted as turbidites by Seilacher & Meischner (1965; S. Egenhoff, personal communication 2006), but this interpretation has been questioned by Bjørlykke (1974, p. 16). He found that they show rapid lateral change in thickness and character (some hundred metres and up to a few kilometres) and occur in an environment characterized by low relief. Instead he suggested they merely represent storm deposits. The sole markings at the base of some of the calcarenites at Huk on Bygdøy (grid reference: NM 938 410–932 412) show current directions from the SW to NE and were interpreted as formed by currents parallel to an assumed coast. This interpretation would allow for a northern source area as proposed by Bjørlykke (1974) on the strength of geochemical similarities between the Elnes Formation and a source from volcanic rocks in the Trøndelag area of mid-Norway. The suggested scenario is not accepted here for the following reasons: (1) The sandy beds present a typical partial Stow sequence for fine-grained turbidites, but lack hummocky cross-stratification, a basal shell-lag, or horizons with rip-up clasts typical for tempestites. The presence of seemingly structureless sand- or siltbeds up to tens of centimetres in thickness (Fig. 7) together with water-escape structures indicating abrupt deposition, are further evidence in favour for the turbidite interpretation. (2) The individual turbidites may not be traceable for more than a kilometre or so but the turbidite intervals can be followed with little change over much larger distances. The proposed rapid lateral change is due to a combination of the strong Caledonian faulting and folding of the Cambro–Silurian rocks in the area, making correlations difficult, and the fact that Bjørlykke appearently was unaware of silty intervals outside the Oslo and Bygdøy area and wrongly thought them to be absent from the Slemmestad

Fig. 6. A. Turbidite with partial Stow sequence showing quartz-rich basal current ripples with north-east dipping cross-stratification succeeded by slightly finer-grained sediments with planar lamination ending with silty mudstones at the top. Thickness of current ripples approximately 2 cm. Vego at Krekling, Eiker-Sandsvær. B. NE-dipping cross-stratification from more than one turbidite. Vego at Krekling, Eiker-Sandsvær. Bars approximately 5 cm.

Fig. 7. Four massive, fine-grained turbidites interbedded with silty mudstone. The upper and lower turbidites are more than 30 cm thick. Vego at Krekling, Eiker-Sandsvær. Bar equals 10 cm.

area. (3) With regard to bottom topography, the Cambro–Silurian rocks of the Oslo Region show marked lateral and vertical facies changes, reflecting pronounced changes in relief unlike anything observed on the main Baltoscandian platform (Bockelie 1978). Thus contrary to the assertion of Bjørlykke (1974), the topography would not be a problem for the formation of turbidites. (4) The generally north-easterly current direction corresponds well with the presence of a westerly Telemark land area supplying the Oslo Region with quartz-rich terrestrial sediments (Fig. 4). This is further corroborated by the clearly nearshore, cross-stratified deposits of the Huk Formation in the southernmost part of the Oslo Region (Nilssen 1985) and the presence of a western change in the Palaeozoic facies-types observed by several authors (e.g. Bockelie 1978; Henningsmoen 1960; Størmer 1967). These observations do not support a north-western source-area located around the Trondheim region, 200 km to the north (Fig. 4).

The upper part of the Elnes Formation demonstrates a gradual return to more marly deposits with the absence of septarian concretions, which are

replaced by smaller, but increasingly more frequent, limestone concretions. In the southern to central Oslo Region this also marks the transition to the succeeding Fossum and Vollen formations, but to the north there is a return to the dark mudstones encountered in the middle part of the formation. A similar development is seen for the contemporaneous Andersö Shales in central Sweden to the north and at the top of the Killeröd Fm. in Scania, southern Sweden (Månsson 2000; Pålsson *et al.* 2002).

The southern Oslo Region

The southern Oslo Region is composed of the Skien-Langesund, Eiker-Sandsvær and Modum districts (Fig. 1). Here the Elnes Formation starts with a thin (1–3.5 m thick) lower marl unit, the Helskjer Member, succeeded by the Heggen Member, which reach a minimum thickness of at least 20 m (Nilssen 1985). The base of the last of these is defined at the first thick mudstone bed. The lower part is characterized by marly mudstones containing moderately frequent horizons of medium-sized carbonate concretions (Owen *et al.* 1990; Wandås 1984). Upwards the marly mudstones gradually change into purer and darker siliciclastic mudstones, interrupted at certain levels by thick (perhaps > 30 m in Eiker-Sandsvær) intervals of sandy turbidite beds. The upper part of the formation shows a gradual return to marlier, concretionary mudstones. The transition to the succeeding Fossum Formation is found where the small carbonate concretions begin to dominate the sediments (Owen *et al.* 1990).

The central Oslo Region

The central Oslo Region consists of the Oslo-Asker and Ringerike districts (Fig. 1). The Elnes Formation in this area – although the stratigraphical knowledge of the Elnes Formation is not well known in the Ringerike district – is characterized by an approximately 1.5 m lower marl unit, the Helskjer Member, the top of which is fairly abrupt. The overlying Sjøstrand Member is composed of dark grey to blackish mudstones with large, sparsely distributed septarian concretions. In the lower part fissile mudstones rich in pyrite are present. Upwards these sediments become gradually lighter, only interrupted by a couple of turbidite intervals. The member is typically rich in graptolites and phosphatic shelled brachiopods. The top of the Sjøstrand Member changes gradually into marly mudstone deposits rich in shelly fossils. The concretions become slightly smaller while changing from fossil-poor septarian concretions to fossiliferous carbonate concretions,

Fig. 8. Shell-lag with erosive base preserved in calcareous concretion. Middle Engervik Member at Elnestangen north of Slemmestad. Bar equals 1.0 cm.

some of which formed in connection with thick shell-lags. The base of the succeeding Engervik Member is defined at the base of the first coherent limestone bed (Owen *et al.* 1990). Except for the lower few metres dominated by dark grey, siliciclastic mudstone, with large (10–20 cm), rounded carbonate concretions, it has a rich shelly fauna found in marly mudstone with numerous, small and often irregular carbonate concretions and several distinct shell-lags (Fig. 8). Certain intervals are rich in large *Chondrites* burrows. The base of the overlaying Håkavik Member is placed at the base of the first calcarenite at which coarse-grained siliciclastic laminae and beds become a dominant feature of the upper Elnes Formation. The coarser beds and lamine may contain cross-stratification, sole-marks, flame structures, fining-upward features, planar and convolute lamination and in extremely rare cases also a basal shell-lag, indicating a possible storm-induced origin for some of the beds. The current direction, as in the earlier Sjøstrand Member, was from the south-west (Seilacher & Meischner 1965; Owen *et al.* 1990; S.O. Egenhoff, personal communication, 2005; this study). Generally, the size of the extrabasinal quartz does not exceed 30 to 40 μm (Bjørlykke 1974, pp. 15–16). The base of the succeeding Vollen Formation is located at the top of the last calcarenite bed, where the lithology becomes dominated by small carbonate concretions in a marly mudstone matrix (Owen *et al.* 1990). The base does not necessarily correspond to the reappearance of *Chondrites*, although this trace fossil does appear close to the boundary in the Slemmestad area.

The northern Oslo Region

The development of the Elnes Formation is largely identical in the northern Hadeland and Mjøsa districts, the latter here refered to as a single district. In Hadeland the base of the formation is characterized

by the Helskjer Member, an approximately 10-m-thick marl and argillaceous limestone unit, the base of which is drawn at the first shale above the compact limestone of the underlying Huk Formation (Owen *et al.* 1990). In the northernmost part of the Mjøsa district the Elnes Formation overlies the Stein Formation, and the base here is defined at the level where the mudstones dominate over the underlying nodular marls and limestones (Skjeseth 1963, p. 71; Rasmussen & Bruton 1994, p. 204). The succeeding succession composed of dark mudstones with horizons of fossiliferous septarian concretions is as yet undefined (Owen *et al.* 1990), but it does resemble the Heggen Member in the south. The main differences are the development of septarian cracks in the concretions, the general lack of turbidites, and the more siliciclastic composition of the mudstones. Due to the overall similarity, the dark mudstones of the lower to middle Elnes Formation in the northern area are here regarded as lateral equivalents to part of the Heggen Member. Above the mudstones the lithology changes and is followed by a few metres of light grey mudstone full of small (2–7 cm) irregular carbonate concretions, which are silicified. This unit, termed 'Feinknolliger Kalk' by Holtedahl (1909), is regarded as a distal equivalent of the Engervik Member. Above this is a blackish mudstone succession containing large (10–20 cm) carbonate and septarian concretions and termed the 'Cephalopod Shale' by Størmer (1953).

Stratigraphical history

The central Oslo Region is the lithologically best studied area with regard to the Elnes Formation having been investigated by many authors (e.g. Kjerulf 1857; Brøgger 1882, 1890; Størmer 1953; Bjørlykke 1965, 1974; Seilacher & Meischner 1965; Skaar 1972; Owen *et al.* 1990; Maletz & Egenhoff 2005) (Fig. 9). The first attempt at a stratigraphic subdivision of the Cambro–Silurian rocks was made by T. Kjerulf (1857, pp. 284–285), who introduced a numerical system of which the Helskjer Member together with the preceding Huk Formation corresponds to 'Etage 3α' (Fig. 9). 'Etage 3β', which is described as non-calcareous black or grey graptolite shale with pyrite, corresponds to the Sjøstrand Member and lower third of the Engervik Member; the upper boundary probably located between 4.7 and 6 m above the base of the Engervik Member. His 'Etage 4' is described as marly grey shale with kidney-sized limestone concretions and limestone beds. This unit includes the rest of the Elnes Formation together with the succeeding Vollen Formation. The system was further elaborated on by W. C. Brøgger (1882, p. 26), who changed the

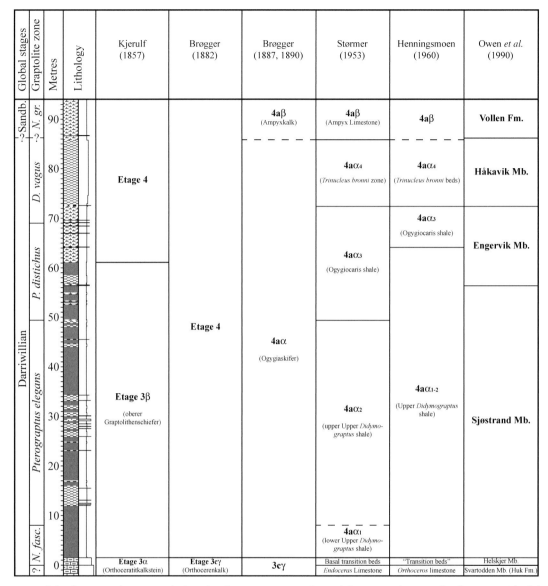

Fig. 9. Historical development of names applied to the Elnes Formation and its subdivision in the central Oslo Region. The lithological log represents a composite log for the Slemmestad area.

'etage' numbering and subdivided them (Fig. 9). Only the lower part of the Elnes Formation was treated. The Helskjer Member was now assigned to the upper part of the 'Orthocerenkalk' or 'Etage 3 cγ', while the succeeding Elnes Formation figured as 'Etage 4'. Just 5 years later he presented his final subdivision of the interval, placing the whole Elnes Formation with the exception of the basal Helskjer Member in his 4aα zone, while the overlying Vollen Formation was given the code 4aβ. For the next half century little attention was given to the stratigraphical record of the central Oslo Region, before interest was revived in the 1950s with the initiation of the 'Middle Ordovician of the Oslo Region' project led by Professor Leif Størmer. This project aimed at a complete redescription of the

stratigraphical and fossil record with a focus more on the biostratigraphy than lithostratigraphy, as reflected in the resulting stratigraphic division based on both lithological and biostratigraphical criteria. The Helskjer Member was identified as a separate unit (zone 3cδ). Brøggers (1890) zone 4aα was divided into the four subzones $4a\alpha_{1-4}$ (Fig. 9). The lower two representing Størmer's 'Upper Didymograptus Shale' (= Sjøstrand Member) were divided based on the presence of the graptolite *Didymograptus bifidus* in the lower part. The boundary could perhaps correspond with the top of the *Nicholsonograptus fasciculatus* graptolite zone, which is characterized by a distinct change in the graptolite fauna (Maletz 1997). Zone $4a\alpha_3$, also called the 'Ogygiocaris Shale', is characterized by grey weathering of the fossiliferous concretions, which in general become more flattened than those found in the preceding Upper Didymograptus Shale. The zone appears to equal the present Engervik Member. The final zone, 4aα₄ or *Trinucleus bronni* Zone, has at its base, the first sand bed and is characterized by calcarenitic beds with frequent fragments of the trilobite *Botrioides* (earlier *Trinucleus*). The *B. bronnii* Zone corresponds completely with the Håkavik Member of Owen *et al.* (1990). In publications after 1953, authors largely agreed with Størmer's division for the Middle Ordovician, but Henningsmoen (1960) drew the boundary between 4aα2 and 4aα3 at the first limestone bed above the dark grey mudstones observed at Engervik near Slemmestad, Oslo-Asker. An examination of this section shows that this boundary is placed much higher than in any of the earlier definitions and well into the concretionary unit (Fig. 9).

Owen *et al.* (1990) replaced the former combined bio- and lithostratigraphy and erected formal formations and members based on differences in the dominant lithology. The Helskjer Member from the underlying unit was transferred to the Elnes Formation, while the former 4aα₁₋₂ zones were combined in the Sjøstrand Member (Fig. 9).

The stratigraphical history of the southern Oslo Region is outlined in Figure 10. Here the Elnes

Formation was first treated stratigraphically by Dahll (1857), who in line with the system for the central Oslo Region established by Kjerulf (1857) used numbers for the different units. Thus he combined the complete Huk and Elnes formations of the Skien-Langesund area in his 'Straten 3', succeeded by 'Straten 4α' (= Fossum Fm.) (Fig. 10). Brøgger (1884) followed this with his own division, erecting an identical system to the one presented for the central Oslo Region. The basal Helskjer Member and the Svartodden Member of the underlying Huk Formation represented the top of his 'Etage 3' or 'Orthocerenkalk', while the succeeding Elnes Formation figured as 'Etage 4α'. Størmer (1953) establish a tripartite subdivision of the formation, where the lower marly Helskjer Member was combined with the massive Endoceras Limestone (3 cγ), followed by the Upper Didymograptus Shale and succeeding Ogygiocaris Beds. The transition between the latter two units is placed at a phosphatic layer in the Skien-Langesund district, but is more obscure in the Eiker-Sandsvær district. Owen *et al.* (1990) reassigned the lower marly unit to the Elnes Formation, while introducing the Helskjer Member at the base overlain by the concretionary mudstones of the Heggen Member. The latter contains both the Ogygiocaris Beds and the Upper Didymograptus Shale of earlier usage.

The history of the stratigraphic subdivision of the Elnes Formation in the northern part of the Oslo Region is summarized in Figure 11. Holtedahl (1909) in his work on the Etage 4 *sensu* Brøgger (1887) erected three zones in this region. What later became termed the Helskjer Member was placed within the top of the Orthocerenkalk and is succeeded by the blackish Ogygia Shale. The shale unit is topped by a 3- to 4-m-thick, light grey nodular unit, the A2 Zone, which he described as extraordinaric unfossiliferous. The nodular unit is overlain by fossiliferous black shale with larger elliptical concretions containing barite crystals, his A3 Zone. Størmer (1953) made an attempt at transferring and incorporating the stratigraphical

Dahll 1857	Brøgger 1884	Størmer 1953	Owen *et al.* 1990 and this study
Straten 4α	Etage 4b	Ampyx Limestone	Fossum Fm.
Straten 3	Etage 4a	Ogygiocaris Beds	Heggen Mb.
		Upper Didymograptus Shale	
	Etage 3	3cγ	Helskjer Mb.
			Huk Fm.

Fig. 10. Historical development of names applied to the Elnes Formation and its subdivisions in the southern Oslo Region during the last 150 years.

Holtedahl 1909	Størmer 1953	Skjeseth 1963	Owen *et al.* 1990	This study
Zone A3 (Shale with cephalopods and gastropods)	Cephalopod Shale	Cephalopod Shale	Undifferentiated	Cephalopod Shale mb.
Zone A2 (micro nodular limestone)	Ogygiocaris Bed	Ogygiocaris Shale + Upper Didymograptus Shale		Engervik Mb.
Zone A1 (Ogygia Shale)	Upper Didymograptus Shale			Heggen Mb.
Orthocerenkalk	3cγ	Helskjer Shale and Limestone	Helskjer Mb.	Helskjer Mb.
		Endoceratid Limestone	Huk Fm.	Huk Fm.

Fig. 11. Historical development of names applied to the Elnes Formation and its subdivisions in the northern Mjøsa district during the last century.

names used for the Oslo-Asker district into the stratigraphical system of Holtedahl (1909) even though no marked difference was found between the supposed Upper Didymograptus Shale and succeeding Ogygiocaris Beds. It is not clear from his description whether he included Holtedahl's A2 Zone or rather regard it as a separate unit dividing the Ogygiocaris Beds from the overlying 'Cephalopod Shale'. Skjeseth (1963) introduced the name Helskjer Shale and Limestone for the transitional marly deposits between the main Huk Formation and the shales of the Elnes Formation proper, but regarded it as a part of the underlying 'Orthoceratite' Limestone. Owen *et al.* (1990) assigned the Helskjer Member to the Elnes Formation but avoided subdividing the main part of the formation. The most recent revision of the lithostratigraphy of the northern Mjøsa district is by Rasmussen & Bruton (1994), who assigned the limestones of the former Huk Formation (*sensu* Owen *et al.* 1990) north of Lake Mjøsa to the Stein Formation, overlain by graptolitic shales and mudstones of the Elnes Formation proper.

Localities

Southern Oslo Region

In the southern Oslo Region, attention was directed to the central Eiker-Sandsvær district approximately 60 km south-west of Oslo (Fig. 1, location 3), known for the classical trilobite study area (Brünnich 1781; Brøgger 1882; Henningsmoen 1960; Owen 1987). The present study concentrated on three localities containing several subprofiles, selected along a west-northwest facing cliff (Fig. 12). The sedimentary rocks exposed in all the sections have been subjected to contact-metamorphism resulting in a fairly hard and brittle rock, where recrystallization has produced a coarser grainsize than the original. The contained carbonate nodules have largely been dissolved, leaving holes in the rock, and the carbonate has probably migrated out into the surrounding mudstones increasing their carbonate content.

Råen. – This locality, which is situated just south-west of lake Fiskum, consists of four profiles distributed on one small slope-profile and three road-cuts along a small forest road (Fig. 13; grid references: NM 453 168–457 169). The four profiles have been numbered from the stratigraphical top downward and were all logged on a centimetre scale, though only profile 1, 4 and parts of profile 3 were sampled in detail. Profile 1 covers around 25 m of the upper half of the Heggen Member, but lacks the top

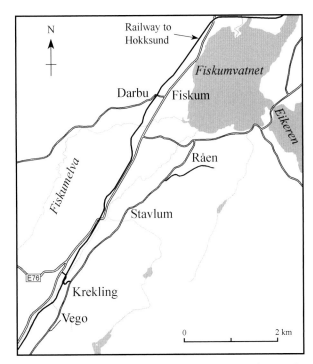

Fig. 12. Map of the central Eiker-Sandsvær district shoving the important localities south of Fiskum town.

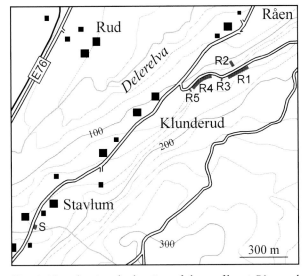

Fig. 13. Map showing the location of the profiles at Råen and Stavlum, Eiker-Sandsvær. Isochore lines in metres. Profiles 1 to 5 at Råen are indicated by line R1 to R5, while the short profile at Stavlum is indicated by an 'S'.

(Fig. 14). When present, the top is characterized by marly mudstones with many small (> 5 cm), round, carbonate concretions. Downwards, the number of concretions decreases and mudstones dominate the lithology (Fig. 15). This, the main part of the profile, is also characterized by a general increase in the abundance of shelly fossils and in the size of the

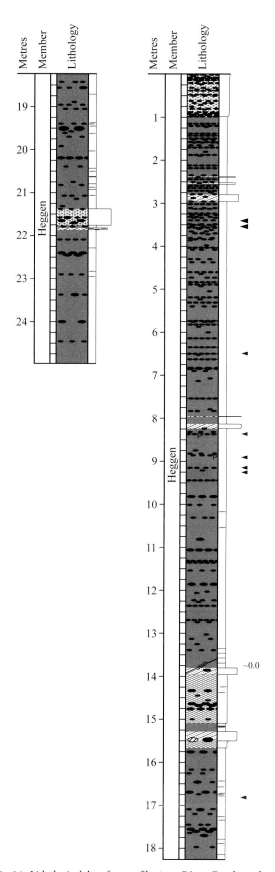

Fig. 15. Upper part of profile 1 at Råen, Eiker-Sandsvær. The calcareous concretions have been removed by weathering.

concretions. The middle and lower part of the profile includes a few sandy beds a few centimetres thick. Profile 2, which contains the upper part of the middle Heggen Member, changes from dark grey mudstones with sparsely distributed carbonate concretions at the top to completely sand-bed-dominated deposits rich in sedimentary structures such as current ripples, water escape structures, planar lamination and fining-upward beds (Fig. 16). Basal shell-lags are absent, and the silty and sandy interval is generally unfossiliferous. The current-ripples for the most part indicate a north-easterly current direction. Profile 3 is no more than 2.6 m thick and is dominated by fine sand beds alternating with dark mudstones with graptolites (Fig. 17A). Profile 4 is an 11-m-thick mudstone succession with moderately large carbonate concretions (Fig. 17B). It is fairly rich in shelly fossils. The profile corresponds to the lower 9 m of profile 1, which is why both profiles are treated together in the section discussing the biostratigraphy. The absence of coarser sediments in this profile may be an artifact of weathering, making silty intervals difficult to distinguish. A fifth profile is found immediately adjacent to Profile 4, but covers more or less the same interval.

Stavlum. – This small road-cut is located midway between Råen and Vego at Krekling (see below) (Fig. 13; grid reference: NM 444 160). This locality contains the lower part of the Helskjer Member together with the underlying Huk Formation and yielded trilobites studied by Wandås (1982, 1984) and herein. The lithology consists of marly mudstones

Fig. 14. Lithological log for profile 1 at Råen. For legend see Figure 26.

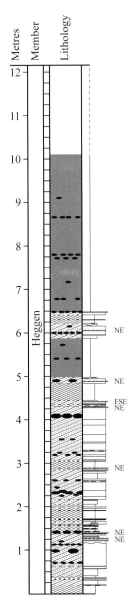

Fig. 16. Lithological log for profile 2 on the slope below the forest road at Råen. Note the dominantly north-easterly current direction.

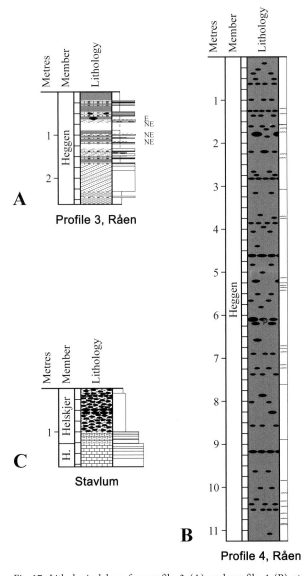

Fig. 17. Lithological logs for profile 3 (A) and profile 4 (B) at Råen and road-cut at Stavlum (C). Same legend as in Figure 26.

completely dominated by small and round carbonate concretions rich in asaphid trilobites (Fig. 17C).

Vego at Krekling. – This locality consists of six sub-profiles situated along a small private forest road just above Vego south of Krekling (Figs 12, 18; grid references: NM 429 137–430 138). Taken together, the exposures show intervals in the middle to upper Heggen Member studied mainly to get an idea of the general changes through this stratigraphic unit (Fig. 19). Fossils are extremely rare in the lower silty and sandy part of the transect, and detailed sampling was restricted to the more fossiliferous mudstone intervals of profile 6. Samples were not collected

from the upper marly part corresponding to the upper Helskjer Member at Råen. In Profile 6 numerous current ripples can be studied in the lower coarser part of the turbidites (Fig. 6). Here current directions are both from the NNE and the SSW (Fig. 19).

Central Oslo Region

In the central Oslo Region, three main exposures were selected for study in the vicinity of Slemmestad approximately 20 km south-west of Oslo (Fig. 20). The Slemmestad area contains a complete Elnes Formation, making this the best study area in the Oslo Region. The estimated thickness of the Elnes Formation is 86 m (see Fig. 9).

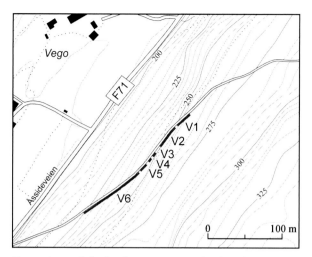

Fig. 18. Map of the locality at Vego south of Krekling, Eiker-Sandsvær. The profiles are numbered from V1 to V6. The figure is based on a technical map from Statens Kartverk, Norge.

The old eternite quarry. – This locality (Fig. 21, Locality 1 and 2; grid reference: NM 839 289) was described by Maletz (1997), and forms the basis for his defined graptolite biostratigraphy for the lower part of the formation (see Maletz *et al.* 2007 for the upper). The quarry exposes the Tøyen and Huk formations and a complete transect through the lower 40.5 m of the Elnes Formation (Fig. 22), accessible in a short south-western and a longer north-western profile cutting nearly perpendicular to the strike. The strike and dip of the strata are 24° and 20° W, respectively. The quarry has been disused for many years, resulting in a fairly strong weathering of the nearly vertical outcrops, and excavation is needed to extract fresh material. The south-western profile was logged from 16 cm below the top of the Svartodden Member of the underlying Huk Formation and collections were made at 1 cm intervals through the overlying Elnes Formation. The base of Elnes Formation is taken as the first thin shale bed above beds with frequent endoceratid cephalopods. The basal marly interval, the Helskjer Member, is 1.49 m thick and becomes finer grained upwards, the transition with the succeeding dark grey to blackish mudstones of the Sjøstrand Member being marked by a more or less distinct drop in marl content. The lower 7 to 8 m of the Sjøstrand Member is characterized by a gradually decreasing grain-size and carbonate content together with an increasing pyrite and organic content ending with a black, fissile mudstone rich in planktonic graptolites and linguliformean brachiopods and lacking in calcium carbonate concretions. The fissile mudstone interval is terminated by the appearance of quartz-silt and fine sand about 10 m above the base of the member and at the top of the south-western section.

The lower 11 m of the north-eastern profile corresponding to the equivalent beds in the south-western section were not logged as efforts were directed towards detailed collecting from here to the top of the profile. This shows alternating intervals of dark grey mudstones and thick silt- or sandstones up to several centimetres thick. The mudstones are generally poorly fossiliferous but do contain graptolites and phosphatic shelled brachiopods. The lack of clear bedding indicates a fairly strong biotubation and unfossiliferous septarian concretions (10–50 cm in diametre) occur at widely spaced intervals. The sandy beds vary from less than 1 cm in thickness to almost 50 cm for the largest combined bed recognized at 29 m above the base of the formation. They consist of light grey quartz-silt or fine sand with some mica. Planar lamination is present and bottom structures showing cross-stratification are best seen in a bed 24.9 m above the base of the formation. The sandy beds contain *Planolites*, but are otherwise poorly fossiliferous.

Beach section north of Djuptrekkodden. – The locality is approximately 600 m north-east of the Old Eternite Quarry (Fig. 21; grid references: NM 842 294–842 295), and at low tide presents a nearly complete exposure from 26.75 m above the base of the Elnes Formation and up into the overlying Vollen Formation (Fig. 23). The only gap in the outcrop is found between 23.18 and 27.56 m above datum corresponding to the uppermost part of the Sjøstrand Member.

This is a classic section for the Elnes Formation and represents the main profile on which Henningsmoen (1960) based his *Ogygiocaris* biostratigraphy. His definition of the boundary between the Sjøstrand and the succeeding Engervik Member differs somewhat from that subsequently given by Owen *et al.* (1990) and is thus located at a limestone bed 7.76 m above the recent boundary (Fig. 9).

The lower part of the profile exhibits a lower interval with silt and fine sands beds, either compact or showing planar lamination or cross-stratification (Fig. 24). This is followed by dark grey mudstones rich in thin layers of pyrite and the scattered horizons of septarian concretions. At about 17 m above the base, the sediments become slightly coarser and there is an increase in marl with a rich and diverse shelly fauna around 22.5 m above the base. Beds with fossiliferous concretions replace those with septarian concretions.

Above the gap in the section the first beds include a 10-cm-thick, fine-grained marl or mud limestone, lacking fossils and forming the base of the succeeding Engervik Member. This bed can be traced over a

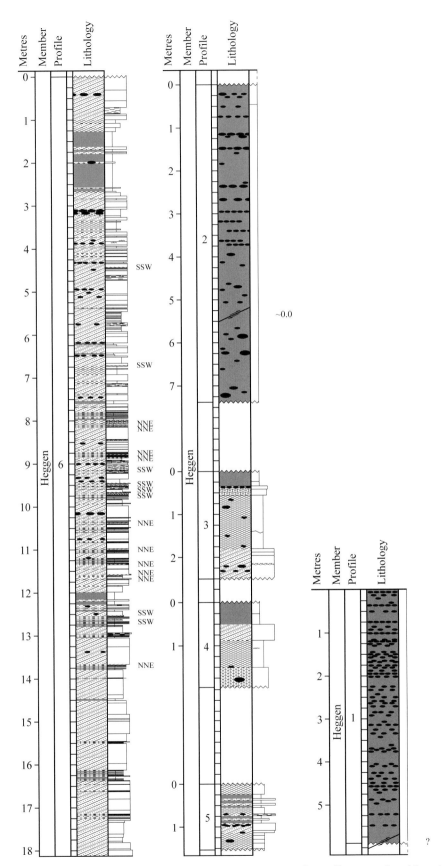

Fig. 19. Lithological log for the six profiles at Vego, Eiker-Sandsvær shown on Figure 18. The profiles are numbered from 1 to 6 from the upper right to lower left corner. Note the cyclically reversing current direction as indicated by the current ripples in profile 6. Legend as in Figure 26.

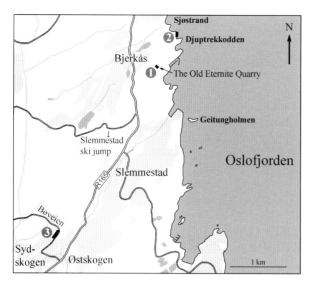

Fig. 20. Map of the Slemmestad area showing some of the localities mentioned in the text. Built-up areas are indicated by light grey shading. 1. The Old Eternite Quarry just north of Slemmestad. 2. Beach profile just north of Djuptrekkodden. 3. Road-cut along Bøveien south of Slemmestad.

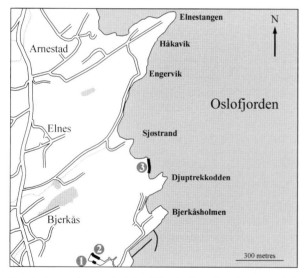

Fig. 21. Map of important localities just north of Slemmestad town. 1. South-western profile in the Old Eternite Quarry. 2. North-eastern profile in the Old Eternite Quarry. 3. Beach-profile just north of Djuptrekkodden.

distance of at least 4.5 km from Elnestangen (grid reference: NM 843 302–845 306) to the profile at Bøveien (see Fig. 20). It is overlain by 2 m of a relatively coarse and marly mudstone rich in pyrite and containing moderately sized, ellipsoidal carbonate concretions with a rich shelly fauna. From 30 to 32 m above the base there is a brief return to darker mudstones containing septarian concretions. The next unit is generally marlier with smaller, flattened carbonate concretions. Large *Chondrites* occur at intervals from 34.65 to 41.10 m above the base, but

the trace fossils are generally absent in intervals rich in pyrite. The Håkavik Member starts at 43.98 m where unfossiliferous beds of silt- and sandstone dominate. The sandstones are structureless or contain planar lamination with fining upwards trend, but in a few cases north-east dipping cross-bedding can be seen. Basal shell-lags have been observed at two levels. The top of the Håkavik Member is characterized by a gradual decrease in the number and thickness of the sand-beds, an increase in marl content and a return of the trace fossil *Chondrites*. The top of the Elnes Formation is placed at the last thin fine sandstone bed 57.88 m above the base of the section (Fig. 25).

The Bøveien profile. – This road-section is located on the western side of the small road from Sydskogen (Fig. 20, location 3; grid reference: NM 821 261). It differs from the two previous localities in having been metamorphosed, making the rocks hard and brittle with the partial removal and recrystallization of carbonate material. This is the type locality for the brachiopod *Cathrynia stoermeri* (Spjeldnæs 1957), but has seldom been examined. The section is cut by several large faults, but shows a more or less undisturbed transect from the lower part of the formation and up into the succeeding Vollen Formation. The present study has only focused on the upper middle part of the profile where beds corresponding to the missing interval at Djuptrekkodden are available for study (Fig. 26) and the fossils logged. The characteristic limestone bed marking the base of the Engervik Member is partly dissolved by contact-metamorphism, but is still identifiable.

Northern Oslo Region

The northern Oslo Region with the northern Mjøsa district has long been known for its rich and diverse trilobite fauna from the lower part of the Elnes Formation which, through time, has yielded several new species not found further south in the Oslo Region (Bruton 1965; Henningsmoen 1960; Nikolaisen 1965; Wandås 1984). The section selected for this study (Fig. 27; grid-reference: PN 105 473 to 106 472) is a relatively new road-cut located on the eastern side of highway E6 just south-west of Nydal town where road Fv 84 between Nydal and Hamar crosses the highway. The rocks here have an average strike and dip of 37° and 55°W respectively, but near the top of the section the beds are disturbed and have been subjected to strong shear stress and some heating, removing all traces of fossils in the mudstone. Fossils are quite abundant in the marls and limestones of the Helskjer Member. The Helskjer Member corresponds to the lower 9.25 m of the Elnes Formation at

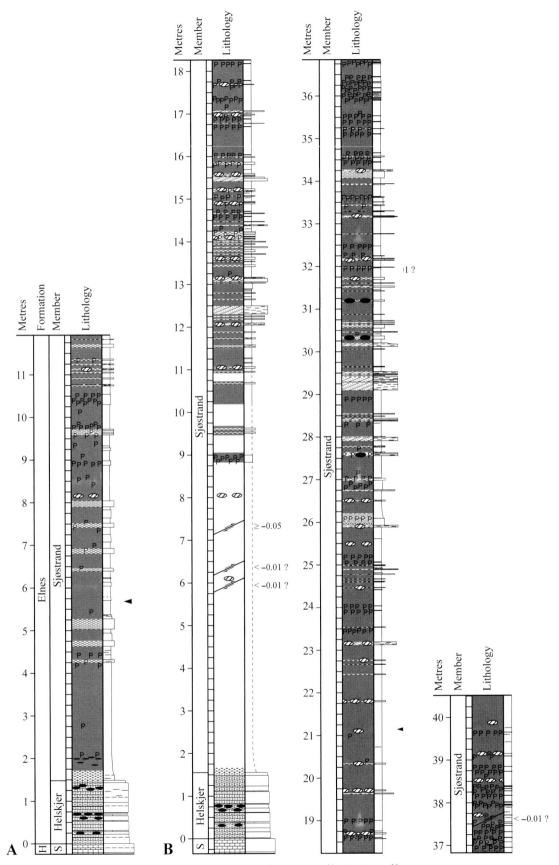

Fig. 22. Lithological logs for the Old Eternite Quarry at Slemmestad. A. SW profile. B. NE profile.

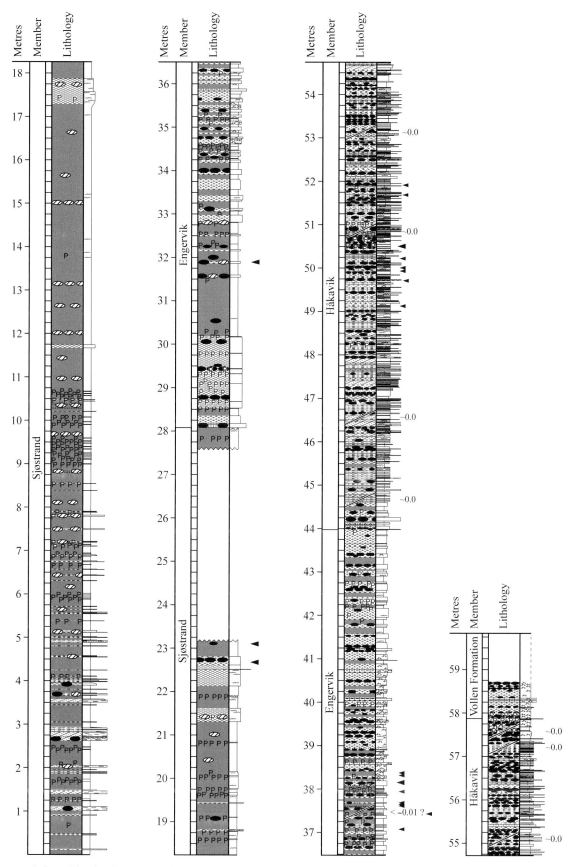

Fig. 23. Lithological log for the beach profile just north of Djuptrekkodden, Oslo-Asker. Legend as in Figure 26.

Fig. 24. Lower part of beach profile at Djuptrekkodden (~5 m above datum) showing fine-grained turbidites in dark grey mudstone. Bar equals 20 cm.

Fig. 25. Håkavik Member and succeeding Vollen Formation just north of Djuptrekkodden, Oslo-Asker. The boundary between the Elnes Fm. and the overlying Vollen Fm. is indicated with a white arrow.

Nydal (Fig. 28). It is succeeded by black mudstones or shales with frequent horizons of small, fossiliferous septarian concretions. This part strongly resembles the Heggen Member to which it is assigned. The middle part of the exposure is cut by two larger faults causing gaps in the sequence. Upwards the mudstones containing septarian concretions changes into more marly deposits containing small (2–7 cm), grey, siliceous carbonate concretions extremely poor in fossils (Fig. 29). This interval has been interpreted as a marginal part of the Engervik Member. The interval is no more than around 6 m thick and is overlain by black mudstones largely identical with the underlying Heggen Member. The upper part of the Elnes Formation is not exposed in the Nydal area.

Additional localities. – In addition to the main localities described above, material from another four exposures covering the basal part of the formation has been included. The profiles consist of a road-cut at Vikersund skijump in Modum (southern Oslo Region; Fig. 1, locality 4; grid-reference: NM 563 453); the Hovodden peninsula on the eastern

shore of Randsfjorden, Hadeland (Fig. 1, locality 7; northern Oslo Region; grid-reference: NM 785 952); a small road-cut just north of Furnes Church, Mjøsa (Fig. 27; northern Oslo Region; grid reference: PN 099 472) and Helskjer on Helgøya, Mjøsa (Fig. 1, locality 8; northern Oslo Region; grid-reference: NN 088 329). The profiles were measured and collected for trilobites by Wandås (1982, 1984). Further profiles to which reference is made are the Stathelle profile from the Skien-Langesund area measured by Nilssen (1985) (Fig. 1, locality 1) and the Rønnings-fossen profile from the Eiker-Sandsvær area (unpublished data by D. L. Bruton and A. W. Owen) (Fig. 1, locality 2), which covers most of the Elnes Formation together with the upper boundary to the succeeding Fossum Formation. From Sweden details of the Andersön profile from Jämtland, were presented by Pålsson *et al.* (2002) (Fig. 1, locality 10).

Bio- and Ecostratigraphy

Even though several comprehensive works have been published on the trilobites and their general stratigraphic distribution, there has never been any serious attempt at a biostratigraphical subdivision of the Elnes Formation (Henningsmoen 1960; Nikolaisen 1963, 1983; Owen 1987). However, Wandås (1982)

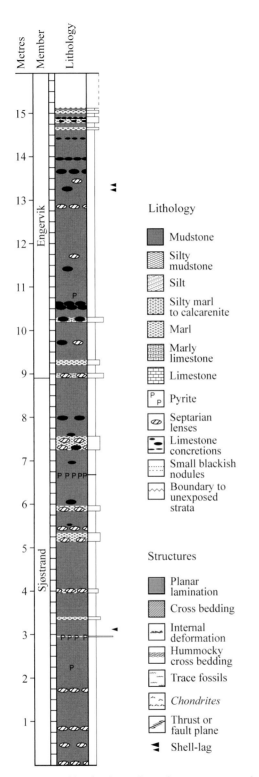

Lithology

- ▓ Mudstone
- ▒ Silty mudstone
- ░ Silt
- Silty marl to calcarenite
- Marl
- Marly limestone
- Limestone
- P / P Pyrite
- ⬭ Septarian lenses
- Limestone concretions
- Small blackish nodules
- ∿ Boundary to unexposed strata

Structures

- ▓ Planar lamination
- ▨ Cross bedding
- Internal deformation
- Hummocky cross bedding
- Trace fossils
- *Chondrites*
- Thrust or fault plane
- ◄ Shell-lag

Fig. 26. Lithological log for the road-cut along Bøveien just above Sydskogen, Oslo-Asker. The base is set at a fault line.

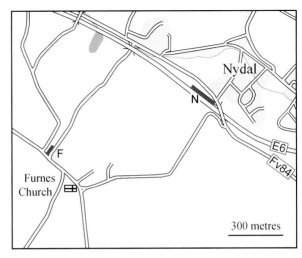

Fig. 27. Map showing the location of the Furnes (F) and Nydal (N) profile. Built-up areas are shaded light grey.

occurrence. Both realized that their main problem was the rather poor knowledge of the stratigraphical distribution of the various species and corresponding difficulties in establishing anything more than a coarse correlation between the various areas within the Oslo Region.

The detailed sampling carried out in this study together with the earlier collections made by Wandås (1982, 1984) has led to a much better understanding of the geographical and stratigraphical distribution of species. The results, which are presented in Figures 30–37, also makes it clear that the main part of the fauna is either extremely sporadic or else more environmentally than stratigraphically controlled. An example of the latter is *Nileus armadillo*, which exhibits some stratigraphically related changes in the width of the pygidial border and marginal terrace-lines. Unfortunately, the changes do not occur at the same stratigraphical level or in the same chronological order in different parts of the Oslo Region, thus highlighting the importance of comparing collections from more than one locality, when studying possible evolutionary trends. Regrettably few trilobite species appear as sound candidates for use in constructing a trilobite biozonation, and those that do are generally only found within a short stratigraphical interval delimited by large undistinguished zones or are only found inside a limited geographical area. For these reasons only the basal part of the Elnes Formation lends itself for a biozonal subdivision based on the appearence of megistaspid species described by Wandås (1982, 1984). He defined three biozones based on the first occurrence of *Megistaspis (Megistaspidella) giganteus runcinatus* n. subsp. [= *M. (M.)* sp. of Wandås (1982, p. 138)], *M. (M.) giganteus giganteus* Wandås, 1984 and *M. (M.) maximus* Wandås, 1984, forming a

found a clear stratigraphic sequence for the megistaspids in the basal part of the formation, while Henningsmoen (1960) suggested that species of *Ogygiocaris* may show real differences in their stratigraphical

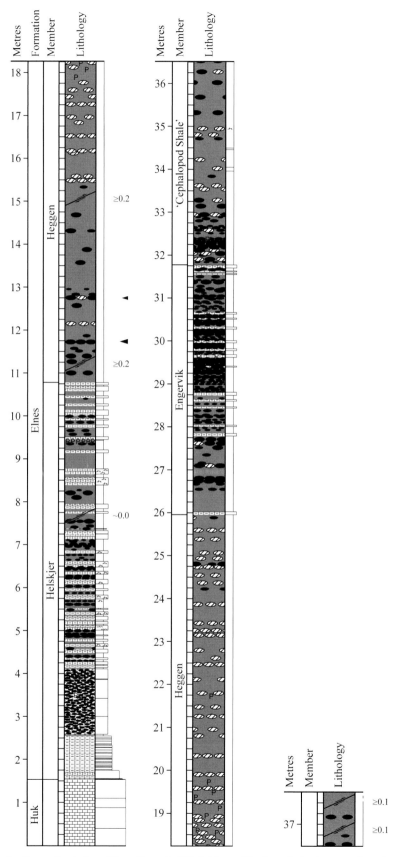

Fig. 28. Lithological log for the road-cut along E6 south-west of Nydal town. The boundaries between the subprofiles are located 11.22 and 15.06 and 36.86 m above datum. Legend as in Figure 26.

Fig. 29. Road-cut along E6 at Nydal, Mjøsa district. The massive limestone to the left represents the Huk Formation. The micro-nodular interval separating the lower Heggen Member from the upper 'Cephalopod Shale' is seen in the lower right corner. Picture taken by Hans Arne Nakrem, 2003.

continuous evolutionary line. *M. (M.) giganteus runcinatus* ranges from the top of the underlying Huk Formation into the basal few metres of the Elnes Formation, the upper boundary apparently corresponding to a level slightly below that of *M. (M.) heroica* Bohlin, 1960 (Fig. 53). The zone is thickest in the northern and western Oslo Region, becoming thin in the Oslo-Asker area. It is succeeded by the relatively brief zone defined by the first occurrence of *M. (M.) giganteus giganteus*. This zone, which possibly never exceeds half a metre in thickness, is generally located above the Helskjer Member, although it is found in the upper part of the member at Furnes in the Mjøsa district (Fig. 53). The *M. (M.) maximus* Zone may possibly extend from the lower middle to upper part of the *Nicholsonograptus fasciculatus* graptolite Zone.

In addition to the megistaspid trilobites, others that may be useful include those described below.

The boundary between the Huk Formation and the overlying Helskjer Member of the basal Elnes Formation corresponds to the transition between *Asaphus sarsi* and the succeeding *Asaphus striatus*, the latter being extremely common in the Helskjer Member of the southern to central Oslo Region (Figs 30, 32, 34–36). An overlap is found within the lower half metre of the Helskjer Member. Both species appear to be good candidates for biozonation as they can be traced into the East-Baltic area (see Ivantsov 2003).

Subasaphus platyurus (Angelin, 1854), is extremely rare in Norway, but is a good index fossil when comparing the Oslo Region succession with that of the Baltoscandian platform, where it occurs in the Vikarby and Segerstad limestones of Sweden and equivalent beds in Estonia and the Doboviki Formation of NW Russia (Angelin 1878; Törnquist 1884; Schmidt 1901; Jaanusson 1953b; Jaanusson 1964). In the Elnes Formation, *Subasaphus platyurus* ranges from the lower Helskjer Member to basal Heggen Member (lower *M. (M.) giganteus runcinatus* Zone to possibly lower *M. (M.) maximus* Zone).

Pseudobasilicus (Pseudobasilicus) truncatus n. sp. and *Volchovites perstriatus* s.l. (Bohlin 1955) are both found in the lower *M. (M.) giganteus runcinatus* Zone (Figs 30, 34–37). *Volchovites perstriatus* also extends into the *M. (M.) giganteus giganteus* Zone and may even reach into the basal *M. (M.) maximus* Zone.

Cybelurus cf. *mirus* (Billings, 1865) is extremely rare, but is an interesting species because of its very strong affinity with taxa from both North America and Spitsbergen (see Whittington 1965a; Fortey 1980). It most closely resembles the specimen described by Fortey (1980) from the lower Darriwilian Profilbekken Member on Svalbard. It has been found in the uppermost part of the Helskjer Member and the Heggen Member in the northern Oslo Region (Fig. 37).

Among the species of *Botrioides*, the *B. efflorescens*–*B. foveolatus* lineage from the upper part of the Elnes Formation is of particular interest. *B. efflorescens* (Hadding 1913) ranges from the very top of the *Pterograptus elegans* graptolite Zone into the lower *Pseudamplexograptus distichus* Zone, where it is succeeded by *B. foveolatus* (Angelin 1854) in the middle *Pseudamplexograptus distichus* Zone. In lithostratigraphical terms the latter species occurs from the middle Engervik Member and up into the lower Håkavik Member (Figs 30–32). Both species can be identified in central and southern Sweden, where the same transition is recognized in the uppermost Killeröd Formation. Only *B. efflorescens* has been recorded in the Lower Anderö Shale member in Jämtland, but the material is poor and discrimination of species difficult.

Botrioides broeggeri (Størmer, 1930) first appears in the upper part of the Elnes Formation, while *B. simplex* Owen, 1987 appears restricted to the lower part (Figs 30, 32, 35).

Species of *Ogygiocaris* are generally not good biostratigraphic indicators as most occur together and have a long stratigraphical range. However, exceptions are the two subspecies of *Ogygiocaris striolata* Henningsmoen, 1960. The older subspecies, *O. striolata corrugata* Henningsmoen, 1960, is found from just below the *Pseudamplexograptus distichus* graptolite Zone of the upper Sjøstrand Member and

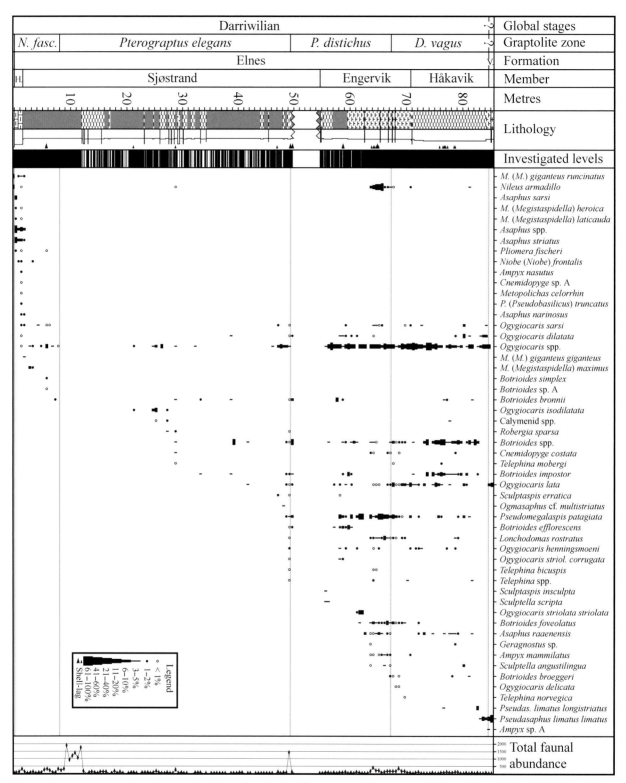

Fig. 30. Composite range chart showing the relative abundance of the various trilobite taxa in relation to the total number of macro-fossils counted for each half metre. Area north of Slemmestad, Oslo-Asker district.

middle Heggen Member, and continues up into the lower Engervik and upper Heggen Member (Figs 30, 32, 34). Here it is succeeded by *O. striolata striolata* Henningsmoen, 1960, which appears soon after in the middle *Pseudamplexograptus distichus* Zone, corresponding to the middle Engervik Member of the Olso-Asker area and the top Heggen Member in the south-western Oslo Region (Figs 30, 32).

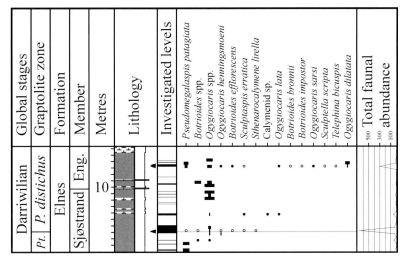

Fig. 31. Range chart for Sjøstrand-Engervik boundary at Bøveien road-cut, Oslo-Asker. Legend as in Figures 26 and 32.

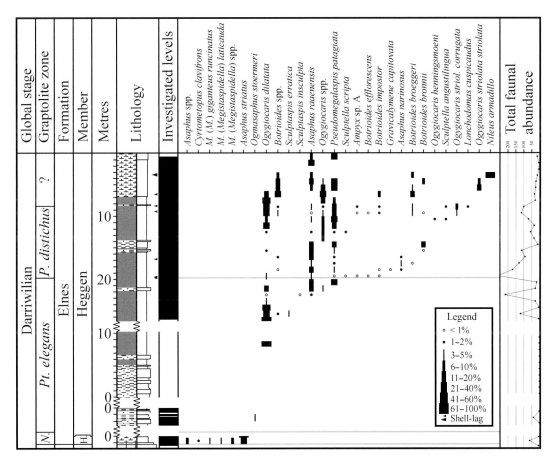

Fig. 32. Range chart for the profiles at Råen and Stavlum near Fiskum, Eiker-Sandsvær. Percentages given in relation to total amount of macrofossils counted for each sub-interval.

Pseudomegalaspis patagiata (Törnquist, 1884) extends up through the whole middle and upper Elnes Formation and continues up into the succeeding Vollen Formation and contemporaneous 'Cephalopod Shale' to the north (Figs 30–34). Even though

P. patagiata has a rather large stratigraphical range, it may be useful for general correlation with the main Baltoscandian platform, where it occurs in the lower Andersö Shale member, the Folkeslunda and Furudal limestones and the Killeröd Formation in central to

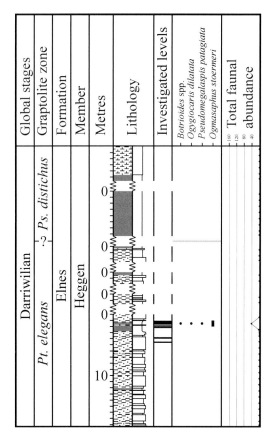

Fig. 33. Range chart for the exposures at Vego near Krekling, Eiker-Sandsvær. Legend as in Figures 26 and 32.

southern Sweden. It is furthermore indicated to occur in the uppermost 2 m of the Stein Formation in Jämtland, Sweden (Karis 1982), suggesting that the top of this formation corresponds to somewhere around the uppermost Sjøstrand to middle Engervik Member. However, this correlation is probably based on misidentification of *Ogmasaphus stoermeri* Henningsmoen recorded from the *Nicholsonograptus fasciculatus* graptolite Zone.

Pseudasaphus limatus Jaanusson, 1953a represents another good and potential stratigraphic indicator as it appears at the very top of the Elnes Formation in the central Oslo Region and near the base of the contemporaneous 'Cephalopod Shale' in the north (Figs 30, 34). Outside Norway, *P. limatus* has a wide distribution having been found in the Folkeslunda Limestone and Dalby Limestone in the Siljan district of central Sweden (Jaanusson 1953a) and in a borehole core at the town of Mukhovtsy in the St. Petersburg district, NW Russia (Balashova 1976).

Pelagic trilobites such as the telephinids and the remopleurids *Sculptella* and *Sculptaspis* are generally poorly represented in the material and several species have a quite variable geographical occurrence up

through the Elnes Formation, making correlations difficult. However, *Sculptaspis erratica* Nikolaisen, 1983 is potentially useful as it occurs from the upper *Pterograptus elegans* Zone into the middle to upper *Pseudamplexograptus distichus* Zone (Figs 30, 31, 34).

Regional correlation

Based mainly on the biostratigraphy of trilobites and aided in some instances by graptolites and other fossil groups, a correlation has been attempted between various sections in the Oslo Region and a single section up in Jämtland, central Sweden (Fig. 38). As already mentioned in the lithological description of sections, the boundary between the Huk Formation and the Elnes Formation (Fig. 38, line 1) is characterized by a change from massive calcarenites rich in endoceratid cephalopods to marls and muddy limestones with shaly intercalations, which constitutes the Helskjer Member. In the north this boundary is less distinct as cephalopods usually are absent, but the boundary is readily recognizable because of the distinct change from massive limestones to more muddy and marly limestones with wavy shale intercalations.

On Figure 38 the boundary between the *Megistaspis* (*Megistaspidella*) *giganteus runcinatus* Zone and the *M.* (*M.*) *giganteus giganteus* Zone is represented by line 2 between Slemmestad and the Vikersund skijump (Vikersundbakken) and indicates the succession at Slemmestad to be strongly condensed. This probably explains why the trilobite fauna of the Helskjer Member in the Oslo-Asker district is so different from the rest of the Oslo Region with its complete dominance of *Asaphus striatus*. The *M.* (*M.*) *maximus* Zone (line 3) can be followed from the northern to the southern Oslo Region. In the Hadeland district the boundary occurs in the upper part of the Helskjer Member 8.3 m above boundary 1, while it occurs 0.65 m from the top of the Helskjer Member at Furnes in the Mjøsa district. Towards the south, the zone boundary can be traced higher above the Helskjer Member, and in the Modum district it occurs nearly 4 m above the member. Line 4 (Fig. 38) indicates the first appearance of the zone fossil *Pterograptus elegans*, while line 5 represent the base of the *Pseudamplexograptus distichus* graptolite Zone located a few centimetres below a distinct shell-lag, which can be traced from the Eiker-Sandsvær district to the Oslo-Asker district and possibly even as far as the Mjøsa district as indicated by the appearance of *Pseudomegalaspis patagiata*, *Ogygiocaris henningsmoeni* and *Ampyx mammilatus* around such a shell-lag in the Nydal profile. Other trilobite species, which seem to have their first appearance around or slightly below this level, are *Botrioides efflorescens* and *Ogygiocaris*

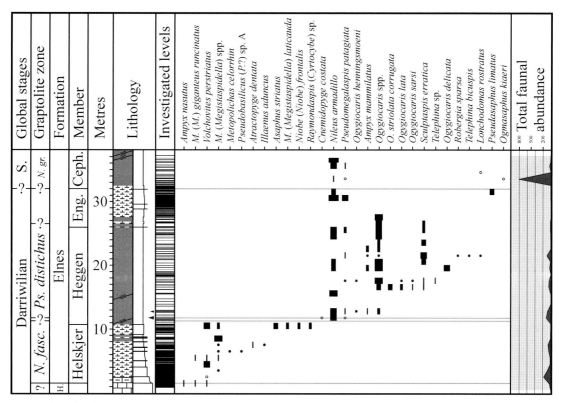

Fig. 34. Range chart for the upper Huk Formation and lower to middle Elnes Formation at Nydal, Mjøsa. Stratigraphy follows that of Figure 11. Legend as in Figures 26 and 32.

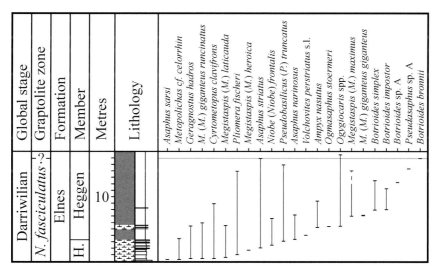

Fig. 35. Range chart for basal part of the Elnes Formation at road-cut below Vikersund skijump. Data based on fossil material sampled by Wandås (1984).

striolata corrugata. Both are seemingly characterized by a fairly short stratigraphical range. *O. striolata corrugata* occurs higher up in the Nydal profile and is therefore a good indication that the shell-lag bed can be equated with the *Pseudamplexograptus distichus* Zone. Line 6 (Fig. 38) between the Bødal and the

Slemmestad profiles is purely lithological and is based on the base of a thick limestone bed found in the Slemmestad area. The distance between this bed and the correlative line 5 suggests that there is a gap in the Slemmestad profile of around 6.5 m just below the Engervik Member. Line 7 (Fig. 38) is based on the

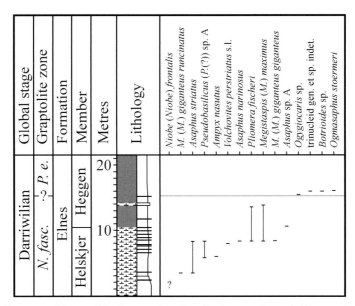

Fig. 36. Range chart for profile at Hovodden, Hadeland. Figure based on fossil material sampled by Wandås (1984).

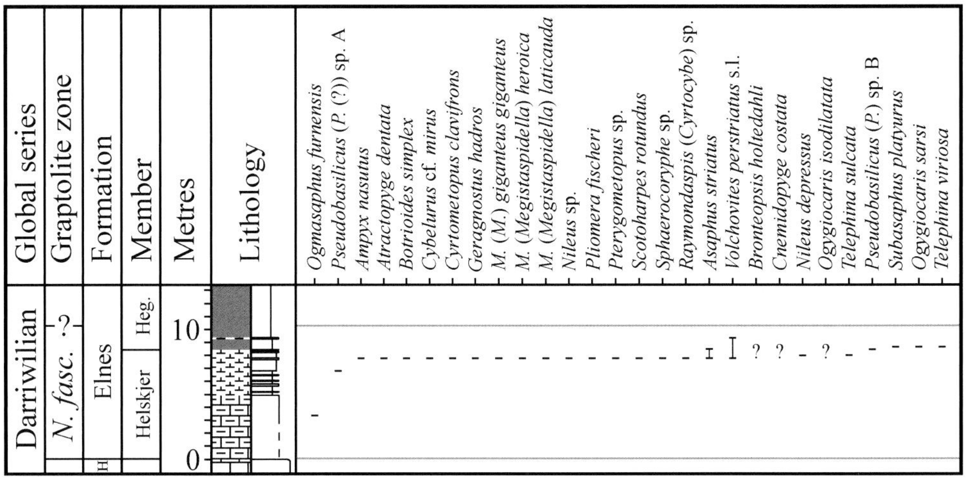

Fig. 37. Range chart for profile at Furnes Church, Mjøsa. Figure based on fossil material sampled by Wandås (1984).

last appearence of *Ogygiocaris striolata corrugata*, a level which is found slightly below a more or less distinct shell-lag in the Oslo-Asker and Eiker-Sandsvær districts. A corresponding layer rich in the trilobites *Ogygiocaris dilatata*, *Pseudomegalaspis patagiata* and the occasional raphiophorid crops out at Rønnings-fossen (grid reference: NM 358 035) in the Eiker

Sandsvær district (unpublished data from D.L. Bruton and A.W. Owen). Line 9 (Fig. 38) represents the first appearance of *Ogygiocaris striolata striolata* and is followed by line 10 representing a short interval in which there is a large increase in the number of *Nileus* specimens in the Oslo-Asker and Eiker-Sandsvær districts.

Fig. 38. Stratigraphical correlation between various lithological logs covering the whole Oslo Region and a single locality from Andersön in central Sweden. 1, Top massive limestones with endoceratid cephalopods. 2, Base *Megistaspis (Megistaspidella) giganteus giganteus* trilobite Zone. 3, Base *M. (M.) maximus* trilobite Zone. 4, Base *Pterograptus elegans* graptolite Zone. 5, Base *Pseudamplexograptus distichus* graptolite Zone. 6, Base of the first thick limestone bed above the black shales of the Sjøstrand Member. 7, Last appearence of *Ogygiocaris striolata corrugata*. 8, Shelly layer rich in the trilobites *Ogygiocaris dilatata*, *Pseudomegalaspis patagiata* and some raphiophorids. 9, First appearance of *Ogygiocaris striolata striolata*. 10, Maximum in short but high frequency of the trilobite *Nileus*. 11, Inferred base of *Protopanderodus graeai* conodont Zone (Rasmussen 2001). 12, Inferred base of *Eoplacognathus reclinatus* conodont Zone (Rasmussen 2001). 13, First appearance of *Pseudomegalaspis patagiata* as based on Karis (1982) and this study.

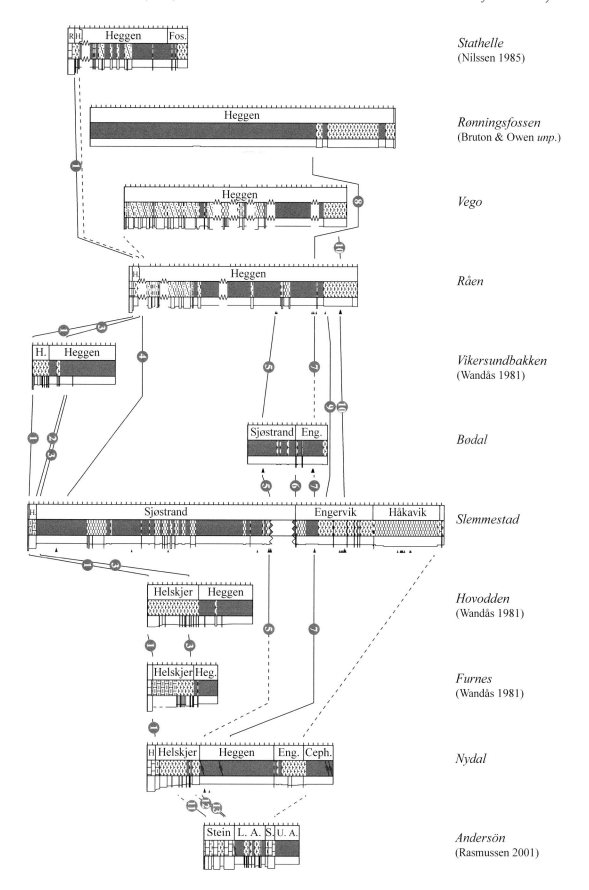

The above observations clearly suggest that the upper boundary of the Helskjer Member in the Mjøsa District correlates to somewhere in the Helskjer Member south of this district and, perhaps, to the basal Heggen Member in the Modum district. The base of the southern Heggen Member corresponds to somewhere in the upper half of the Helskjer Member in the Oslo-Asker district as indicated by the occurrence of *Pseudobasilicus* (*Pseudobasilicus*) *truncatus* n. sp. It may also equate to some level in the lower half of the Helskjer Member. The base of the northern Heggen Member correlates more or less with the base of the Sjøstrand Member in the Oslo-Asker district as indicated by the megistaspids. Higher up in the formation the base of the Engervik Member of the central Oslo Region corresponds to somewhere in the upper half of the Heggen Member to the north and south, whereas the basal Håkavik may correlate to somewhere around the top of the Elnes Formation in the Eiker-Sandsvær and Modum districts to the southwest and possibly the lower boundary of the Engervik Member in the Mjøsa district. The base of the Vollen Formation of the central Oslo Region seems more or less identical to the base of the northern 'Cephalopod Shale', where the first appearance of *Pseudasaphus limatus* and *Ogmasaphus kiaeri* occurs.

If these correlations are followed to the Andersö Shale formation of the central Swedish Jämtland area, it appears that the lower part of the Andersö Shale formation corresponds to the upper half of the Elnes Formation as indicated by the early appearance of *Lonchodomas rostratus* and *Robergia sparsa* (Pålsson *et al.* 2002). Observations by Karis (1982) of *Pseudomegalaspis patagiata* in the uppermost 2 m of the Stein Formation, if true, confirm this interpretation as this would suggest a correlation between the top Stein Formation and the uppermost Sjøstrand to middle Engervik Member of the Oslo-Asker district. The middle Ståltorp Limestone member of the Andersö Shale formation corresponds more or less to the Engervik Member of the Mjøsa district, with the upper boundary correlating to somewhere around the top of the Håkavik Member as indicated by the appearance of *Ampyxoides minor* and *Paraceraurus* sp. (Pålsson *et al.* 2002). Correlation can also be attempted in relation to the Killeröd Formation of Scania, where the change from *Botrioides efflorescens* to the succeeding *B. foveolatus* occurs somewhere between beds J and N (*sensu* Månsson 1995) of the uppermost part of the formation. This corresponds with the lower middle Engervik Member at Slemmestad, making it a particularly interesting correlation as the limestone rich Killeröd Formation is succeeded by dark mudstones and shales instead of shallow water limstones and marls or coarser grained

siliciclastic deposits. The difference in the lithological development could be related to the orogenic uplift of the southern and central Oslo Region in connection with the obduction of a island arc (see Sturesson *et al.* 2005), but another explanation is that it reflects the possible presence of an unconformity or condensed section at the very top of the Killeröd Fm. as the deepening of the Swedish basin in general seems to have occured somewhat later around the beginning of the Late Ordovician.

In short the trilobite biostratigraphic and lithostratigraphic correlations tie the top of the Håkavik Member to the top of the Engervik Member in the Mjøsa district and the corresponding top Ståltorp member in Jämtland, Sweden. The latter has been found to equate with the base of the *Nemagraptus gracilis* Zone and thereby the base of the Upper Ordovician (Pålsson *et al.* 2002). This result agrees well with recent studies on chitinozoans, which suggests that the lithostratigraphical boundary between the Håkavik Member and the overlying Vollen Formation in the Oslo area should correlate to the very top of the Darriwilian (Grahn & Nõlvak 2007).

Palaeoecology

Before any ecological analyses are carried out, transport sorting and sampling bias should be discussed as these factors may have strong influence on the data set. The fossil material was sampled directly from the rocks in the field and it is not possible to estimate even tentatively how many specimens, especially the smaller ones, have avoided detection or become destroyed during the process. A large part of the small calcareous fossil assemblage such as ostracodes and the tiny trilobite growth stages have furthermore been destroyed by later diagenesis adding to the generel bias towards an overrepresentation of larger specimens. This problem though is the same for all the localities and may therefore largely be neglected as long as one is aware that the results do not represent a true life and death assemblage. A similar problem is seen for the graptolites in the Eiker-Sandsvær and Mjøsa district, where the resulting ratio between organic walled and shelly fossils is somewhat distorted, making direct comparisons between the different areas of the Oslo Region problematic with regard to the faunal composition.

With regard to the question whether the fossils are *in situ* or have been transported before deposition, it is quite appearent that there has been little or no transport in the middle part of the formation, whereas the basal and upper part shows that episodic current deposition has taken place due to storm

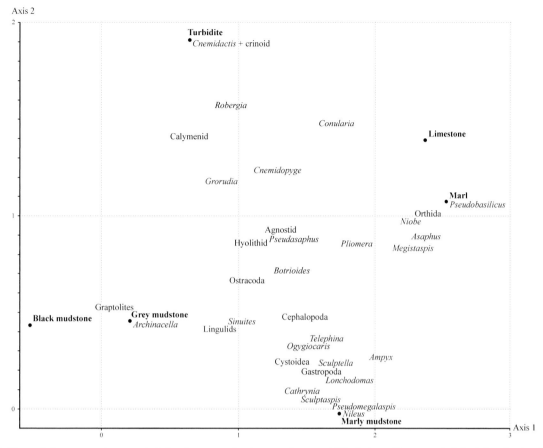

Fig. 39. Detrended correspondence analysis for the connection between fauna and lithofacies at Slemmestad, Oslo-Asker district. The first axis reflects carbonate content/water depth, while the second axis represents stress at the sea-bottom in the form of turbidite frequency.

activity. Even so the general impression is that the fossils found in these shell-lags are locally derived as the faunal composition is more or less the same as that in the fair-weather deposits, albeit with a slightly increased ratio of larger and heavier shell fragments. Thus the palaeo-transport may be ignored.

The palaeoecological facies preferences of the trilobite genera and associated faunal groups encountered in the Elnes Formation have been investigated using multivariable Correspondence and Detrended Correspondence analysis (Figs 39, 40). The resulting plot for the fauna at Slemmestad has a cross-like outline with a general deepening of the basin towards left side of diagram as indicated by the general change from limestones and marls towards the right side of diagram and black to dark grey mudstones to the left (Fig. 39). The second axis represents an upward increase in the benthic stress level with a change from muddy sediments formed in a fairly quiet environment just above storm wave base, to turbidite-dominated deposits formed during periods with frequent disturbance of fine-grained turbidite currents. Most trilobites are found in rocks deposited at or just above storm wave base characterized by marly mudstone with calcareous concretions. The faunal diversity decreases gradually towards the deeper, more dysoxic siliciclastic mudstone environment, the turbidite-dominated environments and towards shallower waters, characterized by limestones deposited close to fair weather wave base. No clear clusters are found due to the rather gradual transition found between the various lithofacies, but *Ampyx mammilatus*, *Lonchodomas rostratus*, *Nileus armadillo*, *Pseudomegalaspis patagiata*, *Sculptaspis* spp., *Sculptella* spp., *Telephina* spp. and to some degree *Ogygiocaris* spp. seem to have a strong preference for marly mudstone facies formed just above storm wave base but with relatively quiet and stable bottom conditions. They may form one distinct trilobite association. The trilobites *Robergia sparsa*, *Cnemidopyge* and some calymenids together with other faunal representatives such as echinoderms and the brachiopod *Alwynella* appear to have been rather tolerant to the effects of turbidites. In the case of *Robergia* this is clearly due to its pelagic life-strategy, and the extremely rare calymenids may simply be

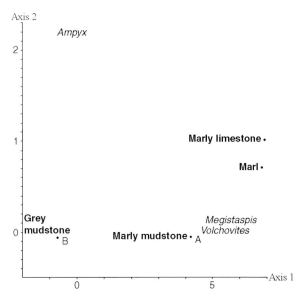

Fig. 40. Correspondence analysis for the relationship between lithofacies type and trilobite fauna at Nydal, Mjøsa district. The limestone facies plots far above the rest with a value of 18.13 on axis 2 and has for clarity been omitted from the diagram. A. *Asaphus striatus, Atractopyge dentata, Illaenus aduncus, Metopolichas* cf. *celorrhin, Niobe (Niobe) frontalis, Pseudasaphus limatus, Pseudobasilicus* sp. and *Turgicephalus* sp. B. *Cnemidopyge costata, Lonchodomas rostratus, Nileus armadillo, Ogygiocaris* spp. *Pseudomegalaspis patagiata, Robergia sparsa, Sculptaspis erratica, Telephina* spp. and *Ogmasaphus kiaeri.*

plotting here due to randomness rather than a true preference. *Asaphus* spp., *Megistaspis (Megistaspidella)* spp., *Niobe (Niobe) frontalis* and *Volchovites perstriatus* form a second trilobite association strongly linked to the marlier and more calcareous facies deposited between storm wave base and fair weather wave base. All trilobites become extremely rare outside the storm wave base, with the exception of *Ogygiocaris isodilatata* n. sp. and some few raphiophorids, trinucleids, remopleuridids and telephinids. The trilobite associations found at Nydal in the Mjøsa district are much more distinctive, clustering close to either the marly mudstone deposits with limestone beds or the transitional lithofacies straddling the border between the marly mudstones and the dark grey, siliciclastic mudstones (Fig. 40). The first association corresponds with the marl-associated trilobite association at Slemmestad, being dominated by *Asaphus striatus, Atractopyge dentata, Megistaspis (Megistaspidella)* spp. and *Volchovites perstriatus*, but also including species such as *Illaenus aduncus, Metopolichas celorrhin, Niobe (Niobe) frontalis, Pseudasaphus limatus* and *Raymondaspis (Cyrtocybe)* sp. The second association found in connection with the transitional lithofacies (Fig. 40) is largely identical with the marly mudstone facies association of Slemmestad, includ-

ing taxa such as *Cnemidopyge costata, Lonchodomas rostratus, Nileus armadillo, Ogygiocaris* spp., *Pseudomegalaspis patagiata* and *Sculptaspis erratica*. Again the limestones have not yelded enough trilobite specimens to define any trilobite associations for the more shallow water facies. The faunal associations at the southern Råen locality, largely correspond with the patterns found at the central and northern localities with their two main clusters around the marl and marly mudstone facies. The results for all the localities are summarized in Figure 41. The main picture for all the localities is that the trilobites are mainly found in connection with the marly mudstone deposits formed in fairly quiet and well-oxygenated waters just above the storm wave base, but well below fair weather wave base. The frequency and diversity decreases strongly below storm wave base, in turbidite-dominated environments and in the limestones and highly calcareous marls deposited well above storm wave base.

Palaeobiogeography

The trilobite fauna of the Oslo Region is dominantly endemic containing very few genera with affinities to other palaeocontinents. Of the more than 90 species identified from the Elnes Formation, only the four species, *Primaspis multispinosa, Porterfieldia humilis*, Elviniid sp. and *Cybelurus* cf. *mirus*, all occurring in the basal to middle part of the formation, seem to have an origin or affinity outside Baltica. The first two genera have a true cosmopolitan distribution, while *Cybelurus* cf. *mirus* is related to a species in the Laurentian fauna on Spitsbergen (Table 1). Elviniid sp. may be cosmopolitan like the first two as it is closely related to the hemipelagic and cosmopolitan species of the genus *Carolinites* discussed by McCormick & Fortey (1999), but is presently only known from a single specimen. The few 'foreign' trilobites clearly indicates a continued isolation of Baltica in the late Darriwilian to earliest Sandbian, although the first exchange of pelagic or depth-tolerating trilobites from Laurentia had already begun, probably in connection with the Darriwilian docking of a volcanic arc along the Caledonian margin (Schovsbo 2003; Sturesson *et al.* 2005). The primarily Laurentian affinity agrees well with the faunal composition found for the early to mid-Darriwilian Otta Conglomerate of mid-Norway, formed in an island arc setting (Bruton & Harper 1981). Avalonia was clearly far removed from Baltica at this time and furthermore, not lying in the path of an oceanic current moving towards Baltica, which might otherwise have transported some Avalonian taxa to Baltica. Another

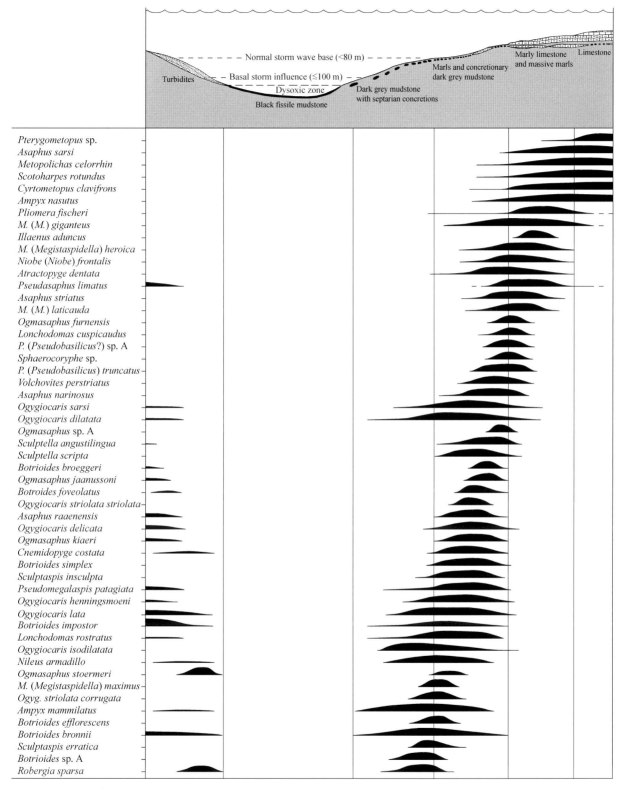

Fig. 41. Summary of the facies relations of selected trilobites from the Elnes and uppermost Huk formations.

interesting aspect with the four species listed above is their complete restriction to the northern Mjøsa district and central Swedish Jämtland area (*Porterfieldia*), even though the southern to central Oslo Region seems more influenced by the associated obduction and uplift of the land area as indicated by the absence of a late Middle to early Late Ordovician deepening of the basin seen for the rest of Baltoscandia.

Table 1. Geographical distribution of the trilobite taxa in the Elnes Formation

Genus	Species	Skien-Langesund	Eiker-Sandsvær	Modum	Oslo-Asker	Ringerike	Hadeland	Mjøsa	Jämtland	SW Scandinavia	East-Baltic/Russia	Outside Baltica
Pseudobasilicus	*truncatus*	X		X	X							
Asaphus	*raaenensis*	X		X	X							
Botrioides	*broeggeri*	X	X	X	X	X						
Ogygiocaris	*striolata striolata*	X	X	X	X	X						
Asaphus	*narinosus*	X	X	X	X		X					
Sculptella	*scripta*	X	X	X	X		X					
Ogmasaphus	*stoermeri*	X	X	X		X	X	X				
Ogygiocaris	*dilatata*	X	X	X		X	X	X				
Megistaspis (Megistaspidella)	*laticauda*	X	X	X	X	X	X	X				
Megistaspis (Megistaspidella)	*giganteus runcinatus*	X	X		X	X	X	X				
Megistaspis (Megistaspidella)	*giganteus giganteus*	X			X	X	X	X				
Megistaspis (Megistaspidella)	*maximus*	X		X	X	X	X	X				
Ogmasaphus	*kiaeri*	X			X		X	X				
Botrioides	*bronnii*	X			X	X	X	X				
Ogygiocaris	*sarsi*	X		X	X	X		X	X			
Ogygiocaris	*lata*	X		X	X			X	X			
Ampyx	*mammilatus*	X			X		X	X	X			
Niobe	*frontalis*	X	X	X	X		X	X		X		
Scotoharpes	*rotundus*	X	X		X		X	X		X		
Botrioides	*foveolatus*	X	X		X					X		
Pseudasaphus	*limatus limatus*	X	X		X	X	X			X	X	
Cyrtometopus	*clavifrons*	X	X		X			X		X	X	
Asaphus	*striatus*	X	X	X	X	X	X	X			X	
Lonchodomas	*rostratus*	X	X		X	X	X	X		X	X	
Megistaspis (Megistaspidella)	*heroica*	X	X	X	X		X	X	X	X	X	
Metopolichas	*celorrhin*	X		X				X		X	X	
Ampyx	sp. A		X									
Gravicalymene	*capitovata*		X									
Sculptella	sp. A		X									
Pseudasaphus	sp. A		X	X								
Sthenarocalymene	*lirella*		X		X							
Trinucleid gen. et sp.indet.			X				X					
Geragnostus	*hadros*		X	X				X				
Lonchodomas	*cuspicaudus*		X		X			X				
Ogygiocaris	*isodilatata*		X		X		X	X				
Ogygiocaris	*striolata corrugata*		X	X				X				
Atractopyge	*dentata*		X		X			X				
Sculptaspis	*erratica*		X		X	X		X				
Ogygiocaris	*henningsmoeni*		X		X		X	X				
Volchovites	*perstriatus s.l.*			X	X		X	X	X			
Botrioides	*impostor*			X	X	X	X			X		
Pseudomegalaspis	*patagiata*		X		X	X			X	X		
Nileus	*armadillo*		X		X		X		X	X		
Botrioides	*efflorescens*		X		X				X	X		
Asaphus	*sarsi*			X	X			X		X		
Asaphus	cf. *striatus*		X			X					X	
Subasaphus	*platyurus*							X	X	X	X	
Megistaspis (Megistaspidella)	*heroica s.l.*			X							X	
Botrioides	sp. A			X								
Ampyx	*nasutus*			X	X	X	X	X	X	X	X	
Pliomera	*fischeri*			X	X		X	X		X	X	

Table 1. Continued.

Genus	Species	Skien-Langesund	Eiker-Sandsvær	Modum	Oslo-Asker	Ringerike	Hadeland	Mjosa	Jämtland	SW Scandinavia	East-Baltic/Russia	Outside Baltica
Botrioides	simplex			x				x				
Cnemidopyge	sp. A				x							
Ogmasaphus	jaanussoni				x							
Pseudasaphus	limatus longistriatus				x							
Pseudasaphus	sp. B				x							
Sculptaspis	insculpta				x							
Sculptaspis	sp. A				x							
Sculptaspis	sp. B				x							
Ogmasaphus	sp. A				x		x					
Ogmasaphus	multistriatus				x		x	x				
Ogygiocaris	delicata				x	x		x				
Pseudobasilicus	sp. A				x		x	x				
Sculptella	angustilingua				x			x				
Telephina	norvegica				x	x	x	x				
Telephina	invisitata				x			x				
Robergia	sparsa				x			x	x			
Telephina	bicuspis				x			x	x			
Telephina	mobergi				x			x	x			
Cnemidopyge	costata				x		x	x	x			
Pseudomegalaspis	cf. formosa				?					?		
Telephina	intermedia					x		x	x			
Asaphus	sp. A						x	x				
Pseudobasilicus	sp.						x					
Telephina	granulata						x	x				
Pseudobasilicus	angustatus						x	x				
Illaenus	aduncus						x	x		x		
Bronteopsis	holtedahli						?	x				
Asaphus	cf. ludibundus							x				
Botrioides	sp. B							x				
Cybellela	sp.							x				
Dionide	jemtlandica							x				
Iceloborgia?	sp.							x				
Ogmasaphus	furnesensis							x				
Pseudobasilicus	sp. B							x				
Pterygometopus	sp.							x				
Sphaerocoryphe	sp.							x				
Telephina	skjesethi							x				
Telephina	sulcata							x				
Telephina	viriosa							x				
Raymondaspis (Turgicephalus)	sp.							x				
Nileus	depressus							x				
Robergia	microphthalma							x	x	x		('C)
Porterfieldia	humilis							x	x		x	('C)
Primaspis	multispinosa							x				S+L
Cybelurus	cf. mirus							x				(L+G)
Elviniid	sp.							x				

S, Spitzbergen; L, Laurentia; G, Gondwana; C, cosmopolitan.
Brackets are used in cases where the closest relative to the species but not the species itself is found outside Baltica.

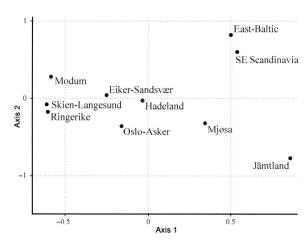

Fig. 42. Principal coordinates analysis with Chord similarity index for the relation between trilobite taxa and their spatial distribution within and outside the Oslo Region. Only trilobite taxa occurring in more than one district have been included. Note the distinct pattern with southern districts plotting to the right and northern districts plotting together with the regions from outside Norway in the left side of the scatter diagram. Ringerike plots far from its real location in relation to the other areas due to the relatively poorer knowledge of the trilobite species occurring in this district.

This, in combination with the strongly varying stratigraphic levels and facies types in which they have been found, suggests that there may have been a direct connection to the Iapetus Ocean in the north. Otherwise one would expect them to be more frequent in the southern Oslo Region with its closer location to the open shelf sea. Consequently the western land area should be regarded as an island chain rather than one unbroken land area shielding the main Baltoscandian platform from the Iapetus Ocean and the approaching island arc.

The biogeographical distribution of the trilobite species has been investigated using Correspondence, Detrended Correspondence, non-metric MDS and Principal Coordinates analysis (Gower, Euclidean and Chord similarity index). The results show only slight differences (Fig. 42). Generally, the individual districts of the Oslo Region plot from south to north (i.e. from left to right on the scatter diagram). This reflects a general change in species with species such as *Ogygiocaris dilatata*, *Pseudobasilicus (Pseudobasilicus) truncatus*, *Botrioides broeggeri* and *Ogmasaphus stoermeri* occuring in the south, while others, such as *Atractopyge dentata*, *Volchovites perstriatus*, *Illaenus aduncus* and *Megistaspis (Megistaspidella) heroica* mainly occur in the north (Table 1). The Ringerike district plots far from its expected location, but this is probably because of poor data. The reason why Eiker-Sandsvær and Modum have changed places in relation to their real geographical association is related to the nearly identical and relatively low diverse

deeper water fauna observed for the Helskjer Member at Eiker-Sandsvær and Slemmestad, whereas Modum is characterized by a more diverse, shallow-water fauna in this part of the formation.

The central Swedish Jämtland area, the south-east Scandinavian area and the East-Baltic area, all plot to the right of the Norwegian districts. The reason for the rather isolated location of the Jämtland area relates to the lack of knowledge of the trilobites occurring in the upper Stein Formation superceeding the Andersö Shale formation. As the uppermost Stein Formation is contemporaneous with the Helskjer and lower Sjøstrand and Heggen members this means a large number of trilobite species from the basal part of the Elnes Formation are excluded from the data for Jämtland. It would be expected to plot much closer to the Mjøsa district than the other two eastern regions. The plot for the three regions outside Norway is a clear indication of a stronger affinity between these and the northern Oslo Region than with the southern and central Oslo Region. This is interesting because south-estern Scandinavia actually lies closer to the southern Oslo Region, where one might expect a closer affinity. The reason for the larger affinity with the northern trilobite associations is to be found in the topographical outline of south-western Scandinavia in Middle Ordovician time, where the Oslo Region was situated on the western margin of a small, southward deepening basin opening out into the western Iapetus Ocean and Tornquist Sea south of the Oslo Region. The Oslo Region was thus partly cut off from the main Baltoscandian platform by deeper and partly dysoxic waters with the only shallow water connection located to the north-east (Fig. 43).

Systematic Palaeontology

The taxonomic revision is mainly based on material stored at the Natural History Museum (Geological Section), University of Oslo, with collection numbers following that of the former Palaeontological Museum (PMO). A few specimens housed at the Department of Geology, University of Lund (LO), Sweden and at the Geological Museum of Copenhagen (MGUH) are included. Type specimens are also quoted from Stockholm Riksmuseum (RM), the Geological Survey of Sweden (SGU) and from the Geological Museum in Uppsala (UM), all Swedish.

The denotation of 'hypotypoid' used for some raphiophorid specimens follows the definition of Størmer (1940) in which it was given to specimens belonging to the original material available when the species in question was described and which includes the original syntype material.

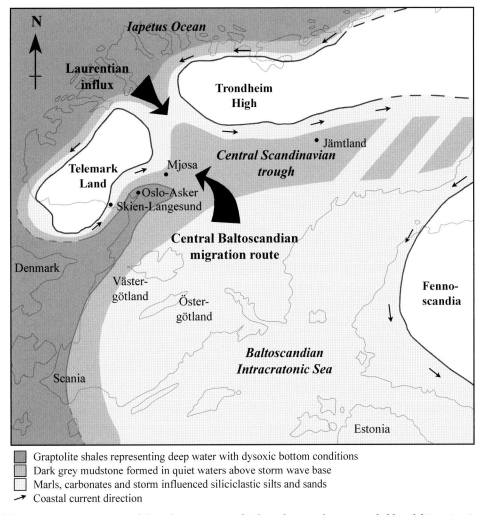

Fig. 43. Model for the north-western part of the palaeocontinent of Baltica showing the most probable trilobite migration routes to the Oslo Region (large arrows). The Caledonian dislocation of the deposits along the continental margin is ignored in the present model. The facies belts are based on data presented by Männil (1966), Nielsen (1995) and Pålsson *et al.* (2002).

Terminology

The terminology used (Fig. 44) is with a few exceptions adopted from the one presented by Whittington & Kelley *in* Kaesler (1997). The trinucleid descriptions follow the terminology of Hughes *et al.* (1975), while the terminology of the agnostid muscle insertion areas follow that of Fortey (1980, Fig. 4). If nothing else is stated the use of length and width follows the guidelines of Whittington & Kelley *in* Kaesler (1997), meaning that the length will be in the sagittal or exsagittal plane, while width is in the transverse plane. Where possibly, hypostomes are described from the ventral side.

Subspecies are here defined as taxa, which are clearly distinguished from the type by a few minor features and which are completely isolated from other subspecies of the same species by time and/or space. This means that if two former subspecies are found together in the same beds, they will either be regarded as separate species or a single species depending on the degree of morphological difference. Local morphological differences observed at a given stratigraphical level in Baltoscandia will not be regarded as more than ecophenotypic variation.

Additional descriptive terms

Ogive. – Term for the V-shaped anterior point of the cranidium, where the facial sutures connects. Synonymous with the anterior beak.

Paradoublural furrow. – Furrow demarcating paradoublural line (Nielsen 1995).

Snout. – Term for the slender, and strongly elongated distal part of the frontal cranidial area seen on some megistaspids (Bohlin 1960).

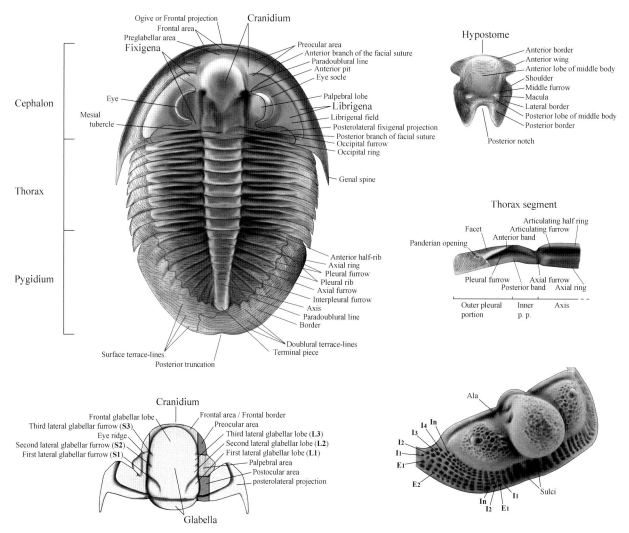

Fig. 44. Trilobite terminology used in the taxonomic descriptions.

Measurements

Unless stated otherwise, all measurements are maximum values for a given distance between two points. No 'life position' measurements have been taken. Some selected measurements are presented in the tables.

Order Asaphida Salter, 1864

Superfamily Asaphoidea Burmeister, 1843

Family Asaphidae Burmeister, 1843

Subfamily Ogygiocaridinae Raymond, 1937

Genus *Ogygiocaris* Angelin, 1854

Type species. – *Trilobus dilatatus* Brünnich, 1781.

Diagnosis. – A genus of Ogygiocaridinae characterized by a facial suture intramarginal in front of glabella; frontal area generally much wider than glabella; occipital ring with distinct, forward pointed V-shaped furrow, which meets the occipital furrow just behind minute mesial tubercle; posterior branch of facial suture distinctly sigmoid resulting in a clearly forward expanded, but relatively short (exsag.) abaxial part of posterior fixigenal projection. Eyes posteriorly, located approximately half their own length in front of posterior cephalic margin and nearly their own length behind front of glabella. Hypostome broadly to elongated ovoid with fairly narrow anterior wings and pointed posterior margin. Thorax with no panderian openings or protuberances; pleural furrows stopping or becoming indistinct 70 to 80% the pleural width from the axial furrow. Pygidium semicircular in outline; pygidial axis occupying between 15 and 25% of pygidial width; pleural ribs well developed, generally 8 to 9 in number; interpleural furrows indistinct or at most moderately developed on abaxial part of pleural rib; inner margin of pygidial doublure and border distinctly wavy (emended from Henningsmoen 1960; Moore 1959).

Remarks. – An extended discussion on the early nomenclatorial history of the genus was given by Henningsmoen (1960) who limited the genus to the Scandinavian taxa. Burskij (1970, pp. 123–124, plate 15, fig. 2, plate 16, fig. 7) in his monograph on Ordovician trilobites from Novaya-Zemlya described a taxon from the middle Lower Ordovician (*Tetragraptus approximatus* Zone), which he termed *Ogygiocaris* aff. *sarsi*. Although the pygidium and thorax resemble the Scandinavian *Ogygiocaris* species the taxon clearly differs in its cephalon, which is characterized by an extremly slender and forward narrowing glabella, large eyes situated extremely close to the glabella and by the deviating outline of its facial suture with its strong forward curve on the posterolateral branch and the angular and adaxial outline of the anterior branch. Though clearly related it should not be assigned to *Ogygiocaris* and neither should the other species tentatively assigned *Ogygiocaris* by Burskij (1970).

Recently Waisfeld & Vaccari (2006) argued for the inclusion of the two genera *Ogygiocarella* Harrington & Leanza and *Araiocaris* Přibyl & Vaněk in *Ogygiocaris*. This was done on the basis of a redescription of *Araiocaris araiorhachis* Harrington & Leanza, 1957 from Argentina, which they found to be characterized by a wavy inner doublural margin on the pygidium similar to that of *Ogygiocaris*, thereby removing one of the major differentiating characters separating the two. The suggested merging of the genera is not supported by the present author for although a close relationship between the three genera is obvious and could at some stage be expressed at subgenus level, the species of both *Araiocaris* and *Ogygiocarella* are still readily separated from the very close-knit group of Scandinavian species belonging to *Ogygiocaris sensu stricto* as follows: (1) Both *Ogygiocarella* and *Araiocaris* are characterized by long (exsag.) postero-lateral fixigenal projections and correspondingly forwardly located eyes, situated close to their own length in front of the posterior cephalic margin, thereby giving them a distinctly *Niobe*-like cranidial appearance. (2) With the exception of *Ogygiocaris striolata* and *O. dilatata*, the two most 'primitive' species of *Ogygiocaris*, the width of the frontal glabellar area on Scandinavian *Ogygiocaris* is markedly wider (tr.) than the glabella (Wglabella/Wfga ≤ 3/5). (3) The hypostome of *Araiocaris* has a very much broader anterior lobe of its middle body, the width corresponding to twice the width described by the maculae. Possible other differentiating characters may be found in the hypostonal outline, but this is too poorly exposed on the specimens figured by Waisfeld & Vaccari (2006) for any sound comparisons. (4) Pygidial axis corresponding to approximately 15% of pygidial width or less on *Araiocaris* and *Ogygiocarella*, while corresponding to around 20% on *Ogygiocaris*. (5) *Ogygiocarella* differs from both *Ogygiocaris* and *Araiocaris* by having 11 to 13 extremely well-defined pleural ribs on the pygidium (see Hughes 1979) and very narrow doublural lines, but corresponds especially with *Ogygiocaris striolata* in the general outline of the hypostome and anterior part of the facial suture and by the configuration and distinctiveness of the lateral glabellar furrows.

Ogygiocaris was revised in detail by Henningsmoen (1960) in connection with his description of the asaphids found in the Middle Ordovician deposits of southern Norway. Based on old PMO material together with new material supplying the biostratigraphic data he showed the genus to contain the following three species and six subspecies:

O. dilatata (Brünnich, 1781)
O. striolata striolata Henningsmoen, 1960
O. striolata corrugata Henningsmoen, 1960
O. sarsi sarsi Angelin, 1878
O. sarsi delicata Henningsmoen, 1960
O. sarsi lata Hadding, 1913

The extensive sampling in the present study has shown that the 'subspecies' of the *sarsi* group co-occur at several stratigraphical levels. This clearly indicates they should be regarded as separate species rather than subspecies as originally proposed by Henningsmoen (1960). This is further supported by the fact that only few if any morphological intermediates have been found even though all three Norwegian members of the *sarsi* group (e.g. *O. sarsi*, *O. regina* and *O. delicata*) show large stratigraphical overlaps. A possible intermediate may be found in a librigena lying together with paratype pygidium PMO 36675 for *O. delicata*. In this example the paradoublural line is located rather close to the eye for an *O. delicata*, but not quite close enough for the typical *O. lata* (see Table 14). The specimens with intermediate features are though as mentioned above rather few and most probably reflect that we are looking at separate skeletal parts instead of a complete specimen, why natural differences seen within a normal population gets much more importance than what it should have. The two subspecies of *O. striolata*, on the other hand, are completely separated stratigraphically

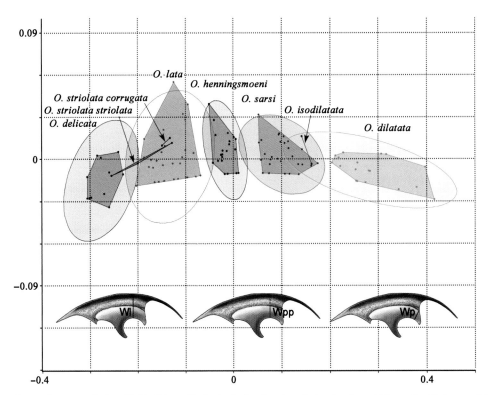

Fig. 45. Principal coordinates analysis with Euclidean Similarity Index for the relative border width on the librigena of *Ogygiocaris* as expressed by the two ratios Wp/Wl and Wpp/Wl. The 95% confidence intervals for each of the eight taxa are shown. The librigenal border width is one of the most important characters when differentiating between *O. delicata*, *O. lata*, *O. henningsmoeni* n. sp. and *O. sarsi*. A similar picture is seen for the cranidium.

with *O. striolata corrugata*, which is also the most *sarsi* like, being the oldest. When this is added to their close morphological similarity it seems most correct to keep them on the subspecies level.

Ogygiocaris regina and *O. lata* were classified as full species by Månsson (2000b; see also Pålsson *et al.* 2002), although no explanations were given. The Jämtland collections stored at the Department of Geology, University of Lund, were examined in the present study in order to verify whether *O. regina* and *O. lata* are discrete taxa as thought by earlier authors. The investigation showed the Norwegian *O. regina* to be conspecific with *O. lata* on strength of the complete match between the by Henningsmoen (1960) selected lectotype librigena for *O. lata* and the librigena of the Norwegian taxon. This result was supported by several other specimens in the Lund collection clearly conspecific with the Norwegian *O. regina*. The majority of the Swedish *Ogygiocaris* material, though, including the juvenile pygidium (LO 2531 T) assigned to *O. lata* by Hadding (1913) belongs to *O. henningsmoeni* n. sp. described herein.

One of the main differentiating characters between the various taxa of the *O. sarsi* group of Henningsmoen (1960) was the relative border width on the cephalon. This character was among other features examined in the present revision of the genus, which

lead to the realization that the genus included two additional species in excess of those previously described by Henningsmoen (1960), namely *O. isodilatata* n. sp. and *O. henningsmoeni* n. sp., both of which are described below. A biometric plot showing the clustering of the specimens due to the border width is presented in Figure 45.

Phylogeny. – A cladistic analysis of the *Ogygiocaris* group (i.e. *Ogygiocaris sensu stricto*, *Araiocaris* and *Ogygiocarella*, including the by Hughes *et al.* (1980) treated Peruvian *O.* cf. *debuchii* (Brongniart in Brongniart & Desmarest (1822)) has been carried out using the diagnostic features listed in Table 2 and, with regard to the measureable characters, illustrated in Figure 46. The Burskij (1970) material from Novaya-Zemlya discussed above was not included due to lack of data for the data matrix. The branch and bound analysis with Wagner optimization of the data matrix yielded a single most parsimonious tree with a length of 56 (consistency index = 0.5) (Fig. 47). *Niobe* (*Niobella*) *imparilimbata* Bohlin, 1955, was used as an out-group as well as the more closely related *Ogyginus corndensis* (Murchison, 1839), which was redescribed by Hughes (1979). The resulting tree suggests a monophyletic origin for the *Ogygiocaris* group, but also clearly indicates that the Scandinavian

Table 2. Diagnostic features and character states defining the different species of *Ogygiocaris* and the related species *Ogygiocarella debuchii* (Brongniart)

		Niobe (Niobella) imparilimbata	*Ogyginus corndensis*	*Araiocaris araiorhachis*	*Ogygiocaris aff. sarsi (Russia)*	*Ogygiocarella cf. debuchii*	*Ogygiocarella angustissima*	*Ogygiocarella debuchii*	*Ogygiocaris striolata striolata*	*Ogygiocaris striolata corrugata*	*Ogygiocaris dilatata*	*Ogygiocaris isodilatata*	*Ogygiocaris sarsi*	*Ogygiocaris henningsmoeni*	*Ogygiocaris lata*	*Ogygiocaris delicata*
Cranidial features	Lb/L ≥ 1/3	–	–	–	–	–	–	–	–	–	+	+	+	–	–	–
	Lb/L ≥ 1/4	–	–	–	+	–	–	–	+	+	+	+	+	+	–	–
	Lb/L ≥ 1/5	–	–	–	+	–	–	–	+	+	+	+	+	+	+	–
	Lbp/L < 60%	+	–	–	–	–	–	–	+	–	+	+	+	–	–	–
	Lbp/L < 70%	+	+	–	+	–	–	+	+	+	+	+	+	+	+	–
	Wo/Wc > 50%	+	+	+	+	+	–	+	+	+	+	–	–	–	–	–
	Lp/Wc > 11%	+	–	+	+	–	–	+	–	–	+	+	+	+	+	+
	Minimum glabellar width ~ 40% of Wc	–	–	–	–	–	–	–	–	–	–	+	+	+	+	+
	Width of preocular area ~ 50% of Wo	–	–	–	–	–	–	–	–	–	–	+	+	+	+	+
	Facial suture marginal anteriorly	–	–	–	+	+	+	–	–	–	–	–	–	–	–	–
	Lf1/L ≥ 20%	+	+	+	+	–	+	+	–	–	+	+	–	–	+	+
	Frontal glabellar lobe with Bertillon pattern terrace–lines	+	–	–	–	+	+	+	+	–	–	–	–	–	–	–
	Eye ≤ 25% of glabellar length from posterior margin	–	–	–	–	–	–	–	+	+	+	+	+	+	+	+
	Niobe type facial suture	+	+	–	–	–	–	–	–	–	–	–	–	–	–	–
	Posterolat. fixigenal field distinctly narrowing adaxially	–	–	–	+	–	–	–	+	+	+	+	+	+	+	+
Librigena	Wp/Wl ≤ 0.34	+	+	–	+	–	–	–	–	–	+	+	+	–	–	–
	Wp/Wl ≤ 0.39	+	+	–	+	–	–	–	–	–	+	+	+	+	–	–
	Wp/Wl ≤ 0.51	+	+	–	+	+	–	+	+	+	+	+	+	+	+	–
	Wb1/Lp ≤ 0.27	–	+	+	–	+	+	+	–	–	–	–	–	–	+	+
Thorax	Waxis/W (thorax) ~ 1/6	–	–	–	+	–	+	+	–	–	–	–	–	–	–	–
Pygidium	Surface terrace-lines coarse	–	+	+	+	+	+	+	–	–	+	+	+	+	+	+
	Doublural terrace-lines coarse	+	+	+	+	+	+	+	–	+	+	+	+	+	+	+
	Waxis/Wpl ≤ 0.72	–	–	+	+	+	–	+	+	+	+	–	–	–	–	–
	Pleural ribs rounded in cross-section	+	+	–	+	–	–	–	+	+	+	+	–	–	–	–
	Interpleural furrows distinct	–	–	–	–	–	+	+	–	–	–	–	–	–	–	–
	Pleural furrows deep and narrow	–	–	–	–	–	+	+	–	–	–	–	–	–	–	–
	Lpa/L ≤ 1/5	–	+	+	–	+	+	+	–	–	–	–	–	–	–	+
	Curvy paradoublural line	–	–	+	+	+	+	+	+	+	+	+	+	+	+	+

Abbreviations used for the measurements are explained in Figure 46.

species should be regarded as a self-contained group. In other words the parsimony analysis clearly supports the conclusion that the genus *Ogygiocaris* should be limited to the Scandinavian species, which among other things are characterized by a strongly S-shaped curve on the posterior branch of the facial suture and by the location of the eyes equal to or less than 25% of the glabellar length from the posterior margin.

The phylogenetic relationship between the various species of *Ogygiocaris* appears to have been turned up side down in the tree with *O. delicata* placed as the most 'primitive' species of the lot. In reality many things such as the order of the stratigraphical first appearances; the very round hypostomes so reminiscent of *Ogygiocarella* and *Araiocaris* found on *O. striolata*, *O. dilatata* and *O. isodilatata* and the Bertillon-type terrace-lines on the frontal glabellar lobe, a cephalic

paradoublural line close to the margin and turning strongly inwards anteriorly at transition to cranidium, a relatively narrow preocular and preglabellar area and the comparatively distinct interpleural furrows on pygidium of both *Ogygiocarella* and *Ogygiocaris striolata* all suggest that *Ogygiocaris striolata* should be regarded as the one morphologically closest to a shared ancestor with *Ogygiocarella* and *Araioracis*, sharply followed by *O. dilatata* and *O. isodilatata*.

O. delicata is favoured as the one closest to *Ogygiocarella* by the marginal location of the paradoublural line on librigena; the narrow pygidial and thoracic axis; the relative length of the pygidial axis and by the flat pleural ribs and their number. Of those the number of pleural ribs is found to vary considerably within the species of *Ogygiocarella* and especially for *O. debuchii* (see Sheldon 1987). *Ogygiocaris* shows a

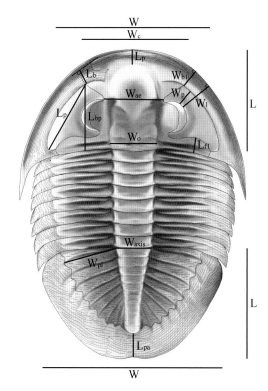

Fig. 46. Measurements on *Ogygiocaris* used in Table 2 and for the cladistic analysis shown in Figure 47.

wide range of cephalic and to a lesser degree pygidial border widths, indicating these characters may be more useful as species characters than on the generic level. This leaves the flat pleural ribs and the axial width as the two best supports for a least derived status of *O. delicata*. *O. dilatata* is mainly characterized by the narrow preocular area, but also shows a long range of similarities with *O. striolata*, the most important being the rounded pleural ribs. Based on characters alone the two species appear largely equal, but when turning to the biostratigraphy *O. dilatata* together with *O. sarsi* and *O. isodilatata* n. sp. turn out to be the stratigraphically oldest group. *Ogygiocaris delicata* and *O. lata* are only found in the middle to upper part of the formation, suggesting the *delicata* extreme with its wide preocular area to be the more derived form.

Ontogeny. – The material investigated for ontogenetically related characters is based on a size range from specimens of approximately 2 cm in length to specimens of around 15 cm of length. *Ogygiocaris* has a number of features that change significantly with growth of which the following cranidial characters are the most pronounced: the relative size of the eye is reduced with age, while the number of terrace-lines on the preglabellar area, the exsagittal length of the occipital ring and the distance from the abaxial end

of paradoublural line to the posterior margin decrease. On the pygidium both the relative pygidial and the axial lengths increase with size, whereas the number of terrace-lines on the doublure at pleural furrow 5 decreases. The most important diagnostic character for differentiation between the various species within the genus, the doublural width, which appears completely unrelated to size, at least for the size range investigated here.

Ogygiocaris dilatata (Brünnich, 1781)

Pl. 1, Figs 1–8; Tables 3–4

v 1781   *Trilobus dilatatus* Brünnich, pp. 393–394.

  1822   *Trilobites dilatatus* – Brongniart, p. 19.

  1822   *Asaphus debuchii* – Brongniart, p. 21 (*partim*) (= only material from Éger [Eiker]).

  1827   *Asaphus dilatatus* – Dalman, pp. 87–88 (272–273), pl. 3, fig. 1 (= lectotype).

  1828   *Asaphus dilatatus* – Dalman, pp. 67–68, pl. 3, fig. 1 (= lectotype).

  1835   *Asaphus dilatatus* Dalm. – Sars, pp. 336–337 (*partim*), *non* pl. 8, fig. 5a–c, pl. 9, fig. 11 (= *O. sarsi*).

  1838   *Tr. dilatatus* Brün. – Boeck *in* Keilhau, p. 141 (mentioned).

  1857   *Asaphus dilatatus* Dalm. – Dahll, p. 309 (listed).

  1857   *Asaphus dilatatus* Dalm. – Kjerulf, p. 285 (listed).

  1864   *O.* [= *Ogygia*] *dilatata*, Brünnich – Salter, pp. 127–128.

  1878   *Ogygiocaris dilatata*. Brünn. – Angelin, p. 92, pl. 42, figs 1a–1c, 2.

  1878   *Ogygiocaris dilatata*. var. Strömi. Ang. – Angelin, p. 96, pl. 42, fig. 2.

non 1878   *Ogygiocaris dilatata*. var. *sarsi*. Ang. – Angelin, p. 96, pl. 42, fig. 1.

non 1913   *Ogygiocaris dilatata* Brünn. var. *strömi* Ang. – Hadding, p. 72, pl. 7, fig. 10 (probably *Ogygiocaris sarsi regina*).

  1930   *Ogygiocaris dilatata*, Brünn. – Reed, pp. 310–311, 313, 315.

  1953   *Ogygiocaris dilatata* forma typica – Størmer, 119–120 (*partim*), pl. 1 only.

  1960   *Ogygiocaris dilatata* (Brünnich, 1781) – Henningsmoen, pp. 217–221, fig. 4, pl. 1, figs 1–7, pl. 2, figs 1–6.

Type stratum and type locality. – Heggen Member (?) of the Elnes Formation. Fossum Iron Works at Fossum, Skien-Langesund district. The lectotype from this locality was selected by Henningsmoen (1960, pp. 217–218) on the basis of the description

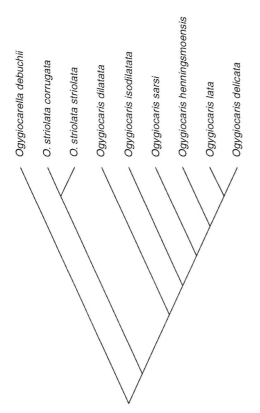

Fig. 47. Phylogenetic analysis showing a most parsimonious tree with length 60 using branch-and-bound algorithm and Wagner optimization. Characters and character states used for the analysis are listed in Table 2.

and figures given by Brünnich (1781). The lectotype, MGUH 3885, is presently deposited at the Geological Museum, University of Copenhagen, Denmark.

Material. – The collections at the NHM contain a large number of more or less fragmentary skeletal parts, mainly from the Eiker-Sandsvær district. A few complete or nearly complete specimens have been found (MGUH 3885, PMO 60436, 60450, 60492, 60495, 67032, 206.958, 206.961).

Description. – A very large species, largest cranidium nearly 8 cm long, largest measurable pygidium reaching 7.0 cm in length.

Cranidium looking somewhat short and broadly rectangular; completely dominated by wide glabella; glabellar width at occipital ring corresponding to

nearly 40% the cranidial width; length of eye less than 20% the cranidial length; anterolateral border width large, paradoublural line generally localized less than half the cranidial length from the posterior margin; maximal length of posterolateral fixigenal projection 20% cranidial length, while shortest length corresponds to between 14 and 18% of cranidial length.

Librigenal border broad, paradoublural line situated 13 to 24% the distance between eye and lateral margin from the eye; doublure with 15 to 17 terrace-lines.

Hypostome subrectangular, slightly longer than wide with narrow borders and broadly V-shaped terrace-lines on central body. Anterior end of terrace-lines turned outwards.

Thoracic segments with moderately long axis (sag.), corresponding to about 30% the axial width. Axis relatively wide, width between 50 and 60% the pleural width or approximately one-fourth the total width.

Pygidial margin with distinct posterior incurvation; pygidium relatively narrow with the length approximating 60 to 70% of the width; axis wide, maximal width between 19 and 24% the pygidial width or nearly as wide as maximal pleural width ($N = 12$; $\mu = 0.94$; min. = 0.71; max. = 1.03); border wide, corresponding to between 20 and 26% the pygidial width at pleural furrow 5 and around 25% at pleural furrow 8 and 9; maximal border width larger than both L4 and L5; pleural field with no more than 7 or 8 pleural ribs in addition to anterior half rib; terrace-lines generally openly spaced, 15 to 22 on doublure and 8 to 14 on border.

Remarks. – *Ogygiocaris dilatata* is mainly found in the Eiker-Sandsvær district, becoming increasingly rarer northwards. It seems to have preferred an oxic to slightly dysoxic environment characterized by fine-grained, but slightly marly sediments deposited well below fair weather wave base.

Occurrence. – Throughout the Elnes Formation except for the basal part of the Helskjer Member. Skien-Langesund: Bø north of Skien, Brevik Station. Eiker-Sandsvær: Krekling, Muggerudkleiva at Sandsvær, Ravalsjøelven, Rønningsfossen, Råen at Fiskum,

Table 3. Cranidial measurements on *Ogygiocaris dilatata* (Brünnich, 1781)

PMO number	L	Lg	Lp	Lo	Lt	Lf1	Lf2	Lae	Le	Lb	Lbp	Wb	W	Wc	Wo	Wgmin	Wae	Outer t.
60451	45.1	39.7	5.4	5.5	4.2	8.6	7.2	19.8	~6,8				~65,6	42.2	25.1	23.9	27.4	
67032	33	29.4	3.6	3.5	2.7	6.3	5.5	~15,1					44.6	~29,4	17.1	17.1	~19,0	
60501	53.3	47.3	6	6	4.7	9.7	7.2	24.1	~9,6	18.3	25.2	35.6		51.9	32		34.1	6
60460	31.1	28.3	2.8	3.4	2.3	6.4	4.6	15.2	5.2				45.9	30.3	18.2		20.1	
206.206b19	25.9	23.4	2.5	3.1	2.2	5	4.4	13.4	5				~37,3	~23,3	15.2		16.6	
4750	43.8	39.9	3.9	5.2	4.6	9.1	7.7	21.9		16.2	21.5	~30,2	~68,6	40.8	27.5		28.2	3
82700	40.8	35.8	5		4.5	3.6	7.9	6.6	19.2	~13,7	~21,3	~27,3	~66,0	~39,6	22.9		25.5	4

Abbreviations as in Table 7.

Table 4. Librigenal measurements on *Ogygiocaris dilatata* (Brünnich, 1781)

PMO number	Le	Lp	Ls	Wbmin	Wp	Wpp	Wl	Wb1	Wb2	Wpm	Terr. Doubl.	Terr. surface
60551	8.5	26.7		14.3	2.7	2.8	18	17.8	15.1	19.7	15	
60451	6.1	16.9	11.6	7.7	2.1	2.3	9.3		8.3	11.1	15	
60451	8.8	26.1		13.4	2.2	2.6	16.8	17.1	14.5			~15
206.290	9.6	25.5			2.1	2.4	11.7	10.6	8.8	17.3	16	
202.341a1					1.3	1.4	5.5					
203.109b15					1.4	1.5	6.6					
203.174	~7,4	23.8		7.6	2.3	2.4	10.6	8.5	7.7		17	
203.229a1					~1,6	1.6	7.6					
203.349					1.4	1.5	6					
203.333	~6,7	24.1		9.6	3.1	3.1	14	12.1	9.6		17	
203.447a9	~5,1			7.1	2.3	2.4	9.8	~9,0				
203.447c8					~2,4	2.4	10.9					
206.276b4	6.6	20.8	9.9	7.1	2	2.1	10	~10,0	7.3	15.3		
206.291	6.1	15.6	~5,5	5.8	1	1	7.5	7.8	5.9		17	
82580	6.8	24.4		11.6	4	4.2	16.4	13.4	11.8	19.9		
82578	6.1	20.6		8.1	2	2.3	11.5	10	8.3	17	17	
4730	6.5	17.7		6.3	1.4	1.4	8.1	8.2	6.4	12	17	
3696					2	2.1	9.9					

Abbreviations as in Table 8.

Table 5. Cranidial measurements on *Ogygiocaris isodilatata* n. sp.

PMO number	L	Lg	Lp	Lo	Lt	Lf1	Lf2	Lae	Le	Lb	Lbp	Wb	W	Wc	Wo	Wgmin	Wae	Outer t.	Inner t.
202.899a1	26.9	22.8	4.1	3.1	1.8	5	4.1	12.6	~6,6	9	15.1	20.5	~42,1	27	13	11	~16,9	3	~16
202.886a/1	14.3	12.4	1.9	1.6	1.2			6.7	2.8	4.3	7.9			12.5	5.6	5.9	7.7		

Abbreviations as in Table 7.

Table 6. Librigenal measurements on *Ogygiocaris isodilatata* n. sp.

PMO number	Le	Lp	Wbmin	Wp	Wpp	Wl	Wb1	Wb2	Wpm	Terr. Doubl.
202.874				~2,5	~2,8	9.3				
202.899C1	6.9	19.4	5.9	2.9	3	10.8	8.5	5.9	14.6	15

Abbreviations as in Table 8.

Vålen at Vestfossen. Oslo-Asker: Bøveien, Djuptrekkodden and Elnestangen at Slemmestad and Bygdøy at Oslo (middle Helskjer (*Nicholsonograptus fasciculatus* Zone) to top Håkavik Member). Modum: Heggen, Vikersund skijump. Ringerike: Gomnes. Hadeland: Brandbu (Heggen Member. Rare). Mjøsa: road-section at Furnes Church (Helskjer Member (?); extremely rare).

Ogygiocaris isodilatata n. sp.

Pl. 1, Figs 9–10; Pl. 2, Figs 1–6; Tables 5–6

Derivation of name. – Referring to the strong resemblance to *O. dilatata* (Brünnich).

1963 *Ogygiocaris* cf. *dilatata* – Skjeseth, p. 63 (occurrence).

Type stratum and type locality. – Sjøstrand Member of the Elnes Formation. NE-Profile in the Old Eternite Quarry at Slemmestad,

Oslo-Asker district. Holotypes and paratypes are all deposited in the PMO collections.

Type material. – PMO 202.899a-c/1 from the middle Sjøstrand Member 27.24 m above base of profile at the Old Eternite Quarry at Slemmestad is selected as holotype. Paratypes are PMO 202.864/1, 202.871a/1, 202.873 c/7, 202.877a, b/1 and 202.886a/1, all from the middle Sjøstrand Member (21.11 to 25.07 m above base of profile) at the Old Eternite Quarry, Slemmestad.

Material. – The examined material includes two cranidia (PMO 202.886a/1, 202.886 c/7), one librigena (PMO 202.874/2), one hypostome (PMO 202.871a/1), some thoracic segments (PMO 202.873 c/7), 19 (20?) pygidia (PMO 3470, 61524 (?), 90494 (two specimens), 202.864/1, 202.872/2, 202.873a/1, 202.873a/2, 202.873 c/7, 202.876/1, 202.877a/2, 202.878, 202.879a/1, 202.880, 202.883a/1, 202.884a/2, 202.884b/3, 202.886b/5, 202.896, 208.644/2) and four articulated but somewhat fragmentary specimens (PMO 33426, 82580, 202.877a, b/1, 202.899a-c/1).

Diagnosis. – Cranidial paradoublural line very posteriorly placed, the abaxial distance to anterior margin approximating one-third the cranidial length. Para-doublural line on librigena situated just over one quarter the distance between eye and margin from eye; width of eye (lat.) corresponding to one quarter the width of the pleural field. Hypostome with outwardly turned terrace-lines in front. Thoracic segments with axial width corresponding to one-fifth the thoracic width. Pygidium semicircular with distinct posterior truncation; nine slightly rounded pleural ribs spaced relatively far apart and a maximal border width and axial width taking up less than one-fifth the pygidial width. Pygidial doublure with coarse pattern of 14 to 16 terrace-lines.

Description. – Largest measured pygidium is 38 mm long, while largest cranidium measures 27 mm in length. Cephalon broadly semicircular in outline with an L/W ratio around 40%.

Cranidial length 64% of the width on PMO 202.899a/1. Glabella slightly contracted at middle, delimited by wide and rather shallow axial and pre-glabellar furrows. Glabellar length corresponding to 85% of the cranidial length. Largest glabellar width is located around 40% the distance between eyes and anterior glabellar margin in front of eyes, while smallest glabellar width between the eyes corresponds to between 40 and 50% the preocular cranidial width. Most anterior part of occipital furrow located just above 10% of cranidial length from posterior margin. Mesial tubercle situated 7 to 8% of the cranidial length from posterior margin. Main part of glabella with four pairs of lateral glabellar furrows or pits; the first, S1, not connected to axial furrow, terminating one-third the glabellar width inside glabella. S2 and S3 very shallow, the last describing a strongly forward convex curve. S4 completely effaced. Posterior lobe, L1, slightly inflated, triangular in outline. L2 and L3 shorter and uninflated, the anterior one strongly curved. Frontal glabellar lobe subcircular to pyriform in outline with broadly rounded anterior margin. Parafrontal band well-developed; clearly part of frontal glabellar lobe. Anterior pits small, located in axial furrow just behind level of largest glabellar width. Preglabellar area is moderately wide, around 15 to 17% of cranidial length at sagittal line. Width of pre-ocular area corresponding to half the glabellar width; anterolateral corners evenly rounded. Preocular area ends just in front of cranidial midline. Paradoublural line curving slightly rearwards from posterolateral margin of preocular field to axial furrow; the initial distance to anterior cephalic margin approximating one-third the cranidial length; distance from posterior cranidial margin to abaxial end of paradoublural line corresponding to just over half the cranidial length. Palpebral lobes small, exsagittal length equiv-

alent to between 20 and 25% the cranidial length, situated about one-third the cranidial length from the posterior border or outside L2. Posterior fixigena moderately broad, maximum exsagittal length of posterolateral projection close to 20% of the cranidial length, transected by moderately deep posterior border furrow. Inner shortened part of posterolateral projection on fixigena corresponding to about 80% the length of the abaxial part. Cranidial surface smooth except for a few terrace-lines on the preglabellar field.

Length of eye corresponding to 36% of length of librigenal field from posterolateral to anterolateral corner on specimen PMO 202.899 c/1; width of eye (tr.) corresponding to 27% the width of the pleural field perpendicular to axis on specimens PMO 202.899 c/1 and PMO 202.874/2. Paradoublural furrow accentuated by raised border located slightly more than one-fourth the distance between eye socle and lateral margin from eye socle. Course of para-doublural furrow largely sub-parallel to border margin; anteriorly curving inwards and finally, when crossing over onto cranidial shield, turning strongly rearwards along cranidial margin. Lateral border wide. Genal spine not preserved. Surface of genal field smooth; border covered by terrace-lines running parallel to the margin. Doublure with around 15 terrace-lines on specimen PMO 202.899 c/1.

Only one fragmentary hypostome (PMO 202.871a/1) preserved. Hypostome subrectangular, only slightly longer than wide, largest width across anterior wing about 40% from anterior margin. Anterior lobe of middle body subcircular in outline, reaching the curved anterior margin. Posterior lobe of middle body forward concave, width nearly equivalent to half the maximum width. Maculae eye-shaped, situated on anterolateral flanks of posterior lobe. Anterior border exsagittally narrow, continuing out into short, evenly rounded wings. Lateral border elongated subtriangular, length corresponding to slightly more than half the length of hypostome, terminating rearwards outside posterior flank of posterior lobe. Posterior border rounded subtriangular. Surface of hypostome covered by coarse terrace-line pattern describing a sharp rearwards pointing V-pattern on the central body. Forward end of terrace-lines turned outwards.

Thoracic axis occupying 20% of the thoracic width, length (sag.) approximates 30% of the width. Pleurae short (exsag.) and wide; transition between inner and outer pleural portion located nearly half the pleural width from axial furrow. Pleural terrace-lines few, parallel to pleural ridges.

Pygidium semicircular in outline with distinct posterior truncation; length between 50 and 60% the width. Axis long and moderately slender, terminating more than 25% the pygidial length from the posterior margin. Axial width varies from 17 to 19% of the pygidial width or between 55 and 68% of the pleural field width. There are 10 to 12 axial rings in addition to the terminal piece and articulating half ring. Pleural field carries nine pleural ribs in addition to anterior half rib, the posterior one rather effaced. Ribs slightly rounded in cross-section, describing a straight line out to shortly before the border, where it turns rearwards, terminating on the inner half of the border. Interpleural furrow normally nearly absent, situated anteriorly on the distal part of the pleural rib. The furrow widens and deepens somewhat at the transition to the border. Pleural furrows moderately wide. Border only enfolding the posterior tip of the terminal axial piece; border width for adult specimens between 14 and 17% pygidial width at pleural furrow 5 and 8; outer surface in cross-section slightly concave, doublure concave. The width at pleural furrow 8 corresponds to between 81 and 106% the abaxial distance between pleural furrow 5 and 9.

Outer surface of pygidium, except for the pleural furrows, covered by relatively coarse terrace-line pattern. Each axial ring with one clearly marked terrace-line curving only slightly convex forward on posterior half; a few less distinct and broken ones may surround it. Terrace-lines on pleural ribs and border directed perpendicular to axial line, on some turning more anterolaterally at pygidial corners. A fictive line perpendicular to the border crossed 11 and 14 terrace-lines respectively on two examined specimens. Pygidial doublure covered by 14 to 17 evenly spaced terrace-lines arranged subparallel to the margin; becoming more closely spaced at inner margin.

Remarks. – *Ogygiocaris isodilatata* n. sp. is morphologically very close to *O. dilatata*, but differs by its longer (sag.) preglabellar field; larger eyes; the paradoublural line on librigena nearer to the margin; the pygidial length/width ratio is less than 70% and the pygidial border and axis are narrower. It also shows some similarities with *O. sarsi* (see below) such as the border width on pygidium and librigena and the nine somewhat flattened pleural ribs. This suggests a close relationship between the two and *O. isodilatata* n. sp. may thus form a link between *O. dilatata* and *O. sarsi*, though probably not a direct lineage as both *O. sarsi* and *O. dilatata* occur down to the basal part of the Elnes Formation.

Occurrence. – Upper Helskjer Member to middle Sjøstrand Member of the Elnes Formation. Eiker-Sandsvær: Vego at Krekling, where it was sampled from level 1.57 m in profile 6. Oslo-Asker:

The Old Eternite Quarry at Slemmestad, where it has been found from 21.11 to 27.24 m above base of formation (middle Sjøstrand Member), road cut above Lille Frøen in Oslo (Sjøstrand Member). Hadeland: Grinaker at Gran (Heggen Member). Mjøsa: Road cut at Nydal-Furnes Church (uppermost Helskjer Member.).

Ogygiocaris sarsi Angelin, 1878

Pl. 2, Figs 7–11; Pl. 3, Fig. 1; Tables 7–8

1835 *Asaphus dilatatus* Dalm. – Sars, pp. 336–337, 342, pl. 8, fig. 5a–b, pl. 9, fig. 11 only.
1838 *Tr. dilatatus* Brün. – Boeck *in* Keilhau, p. 141.
1854 *Ogygiocaris dilatata.* Brünn. – Angelin, p. 92 (*Trilob. dilatatus* in the synonymy list).
1878 *Ogygiocaris dilatata.* var. *sarsi.* Ang. – Angelin, p. 96, pl. XLII, fig. 1, fig. 1a–c.
1886 *Ogygia* (*Ogygiocaris*, Ang.) *dilatata*, Brünnich, var. *sarsi*, Ang. – Brøgger, p. 53, pl. III, figs 38, 38a.
1953 *Ogygiocaris dilatata sarsi* – Størmer, pp. 56–58, 61, 119–121 (lists and occurrence).
1960 *Ogygiocaris sarsi sarsi* Angelin, 1878 – Henningsmoen, pp. 215, 216, 225–227, 258, fig. 3, pl. 4, figs 1–2, 5–10.
1984 *O. sarsi sarsi* – Wandås, p. 232, pl. 10 K, O–Q.

Type stratum and type locality. – The type has never been located, but a neotype, PMO 20287, collected by M. Sars and which may be the original of Angelin's (1878) drawing was selected by Henningsmoen (1960). The neotype is from the Engervik Member of the Elnes Formation at Hjortnestangen in Oslo, Oslo-Asker district.

Norwegian material. – The examined material consists of one complete cephalon (PMO 206.156b12), 18 cranidia (PMO H393, S.1740, 3311, 3468, 3584, 3996, 56299, 56299a, 56314b + c, 61134, 82199, 82873, 82874, 82910, 90491, 142.334, 202.182B2) and 30 librigenae (PMO 3696, 20288, 36761, 56314d, 82206, 82576, 82577, 202.281b8, 203.067b6, 203.314b8, 203.323b2, 203.327c14, 203.351, 203.353b2, 203.354c6, 203.362, 203.378d42, 203.415a1, 203.416, 203.421, 203.447a9, 203.464, 205.022, 205.242c4, 205.962, 206.086, 206.131, 206.156b12, 206.210a4, 206.210b15). The thoracic segments and pygidia are generally too similar to the other species of the *sarsi* group (especially *O. henningsmoeni* n. sp.) to be assigned to a single species, but the following six pygidia may with some confidence be assigned to *O. sarsi*: PMO 82469 (two specimens), 90490, 202.035, 202.060, and 202.130. No complete articulated specimens are known, but five nearly complete but partly disarticulated specimens are in the NHM collections (PMO S.1744, 3689, 20287, 56311 and 82469).

Diagnosis. – Species of *Ogygiocaris* with paradoublural line on cranidium situated far back, posterolateral distance to cranidial margin between 30 and 32% of cranidial length. Paradoublural line on librigena situated between 25 and 33% of the distance between margin and eye from eye; width of eye (lat.) corresponding to one-third the pleural field width. Hypostome narrow with anterior terrace-line terminations directed nearly parallel to sagittal line, deep frontal

Table 7. Cranidial measurements on *Ogygiocaris sarsi* Angelin, 1878

PMO number	L	Lg	Lp	Lo	Lt	Lf1	Lf2	Lae	Le	Lb	Lbp	Wb	W	Wc	Wo	Wgmin	Wae	Outer t.	Inner t.
3996	16.7	13.1	3.6	1.8	1.2		2.1	~8,4		9.6		12.7		~17,8	7.1			6	
3647	32.6	26.5	6.3	3.2	2.4	4.5	2.9	~13,6		10.2	17.1	~23,8	~52,1	35.1	16.1	~15,6		5	
3584	33.6	26.4	7.2	3.4	2.2	5.7	3.9	14.7		10.6	17.7	~25,2	~54,3	36.2	17.1	15.5	20.1	4	
20288	25.5	20	~5,5	2.6	2	4	3.2	10.5	4	7.6	13.8	15.3	~33,9	21.9	9.8	9.6	12.9	5	
206.156b12	~18,8	~15,2	~3,6			3.2	2.3	9.1	~3,8	5.6	11.1	12.1	~26,0	16.8	7.5	7	9.4		16
61134	34.4	25.5	8.9	3.7	2.8	5.5	4	16.8	~6,4	10.7	18.8	23.5		32.5	13.2		19.4	6	

L, maximal cranidial length; Lg, length of glabella; Lp, Length of preglabellar area; Lo, maximal length (exsag.) of occipital ring.
Lt, distance between mesial occipital tubercle and posterior cranidial margin.
Lf1, maximal length (exsag.) of posterolateral fixigenal projection; Lf2, minimum length of posterolateral fixigenal projection found on adaxial half.
Lae, distance from posterior cranidial margin to narrowest part of cranidium just in front of palpebral lobes; Le, length of palpebral lobe.
Lb, distance from distal cranidial termination of paradoublural line and out to anterolateral margin of cranidium semiparallelling outer cephalic margin.
Lbp, distance from distal cranidial termination of paradoublural line to posterior cranidial margin; Wb, distance between distal cranidial terminations of paradoublural line.
W, maximal cranidial width; Wc, maximal preocular cranidial width; Wo, width of occipital ring; Wgmin, smallest glabellar width found between the palpebral lobes.
Wae, smallest cranidial width found just in front of palpebral lobes; Outer t., number of terrace-lines counted on test surface on preocular area; Inner t., number of terrace-lines counted on rostral doublure.

Table 8. Librigenal measurements on *Ogygiocaris sarsi* Angelin, 1878

PMO number	Le	Lp	Ls	Wbmin	Wp	Wpp	Wl	Wb1	Wb2	Wpm	Terr. doubl.	Terr. surface
202.281b8					1.5	1.5	5.2					
203.067b6	5.9		3.8		2.7	2.8	8.5	5.7	4.5	13.6	17	
206.210a4					2	2.1	6.2					
206.210b15					3.4	3.7	10.6					
206.156b12	5.4	13		4.3	2.1	2.3	7.3	5.9	4.5	9	16	
203.314b8	3.3	~7.3			1.05	1.05	3.5					
203.351					3.4	3.5	12.6					
203.323b2					1.7	1.85	5.8					
203.327c14	5.5			7.4	3.3	3.4	11.8	8.8	7.5	11.4	16	
203.354c6		~13,6	8.8	5.5	2.8	2.8	8.6	6.4	6		15	
203.362				8.6	3.9	4.1	13.5	10.6			15	
203.353b2					4.1	4.2	12.8					
203.378d42					2.3	2.4	7.3					
203.415a1					2.9	3	9.5					
203.421					3.7	4.1	12.7					
203.447a9					2.6	2.8	9.5					
203.416					2.6	2.9	8.8					
203.431b1					3.3	3.4	11.6					
203.464		17.2		6	2.7	3	9.1	7.2	6.1		16	
205.022	6.5			8.3	4	4.3	13.8	~11.7				
205.242c4					2.5	2.6	9.6					
205.962					2.5	2.6	8.1					
206.086					4.4	4.8	14.8					
206.131	5.4			4.8	2.2	2.2	7.2	5.2			17	
20288	6.3	16.4	6.7		3.1	3.6	9.9	9.1	6.4	11.9		16
82577	~5.7	15.6		5	2.7	2.7	8.1		5.1			
82576	8.5	21.9	11.4		2.9	3	8.8	6.1		17	18	
82357					1.75	2	6.5					
82206	4.95	13.05	~5.9	5.45	1.9	2.15	7.75	7.1	5.45			
36880					2.9	3.15	8.9					
36761	6.4	~19.7	~7.1	6.7	3.7	4	11.4	~8.7	6.9	~13.6		8

Le, length of eye; Lp, length of librigenal field from posterolateral to anterior corner; Ls, length of spine; Wbmin, narrowest width of doublure.
Wp, shortest distance between eye and doublure.
Wpp, distance between doublure and eye along shortest distance between eye and outer librigenal margin along a line oriented perpendicular to outer margin.
Wl, shortest distance between outer librigenal margin and eye; Wb1, border width anteriorly; Wb2, border width posteriorly.
Wpm, librigenal field width at widest point along a line perpendicular to paradoublural line; We, width of eye.
Terr. doubl., number of terrace-lines on doublure; Terr. surface, number of terrace-lines on outer surface of border.

notch in anterior lobe. Axial width of thoracic segments corresponding to one-fifth the thoracic width. Pygidium semicircular to subtriangular without any posterior truncation; pleural ribs flattened, about nine in number; maximal border width and axial width taking up nearly one-fifth the pygidial width. Pygidial doublure with about 19 to 20 terrace-lines (emended from Henningsmoen 1960).

Description. – An *Ogygiocaris* of intermediate size; largest pygidium about 37 mm long; largest cranidium measures 33 mm. Cephalon slightly parabolic to semicircular (large specimen) in outline with a length corresponding to between 45 and 60% of the width, but the higher values may largely be explained by later distortion.

Cranidial length corresponding to between 62 and 75% of the width on four measured specimens. Glabella slightly contracted at middle, bordered by wide and rather shallow axial and preglabellar furrows. Glabellar length taking up 74 to 81% of the cranidial length. Largest glabellar width is located approximately one-third the distance between eyes and anterior glabellar margin in front of eyes. Minimum glabellar width is found between the eyes and corresponds to just above 40% of the preocular cranidial width. Anteriormost part of occipital furrow located 10 to 11% of cranidial length from posterior margin. Mesial tubercle situated between 7 and 8% of the cranidial length from posterior margin. Glabella with four pairs of lateral glabellar furrows or pits; S1 moderately deep and elongated pit-like, only partly connected to the axial furrow, adaxially terminating around one-third the glabellar width inside glabella. S2 moderately shallow, not connected to axial furrow. S3 forwardly convex, connected to axial furrow. S4 short and generally less pronounced than S1–S3, directed strongly forward adaxially. Anterior pit is shallow or missing. Posterior lobe, L1, slightly inflated, triangular in outline. L2 is subrectangular in outline, largely uninflated. L3 and L4 effaced. Frontal glabellar lobe pyriform with rounded to slightly pointed front. Parafrontal band generally well developed and set apart from frontal glabellar lobe. Preglabellar area wide, between 19 and 26% of cranidial length at sagittal line. Preocular area broader, the width corresponding to more than half the glabellar width; anterolateral corners evenly but relatively sharply rounded. Preocular area ends posteriorly around 40 to 50% of the cranidial length from posterior cephalic margin. Paradoublural line on cranidium mostly directed forward adaxially; adaxial end turning very slightly posteriorly; the abaxial distance to anterior cephalic margin corresponding to 30 to 32% of the cranidial length; distance from posterior

cranidial margin to abaxial end of paradoublural line corresponding to between 50 and 60% of the cranidial length. Palpebral lobes small, exsagittal length equivalent to between 16 and 20% of cranidial length, situated outside L2 less than one-third the cranidial length from the posterior margin. Posterior fixigena with a maximum exsagittal length of posterolateral projection between 14 and 17% of the cranidial length, transected by moderately deep posterior border furrow. Shortened inner part of posterolateral projection corresponding to between 9 and 13% of the cranidial length. Cranidial surface nearly smooth but for some terrace-lines on the preglabellar and preocular fields and in some cases a few wrinkle furrows in front of the mesial tubercle.

Length of eye between 32 and 42% of librigenal field-length from posterolateral to anterolateral corner; width of eye (tr.) corresponding to one-third the pleural field width perpendicular to axis. Eye socle surrounded by broad, shallow furrow, width corresponding to between 25 and 33% of the eye length. Paradoublural furrow marked by raised border situated between 25 and 33% (27 specimens) of the distance between eye socle and lateral margin from eye socle. Course of paradoublural furrow convex to border or diverging forward from border margin; anteriorly turning inwards but generally not rearwards before intersecting the cranidial parafrontal band. Lateral border relatively broad, clearly widening anteriorly. Genal spines compressed oval in cross-section, reaching up to at least two-thirds the cranidial length rearwards. Posterior border narrow, width on librigena less than one-third the width of lateral border. Surface of genal field smooth; doublure covered by 15 to 18 terrace-lines. Border surface generally bearing much fewer, limited to the outer half.

Only one hypostome (PMO 20287) may without doubt be assigned to this species. Hypostome subrectangular, longer than wide with a width/length ratio of nearly 75%, though the hypostome is somewhat compressed laterally and therefore most probably should have a L/W ratio between 75 and 80% as found for several loose ones. Largest width across anterior wing slightly more than one-third the hypostomal length from anterior margin. Anterior lobe of middle body sub ovoid with deep frontal notch and delimited by a very short anterior border, narrow lateral borders and a laterally deep middle furrow posteriorly. Posterior lobe of middle body moderately effaced, forward concave, width approximating half the largest hypostomal width. Maculae sitting on well-developed angular highs placed on posterior side of middle furrow, resulting in a sharp rearwards bend in the furrow and a sub-rectangular posterior end of anterior lobe. Anterior border sagittally very short; laterally continuing out into short, evenly rounded, lateral directed wings with short pointed ends posteriorly. Lateral border narrow, elongated subtriangular, exsagittal length corresponding to half the length of hypostome, posteriorly continuing evenly out into subtriangular posterior border. The surface of the hypostome is covered by coarse terrace-line pattern describing a sharp rearwards pointing V, but turning slightly more forward in front. Central part of middle body smooth, but this probably relates to preburial abrasion.

Thorax only known from disarticulated segments. Axis occupies 20% of the thoracic width; length (sag.) approximating one-third the width. Pleurae short (exsag.) and wide. Transition between inner and outer pleura located between 40 and 50% of the pleural width from axial furrow. Pleural area crossed by moderately wide and deep pleural furrow, terminating close to 70% of the pleural width from the axial furrow. Axial segments bearing few and weak, slightly forward convex terrace-lines. Pleural terrace-lines few, parallel to pleural ridges. Doublural terrace-lines directed forwards and slightly abaxially.

Pygidium semicircular to subtriangular in outline with no posterior truncation; length approximating 60% of the width. Axis long and slender, terminating around 25% the pygidial length from the posterior margin. Width of axis about 16 and 18% of pygidial width and about 60% of pleural field width. There are 11 to 12 (?) axial rings in addition to the terminal piece and articulating half ring. Pleural field with nine to ten pleural ribs in addition to anterior half-rib; posterior one somewhat effaced. Ribs flattened in cross-section. Interpleural furrow shallow to nearly absent. Pleural furrows moderately narrow. Border enfolding the posterior tip of the terminal axial piece; border width at pleural furrow 5 and 8 nearly 20% of the pygidial width; outer surface and doublure slightly concave in cross-section.

Outer surface of pygidium, except for the pleural furrows, covered by relatively coarse terrace-lines. Each axial segment with one distinct and flatly convex terrace-line on middle part followed posteriorly by two to four less distinct and shorter ones; anterior half generally smooth or with one short terrace-line just in front of the main one. Terrace-lines on pleural ribs directed perpendicular to axial line, becoming more subparallel to pygidial margin on border; anteriorly bending to a distinctly lateral direction. Number of terrace-lines crossed by a fictive line perpendicular to border 16 and 21 on two examined specimens. Pygidial doublure covered by 15 to 21 evenly spaced terrace-lines subparallel to margin and becoming more closely spaced at inner margin.

Remarks. – *Ogygiocaris sarsi* is stratigraphically as well as geographically one of the most widespread *Ogygiocaris* species and has been found in the whole Oslo Region as well as central Sweden and occurs stratigraphically up through the whole Elnes Formation.

Occurrence. – Middle Helskjer Member to top Håkavik Member of the Elnes Formation. Skien-Langesund: Trosvik at Brevik. Oslo-Asker: The Old Eternite Quarry at Eternitveien, Slemmestad (1.03 to 6.48 m above the Huk Fm.); the beach exposure between Djuptrekkodden and Sjøstrand at Slemmestad (22.41 to 56.54 m above base of profile); Elnestangen at Slemmestad; the Bøveien profile above Solliveien, Bødalen (3.07 to 13.42 m above base of measured section); at Odden just east of Bekkebukta, Bygdøy (referred to as Vekkopp by Henningsmoen (1960)); Huk, Bygdøy; Hjortnæstangen at Hjortnes Havn, Oslo; Tørtberg at Frøen, Oslo. Mjøsa: Helskjær on Helgøya; road-cut at Furnes Church (basal Heggen Member, 8.6 m above the Huk Formation).

Outside Norway it has been found in the lower Andersö Shale formation of Jämtland, Sweden.

Ogygiocaris henningsmoeni n. sp.

Pl. 3, Figs 2–8; Text-Fig. 48; Tables 9–10

Derivation of name. – In honour of G. Henningsmoen, who made the first detailed study of this genus.

1913 *Ogygiocaris dilatata* Brünn. var. *sarsi* Ang. – Hadding, pp. 70–72, fig. 23, pl. 7, figs 1–7.
1913 *Ogygiocaris dilatata* Brünn. var. *lata* n. var. Hadding, p. 72, pl. 7, fig. 9a–c.
1960 *Ogygiocaris sarsi regina* n. subsp. Henningsmoen, p. 228 (*partim*), pl. 5, fig. 1.
1960 ? *Ogygiocaris sarsi lata* Hadding, 1913 – Henningsmoen, pp. 229–231 (*partim*), pl. 5, fig. 9.
1998 *Ogygiocaris sarsi lata* Hadding – Karis & Strömberg, p. 25, fig. 13.
2000 *Ogygiocaris lata* Hadding, 1913 – Månsson, p. 12, fig. 8i–j.
2000 *Ogygiocaris regina* Henningsmoen, 1960 – Månsson, p. 12 (*partim*), figs ?g, h, ?k.
2002 *Ogygiocaris lata* Hadding, 1913 – Pålsson *et al.*, p. 42, fig. 8d–e.
2002 *Ogygiocaris regina* Henningsmoen, 1960 – Pålsson *et al.*, p. 42 (*partim*), ?fig. 8g–h.

Type stratum and type locality. – Upper Sjøstrand to middle Håkavik Member of the Elnes Formation. Type section at Bøveien, south of Slemmestad. Paratypes are all from the beach exposure between Djuptrekkodden and Sjøstrand at Slemmestad, Oslo-Asker district.

Type material. – The holotype, PMO 206.159/1, was collected from the uppermost Sjøstrand Member, 2.99 m above base of measured section at Bøveien, south of Slemmestad. All paratypes have been collected from the exposure between Djuptrekkodden and Sjøstrand, just north of Slemmestad. Paratype PMO 203.058/4

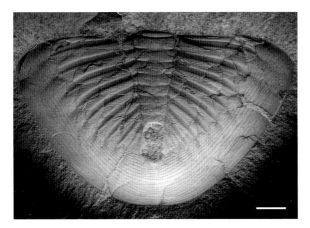

Fig. 48. Pygidium of large specimen of *Ogygiocaris henningsmoeni* n. sp. from Mjøsa district. PMO 72117. Scale bar 1.0 cm.

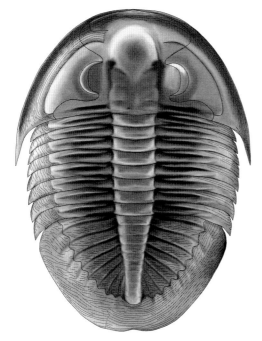

Fig. 49. Reconstruction of a complete specimen of *Ogygiocaris lata* showing the doublural terrace-lines on the left librigena and thorax and right side of the pygidium. Length of specimen around 7 cm.

was found in the uppermost Sjøstrand Member, 22.41 m above base. Paratype PMO 203.067b/8 is from the uppermost Sjøstrand Member 22.46 m above base of profile. Paratype PMO 209.545 was most likely collected from the lower half of profile profile 1 at Råen, Eiker-Sandsvær. Paratype PMO 203.166a-c/3 + 5 is from the lowermost Engervik Member, 31.56 m above base of profile.

Norwegian material. – The examined material contains a cephalon (PMO 33604), six cranidia (PMO 33066, 33368, 203.058/4, 203.109b/12, 206.218b10, 207.433/6), 20 librigena (PMO 203.019/1, 203.046a/1, 203.058/3, 203.061a/1, 203.067b/8, 203.081a/2, 203.109b/14, 203.109b/16, 203.110/1, 203.114/1, 203.166b/3, 203.373/1, 203.462a/4, 205.227b/1, 205.960b/1, 206.042a/1, 206.156d/20, 206.159/1, 206.162/1, 206.217a/3, 207.416/2), three pygidia (PMO 36901, 72117, 203.058/2) and three partly articulated specimens (PMO 203.166a-c/3 + 5, 206.159/1, 209.545).

Diagnosis. – Paradoublural line on cranidium anteriorly placed, posterolaterally situated between 23 and 28% of cranidial length from cranidial border. Paradoublural line on librigena situated nearly two-fifths the distance between eye and margin from eye; width of eye (lat.) corresponding to more than one quarter the width of the pleural field. Thoracic segments with axial width corresponding to one-fifth the thoracic width. Pygidium with no posterior truncation; outline semicircular to rounded subtriangular or in the case of the largest specimens distinctly triangular; pleural ribs flattened, nine to ten in number; maximal border width and axial width taking up less than one-fifth the pygidial width. Pygidial doublure with 18 to 20 terrace-lines.

Description. – One of the largest species of *Ogygiocaris*, although the Norwegian specimens generally only reach a small to moderate size. Largest Norwegian pygidium (PMO 72117) approximately 58 mm long, while largest cranidium is around 28 mm long. Cephalon broadly semicircular in outline with an L/W ratio around 0.4.

Cranidial length approximating two-thirds the width. Glabellar length corresponding to around four-fifths the cranidial length. Lateral glabellar furrows and lobes and occipital furrow similar to

Table 9. Cranidial measurements on *Ogygiocaris henningsmoeni* n. sp.

PMO number	L	Lg	Lp	Lo	Lt	Lf1	Lf2	Lae	Le	Lb	Lbp	Wb	W	Wc	Wo	Wgmin	Wae	Outer t.	Inner t.
203.166bc3	~25.0	~21.,1	3.9			~4.5	3.4		~6.7	7	~16.6	21.5		~27.2	11.3			4	19
203.126a5	11.8	9.9	1.9	1.4	1	2	1.5	5.8		3.2	7.25	~9.2	~19.4	~12.5	4.6	5		5	
206.218b10	16.3	13.1	3.2	1.8	1.5	2.7	2.2	8.3	3.6	3.7	9.9	11.8	~24.5	15.4	6.8	~6.4	9	5	
203.109b12	~18.0	~15.4	2.6			3.8	2.7	8.5		4.7	11.3	16.6		20.8			12.4	3	
33066	16.2	13.3	2.9	1.8	1.6	2.7	2.1	7.7		4.4	10.8	~13.9		~17.7	6.9		9.5		
Exhebition	15.4	12.9	2.5	1.5	1.25	2.6	2	7.5		3.75	9.5	12.6	24	16.5	6.5	5.3	9	4	
82365	14.85	11.8	1.85							3.55	8.8	12.55	~25.2	15.65					
36840	22.75									5.6	~13.6	20.45		24.75					
36857_2	19.4	15.6	3.8							5.3	11.45								
36857_1	22.4	~18.65	~3.75							5.7	~13.0	20.5		25.1					

Abbreviations: see Table 7.

Table 10. Librigenal measurements on *Ogygiocaris henningsmoeni* n. sp.

PMO number	Le	Lp	Ls	Wbmin	Wp	Wpp	Wl	Wb1	Wb2	Wpm	Terr. doubl.	Terr. surface
206.042a1	6.6	17.5	6.8	5.1	3.7	3.8	9.4	6.2	~5.5			
205.972					2.2	2.3	5.8					
205.960b1	7.1	18.6		5.4	3.7	3.9	10	7.6	5.9	12.2	18	13
205.953/1					3.2	3.3	8.4					
205.227b1	~6.5	24		6.6	4	4	10.9	9.4	6.7		19	
203.462a4					~4.3	4.5	~11.4					
203.373		13.7		4.9	3.1	3.1	8.1	6.1	5.2		15	
203.166b3	7.6	23.3		~4.8	3.4	3.6	9.5	8	~4.9	~15.1	16	
206.156d20	3.1	7.8		2.1	1.2	1.3	3.4	2.6		4.2	16	
206.217a3					1.2	1.2	3.3					
203.109b16	3.6	7.4		1.4	1.25	1.3	3.3	2.6	1.6	5.2	13	
206.159					1.8	1.9	4.8					
206.162					3.2	3.3	~8.3					
203.114					2.9	3.4	7.8					
203.109b14	6	~15.3			2.8	3	7.7	5.1	3.9	~10.1	15	
203.081a2	6	16.7	~7.4		3.3	3.3	8.7	6	5.2	11		
203.067b8	~7.3	23.1	11.8	5	3.8	4.3	10.2	7	5	17.7	17	
203.061a1	7.2				3.8	4.1	10.5	8.6	5.8		19	
203.058	7.3	25.3	10.4		4.3	4.7	11.4	7.6	5.8			
203.046a1	~5.6	18.8			3.3	3.6	9.3		4.9		16	
203.019					4	4.4	11.1					

Abbreviations: see Table 8.

O. sarsi. Preglabellar area long, around 14 to 20% of cranidial length at sagittal line. Width of preocular area corresponding to half the glabellar width. Abaxial end of cranidial paradoublural line situated between 23 and 28% of cranidial length from anterior cranidial margin; distance from posterior cranidial margin to abaxial end of paradoublural line corresponding to between 58 and 67% of the cranidial length (10 specimens). Posterior fixigena moderately broad, largest exsagittal length of posterolateral projection approximating 20% of the cranidial length. Cranidial surface nearly smooth but for 17 to 18 terrace-lines on the preglabellar and preocular field.

Length of eye between 27 and 49% of librigenal field from posterolateral to anterolateral corner; width of eye (lat.) corresponding to more than one quarter the width of the pleural field perpendicular to axis. Paradoublural furrow located between 35 and 39% the distance between eye socle and lateral margin from eye socle. Lateral border moderately narrow. Genal spine reaches between 40 and 50% of the genal field behind librigena. Border covered by terrace-lines running parallel to the margin. Doublure with 13 to 19 terrace-lines running parallel to margin.

Hypostome unknown.

Thoracic axis occupying 20% of the thoracic width, length (sag.) approximately one-third the width. Thorax largely similar to that of *O. sarsi*, although the doublure may be slightly narrower (tr.).

Pygidium largely similar to *O. sarsi*; outline semicircular to rounded subtriangular or, in the case of the larger specimens, distinctly subtriangular; posterior end at the highest with slight trace of truncation.

Width of pygidial axis varies between 48 and 65% of the pleural field width. There are 11 to 12 axial rings in addition to the terminal piece and articulating half ring. Border width for adult specimens between 14 and 18% of pygidial width at pleural furrow 5 and 8, widest at 8. Outer surface of pygidium, except for the pleural furrows, covered by relatively coarse terrace-line pattern. Terrace-lines on axis moderately dense. Number of terrace-lines crossed by a fictive line perpendicular to the border counted to between 15 and 17 on three examined specimens. Pygidial doublure covered by 18 to 20 terrace-lines.

Remarks. – The pygidial outline of *O. henningsmoeni* n. sp. seems to become more triangular with increasing size, but a geographical variation is also seen in that the northern specimens generally have more triangular pygidia than the southern ones and also have a slightly wider or shorter cranidium. The geographical variation seen for the last character is also found within *Ogygiocaris lata*, suggesting it is ecologically controlled.

Ogygiocaris henningsmoeni n. sp. is morphologically very close to *O. sarsi*, but differs on the cranidium by its longer and more anteriorly placed paradoublural line and on librigena by a narrower border. The pygidium may be nearly identical, the main differences possibly being a largely more triangular outline and a generally slightly narrower doublure. The species also shows strong similarities with *O. lata* (see below), but differs in its wider librigenal border; the lack of a posterior truncation on the pygidium and by the generally more triangular pygidium.

Occurrence. – Middle to upper Heggen Member and middle Sjøstrand to middle Håkavik Member of the Elnes Formation. Eiker-Sandsvær: profile 1 at Råen (10.20 m below datum). Oslo-Asker: The beach exposure between Djuptrekkodden and Sjøstrand at Slemmestad, where it has been found from 20.50 to 50.33 m above base of exposure and at Bøveien south of Slemmestad, where it occurs 13.35 to 13.43 m above base of measured road-section. Hadeland: East of Grinaker. Mjøsa: E6 road profile at Nydal (Heggen Member).

Outside Norway it occurs as the most common *Ogygiocaris* in the lower to middle Andersö Shale formation in Jämtland, central Sweden.

Ogygiocaris lata Hadding, 1913

Pl. 3, Figs 9–11; Pl. 4, Figs 1–5; Text-Fig. 49; Tables 11–12

1909   *Ogygia dilatata*, Brünn., var. *sarsi*, Ang. – Holtedahl, p. 28.
1913   *Ogygiocaris dilatata* Brünn. var. *lata* n. var. Hadding, p. 72, pl. 7, fig. 8.
1913   *Ogygiocaris dilatata* Brünn. var. *strömi* Ang. – Hadding, p. 72, pl. 7, fig. 10.
1953   *Ogygiocaris dilatata sarsi* – Størmer, pp. 56–58 (*partim*), ?61, 97, 103, 109, 120, pl. 2.
1960   *Ogygiocaris sarsi regina* n. subsp. Henningsmoen, pp. 216, 227–228, 258, pl. 5, figs 2–8, pl. 6, fig. 5.
1963   *Ogygiocaris sarsi regina* – Skjeseth, p. 64.
1984   *O. sarsi regina* Henningsmoen, 1960 – Wandås, p. 232, pl. 10J, N.
2000b *Ogygiocaris regina* Henningsmoen, 1960 – Månsson, pp. 12, 16, fig. 8l.
2000b *Ogygiocaris lata* Hadding, 1913 – Månsson, pp. 12, 16.
2002   *Ogygiocaris regina* Henningsmoen, 1960 – Pålsson *et al.*, p. 42.

Type stratum and type locality. – Andersö Shale formation. The exact level is not known. Holotype librigena (LO 2523 T) collected from Andersön in Jämtland, central Sweden.

Norwegian material. – The examined material contains 11 cranidia (PMO 56299 c, 203.096, 203.107/6, 203.109/2, 205.096/3, 205.109/3, 206.146, 206.156/15, 206.198/2, 207.423/2, 207.424/1), 38 librigenae (PMO 56299 c, 61134, 203.001, 203.043, 203.071/1, 203.116/8, 203.130/14, 203.154/1, 203.190/3, 203.218/1, 203.334/2, 203.344, 205.083/9, 205.095/2, 205.117/4, 205.121/2, 205.129/1, 205.131/5, 205.136/2, 205.139/3, 205.174/2, 205.179/1, 205.182/2, 205.182/3, 205.186/1, 205.187, 205.188/1, 205.189/1, 205.190/2, 205.194/1, 205.200/2, 205.986/2, 206.012/1, 206.207/6, 206.217/1, 207.423/3, 207.423/3, 207.436/3), one hypostomes (PMO 72099) and 59 pygidia (PMO 203.109/1, 203.166/8, 203.190/9, 203.190/11, 203.348/1, 203.424/3, 203.461/5, 205.108, 205.117/2, 205.118/6, 205.119/2, 205.121/1, 205.124/2, 205.126/1, 205.130, 205.131/6, 205.136/1, 205.137/2, 205.140/5, 205.157/4, 205.182/1, 205.183, 205.184, 205.186/5, 205.192/2, 205.197/5, 205.198/1, 205.199/2, 205.200/1, 205.201, 205.208/1, 205.210/2, 205.215, 205.217/3, 205.222/1, 205.232/1, 205.233/1, 205.234, 205.238, 205.239/1, 205.241/2, 205.242/1, 205.952, 205.954/2, 205.958/2, 205.967/2, 205.976/1, 205.991/2, 205.993, 205.994, 206.000/2, 206.005/1, 206.046, 206.052/1, 206.089/1, 206.096/2, 206.147, 206.151). The

complete or partly articulated specimens are: PMO 33335, 36644–46 (one specimen), 72090 (counterpiece 72092), 72098, 82469, 82557 and 205.159/5.

Diagnosis. – Species of *Ogygiocaris* with paradoublural line on cranidium anteriorly placed, posterolateral distance to cranidial margin approximating one-fifth of cranidial length. Anterior pits and lateral glabellar furrow S4 connected, forming a moderate to deep elongated pit in glabellar flanks. Paradoublural line on librigena situated between 40 and 51% of the distance between eye and margin from eye; width of eye (lat.) corresponding to between 30 and 34% of the width of the pleural field. Hypostome narrow with inward curving anterior terrace-lines and a deep frontal notch in the anterior lobe. Axial width of thoracic segments corresponding to one-fifth the thoracic width. Pygidium with small to moderate posterior truncation; pleural ribs flattened, eight to nine in number; maximal border width and axial width taking up nearly one-fifth the pygidial width; less than one tenth in juveniles. Pygidial doublure generally with 16 to 18 terrace-lines (emended from Henningsmoen 1960).

Remarks. – *Ogygiocaris lata* is morphologically very close to *O. sarsi*, *O. henningsmoeni* n. sp. and *O. delicata*, but is easily distinguished by the posterior truncation of the pygidium; the location of the doublural line, which on the cranidium meets the margin 61 to 68% of the cranidial length from the posterior cranidial margin or 16 to 23% of the cranidial length from the anterolateral margin and between 40 and 51% of the distance between the eye and outer librigenal margin from the librigenal margin on librigena and by the generally short preglabellar area (~15 to 20%), though this may vary a bit.

Ogygiocaris lata is perhaps the geographically most widespread and frequently found *Ogygiocaris* species and has been registered from the Eiker-Sandsvær district in the south and up into Jämtland to the north (Henningsmoen 1960; this study). It seems to be more common in the northern districts of the Oslo Region than in the more southern ones and is thus very rare in the Eiker-Sandsvær district while becoming one of the dominating trilobite species in the Mjøsa area. It is less frequent in the Andersö Shale deposits of central Sweden.

Occurrence. – From the middle Sjøstrand Member and up into the Vollen Formation of the Oslo Region. Skien-Langesund: Trosvik at Brevik. Oslo-Asker: Bødalen, Djuptrekkodden (upper Sjøstrand Member to Vollen Fm.), Elnestangen, Bygdøy, Fornebu, central Oslo and Ildjernet. Modum: Melå. Hadeland: Haugalandet, Hovstangen and north of Juv in Gran. Mjøsa: Dyste Bridge in Toten, Hovindsholm on Helgøya and along E6 at Nydal north of Hamar.

Table 11. Cranidial measurements on *Ogygiocaris lata* Hadding, 1913

PMO number	L	Lg	Lp	Lo	Lt	Lf1	Lf2	Lae	Le	Lb	Lbp	Wb	W	Wc	Wo	Wgmin	Wae	Outer t.	Inner t.
206.198b2	17.5	13.8	3.7	2	1.2	3.3	2.4	8.8		3.8	10.9	18.5	~35.2	20.9	~9.6	8.9	~12.2	5	
33270	15	12.9	2.1	1.8	1.4	2.6	2	7.5		3.3	9.6	12.4	~23.3	~14.3	7		7.9	4	
72082	28.9	23.5	5.4	2.9	2.9	4.7	3.6	13.3	6.1	6.2	17.8	23.1	~41.7	28.6	13.4	12.9	16.9	2	
36722	17.6	15	2.6	1.9	1.6	2.9	2	8.1	~4.2	4	11.3	15.6	26.3	19.1	7.1	6.8	10.4	6	15
72090	22.2	18.7	3.5	2.6	1.5	3.6	2.7	10.7	6.7	4.5	14.4	21.9	34.6	24.4	10.6	10	14.1	2	
36790	32	26.5	5.5	3.3	2.3	5.6	4.4	15.2		6.5	20	26.8	~51.2	33.1	13.5	14.3	19.5	2	
205.991a1	13.9	11.1	2.8	1.6	1.3			6.4		~2.3	8.5	~11.7		~14.2	~6.3		~6.8		
206.156d19	9	7.4	1.6	1		1.6	1.1	4.2	2.1	1.7	5.7	~6.5	~13.5	~8.1	3.5		4.7	5	
206.156b15	5.1	4.6	0.5	0.6	0.4			2.5	1.5	0.9	3.4	3.8		4.3	2.2		~2.8		
72099	14.5	11.7	2.8	1.6	1.3	2.7	2.2	7.2		2.9	9.7	12.7	23.9	14.1	6	6.1	8.2	4	
33342	17.6	14.5	3.1	1.8	1.6	3.1	1.8	8		3.6	11.5	16	~29.0	19.3	8.4		10.7	4	
33341	16.5	13.3	3.2	1.8	1.2	2.9	1.9	7.8	3.1	3.3	10.9	15.4	~28.0	17.8	7.4		9.7	4	
33103	19.7	15.3	4.4	2.1	1.6	3.3	2.4	8.8		3.7	13.1	~17.4	~30.8	~19.6	8.5		10.7	2	
33103/2	10.3	8.4	1.9	1	0.8	1.7	1.4	4.9	2.2	1.9	6.8	9.3	?	10.8	4.6		5.8	3	
72675	16.9	14.5	2.4	2	1.5	2.9	2	8.2	3.3	2.7	10.9	15.5	~25.9	17.7	8.4		9.7		
82576	27.3	24.5	2.8	3.4	1.8	5.6	4.8	14.6	8.4	5.2	17.4	26.3	~44.8	29.4	13.9		17.3		19
33255	18.8	15	3.8	1.6	1.3	2.8	2	8.3	3.1	3.6	11.7			~8.4			8.4	1	
203.107a6	9	7.8	1.2	1.1	0.8	1.4	1.3	4.2	2.1	1.7	5.6			4.8					5
S1802	9.4	7.9	1.5	1.1	0.6	1.6	1.2	5		2.1	5.9	7.7	14.2	9.6	4.2	4.1	5.4	4	
82360	17.9	14.5	3.4	2	1.45	3	2.4	8.4		3.85	11.2	15.7	~27.7	18.6	8.9	8.3	10.5	1	3
82367	15.1	12.7	2.4							3	10.25	13.35	~25.9	~15.8					
82363	15.25	13.6	1.65							2.7	~9.9								

Abbreviations: see Table 7.

Table 12. Librigenal measurements on *Ogygiocaris lata* Hadding, 1913

PMO number	Le	Lp	Ls	Wbmin	Wp	Wpp	Wl	Wb1	Wb2	Wpm	We	Terr. doubl.	Terr. surface
203.001	2.2	5	2.2	1	0.9	1.1	2.2	1.3	1	3.6	1.3	13	
203.043		9.3			~1.8	~2.1	4.5	3.2	2.4	6		13	
203.344	5.9	19.1		5.4	3.8	3.8	8.1	4.9	6			15	
203.347	1.8	~3.9		0.7	0.7	0.8	1.7	1	0.7	~2.7			
203.154	6.7	?		5.2	4.5	4.7	10.4	6.7				16	
203.166d	6	~18.8		4.2	4.2	4.5	9.3	~5.6	~4.2				
203.230					4.2	4.2	10.1						
205.083		9.5		2.5	2.1	2.1	4.8	2.7	2.6			14	
205.095b2				2.2	2.1	2.1	4.4	~2.3					8
205.131b5	~7.0	15.1		3.7	3.2	3.3	7.3	4.2	3.7	8.5	0.29		
205.136	~3.5		4	1.9	2	2	3.9	2.4	2.2	5.9			
205.139b3		~12.7		3.2	3.1	3.2	6.4	~3.6	3.3				
205.174	3.9				2.2	2.3	4.6					10	
205.179	3.8	~9.1		2.5	2.3	2.4	4.9	2.7	2.6	~6.2			
205.182b2		12.1		3.8	3	3	6.9	4.4	3.9			13	
205.182b3	3.8	~11.6		2.8	2.3	2.6	5.6	3.4	3.1			16	
205.187	5.6	15.9		~4.6	3.7	4.1	8.3	5.4				16	
205.189a1				3.8	3.4	3.5	7.1						
205.194	6.2			5.1	3.9	4.1	9.4	5.7				17	
205.249					~1.3	1.3	2.9						
205.972					3.2	3.3	6.9						
205.986					1.7	1.8	4						
206.012	3.9			2.3	2	2	4.4	2.7				13	
36786	6.2	14.1		3.5	3.1	3.2	6.9	4.5	3.8	9.4	3	13	
36722	5.3	11.6	8.8	3.4	3.4	3.6	7.4	4.1	3.4	9.8	3	15	
72090	7.4	16.4	~5.5	3.8	3.8	4	7.8	4.3	4.2	12.8	3.7	15	
33269/1	4.3			2.5	2.4	2.5	5.1	2.9	2.6	~6.6		12	7
36758	5.5	14.7		3.6	3	3	7.1	4.7	4			18	
72675	5.5	13		3	2.9	2.9	6	3.6	3.3	8.6		13	13
72098	3.5			2.2	2.2	2.4	4.6	2.5				12	
36630_1					2.5	2.65	5.15						

Abbreviations: see Table 8.

Outside Norway it occurs from the middle Lower Shale member up into the Upper Shale member of the Middle to Upper Ordovician Andersö Shale formation of Jämtland, Central Sweden (Månsson 2000b; Pålsson *et al.* 2002).

Ogygiocaris delicata Henningsmoen, 1960

Pl. 4, Figs 6–12; Tables 13–14

1953 *Ogygiocaris dilatata sarsi* – Størmer (*partim*), pp. 97(?), 103, 109(?), 120 (occurrence).
1960 *Ogygiocaris sarsi delicata* n. subsp. – Henningsmoen, pp. 216, 228–229, 258, pl. 6, figs 1–4, 7.
1963 *Ogygiocaris sarsi delicata* – Skjeseth, p. 64 (listed).

Type stratum and type locality. – Holotype, PMO 36673, was collected from the Elnes Formation on Helgøya, Mjøsa districts; exact level unknown.

Material. – The examined material contains one cephalon (PMO 82904), 19 cranidia (PMO 33067, 33069, 33118, 33446 (two specimens), 36634, 36673 (two specimens), 36792, 36879, 36885, 70555, 72091, 72733, 82362, 82572, 82904, 207.443/3, 207.443/ 11), 17 librigenae (PMO 36673, 36675 (three specimens), 36686 (three specimens), 36783, 36838, 36851, 36879, 72667, 72733, 82372, 82904, 205.150/1, 207.443/1), one hypostome (PMO 36879), 13 pygidia (PMO 36673, 36675 (two specimens), 36684, 36686, 36851 (two specimens), 36879, 67189 (two specimens), 82372, 207.443/2, 207.443/10) and two partly articulated specimen (PMO 72101, 82362).

Diagnosis. – A small *Ogygiocaris* with short preglabellar area; paradoublural line close to cranidial margin, distance at anterolateral corners on cranidium

Table 13. Cranidial measurements on *Ogygiocaris delicata* Henningsmoen, 1960

PMO number	L	Lg	Lp	Lo	Lt	Lf1	Lf2	Lae	Le	Lb	Lbp	Wb	W	Wc	Wo	Wgmin	Wae	Outer t.	Inner t.
72091	11.2	9.5	1.7	1.3	0.9	1.9	1.3	5.6	2.1	1.6	7.9	11.4	~19.3	13.1	5.2		7.1		
36673	16.9	14.4	2.5	1.9	1.8	3.2	2.3	8.4		2.2	11.8	17.2	29	18	8.7	7.5	10		
70555	21.5	18.5	3	2.2	1.8	3.5	2.5	10.9		2.5	15.2	21.8	~34.8	23.3	10	9.3	13.7	4	
82904	12.2	10	2.2	1.4	1.1	2.3	1.7	6.4	3.4	1.8	8.8	~11.9	~19.4	~13.9	6	5.2	7.4		
33118	13.5	11.6	1.9	1.3	0.7			6.8		1.8	9.4	12.9			13.6	5.8		7.7	4
33069	7.7	6.5	1.2	0.9	0.7	1.4	1.2	3.8	1.7	1.3	5.4	~7.6		~8.2	3.2		4.5	4	
33446/1	8.8	7.2	1.6	1	0.9	1.8	1.2	4.5		1.1	6.4	9	14.9	9.7	4.3		5.4	3	
33446/2	15.7	14	1.7	1.8	1.6	3.1	2.3	7.9		2.4	11.4	16.9	~27.5	18.4	7.7		10.7	4	
36885	15.5	13.4	2.1	1.8	1.5		1.9	7.9		2.1	10.7	16.1		17.3	7.1		10.5	6	
36792	4.4	3.9	0.5	1.25	0.6	0.8	0.55	2		0.55	3.4	4.35	~6.7	4.4	1.9		2.65		
36634	18.3	15.6	2.7	2		1.6		6.8	~3.5	2.4	12.9	17.5		19.5	9.1		11.4		
207.443/3	9.4	8.1	1.3	1.2	0.8	1.7	1.3	4.8		1.2	7.1	~11.0	~17.4	~12.6	5.8	4.5	6.2	4	
207.443/11	17.4	14.6	2.8	2.3	1.7	3.2	2.3	8.7	~3.3	2.1	12.3	16.5	~28.4	17.4	7.9	7.4	10.4	4	
82572	20.4	17	3.4				2.1	9.6	4.4	2.7	13.6	18.4		21			~12.6	4	
72100/1	19.1	16.9	2.2	2.1	~1.7		2.1	7.6		3	~12.8	~18.4		~20.2	9	8	10.9	4	
72100/2	18.9	15.8	3.1	2	1.8			9.5		2.9	12.8	~19.0		~22.1	9.5	8.3	12.1	3	
72100/3	15.2	~12.9	~2.3	1.7	1.3			7.9		2.2	10.3	16.1		16.9	~7.3	~6.6	~9.8	3	
33118	14.2	11.9	2.3	1.6	0.9			7.2		2	9.7	13.2		14	6.1	6	8.2	4	9
33069	7.6	6.5	1.1	0.8	0.6	1.5	1.2	3.9	1.7	1.25	5.5	~7.7		~8.4	3.9	3.2	4.4	4	10
36629	9.4	8.5	0.9							1.5	6.5								
82359	18.4	15.3	3.1	1.9	1.65			8.8		2.4	~13.3	17.95		19.3	8.95	8.05	10.95	2	6
33067	2.6	2.3	0.3	0.35	0.35	0.5	0.4	1.4	0.65				5.1	~3.0	1.2		~2.0		

Abbreviations: see Table 7.

Table 14. Librigenal measurements on *Ogygiocaris delicata* Henningsmoen, 1960

PMO number	Le	Lp	Ls	Wbmin	Wp	Wpp	Wl	Wb1	Wb2	Wpm	We	Terr. doubl.	Terr. surface
36630_2					2.8	2.85	5.1						
36629					2.4	2.4	4.3						
205.150	~2.8	7.2	3.6	1.7	1.85	1.9	3.5	1.9	1.9	~5.0		11	
36673	4.5	12.9	3.7	2.2	3	3.2	5.4	2.6	2.5	7.8	2.6	11	
36675	7.9	16.7		4.1	5.8	5.8	9.7	4.3	4.3	11.7	4.5	11	
36675	4.3	11.9		2.1	2.8	3	5	2.3	2.1			11	
36783		17.9		3.2	4.2	4.5	8	4.4	3.6				9
36838	4.5	10.2		2	2.5	2.5	4.4	2.1	2	6.1		11	7
36879	5.4	13.5		2.9	3.8	3.7	6.7	3.1	2.9	9.8	3.5	14	
72100	4.4			1.8	2.9	2.9	4.9		1.8	8.3	2.9		~8
72101	3.1	6.6		1.1	1.9	1.9	3.2	1.4	1.1		1.6		~12
72667	4.8	12	~4.6	2.3	2.8	2.9	5.1	2.5	2.5	9.3			7
82904	4.1	9.3			2.7	2.8	4.6	2	2	8.2			
36675		9		2.3				2.4	2.3	6.9		11	12

Abbreviations: see Table 8.

corresponding to between 11 and 17% of cranidial length. Anterior pit in axial furrow highly effaced, more pronounced on juveniles. Paradoublural line on librigena situated between 53 and 60% of the distance between eye and margin from eye; width of eye (lat.) corresponding to around one-third the width of the pleural field. Axial width of thoracic segments corresponding to just less than one-fifth the thoracic width. Pygidium with small posterior truncation; eight pleural ribs flattened without interpleural furrow; maximal border width close to one-fifth the pygidial width; axial width taking up nearly one-fifth the pygidial width. Pygidial doublure bearing around 18 terrace-lines (emended from Henningsmoen 1960).

Remarks. – *Ogygiocaris delicata* is morphologically very close to *O. lata* and it could perhaps be argued that it merely should be regarded as a subspecies or perhaps ecological phenotype of this, though they clearly overlap geographically as well as stratigraphically. For these reasons it is here regarded as a full species largely restricted to the latest Darriwilian (*Pseudamplexograptus distichus* to *Dicellograptus vagus* graptolite Zone) deposits of the northern Oslo Region. It is characterized from *O. lata* by its smaller size (largest measured cranidium 21.5 mm long), the narrow cephalic doublure, with the paradoublural line closer to the outer margin than to the eye; by the generally smaller exoskeletal size and in the only very slightly developed or absent posterior pygidial notch.

Occurrence. – Uppermost Engervik Member in the Oslo area and the upper part of the shales below the Engervik Member to the north. Oslo-Asker: Djuptrekkodden (uppermost Engervik Member) and Elnestangen (Engervik or Håkavik Member) at Slemmestad. Ringerike: Main road north of Hval. Mjøsa: Flakstadelva at Hedmarken, Hovindsholm on Helgøya, Furnes Church, E6 at Nydal (upper part of shales below Engervik Member).

Ogygiocaris striolata Henningsmoen, 1960

Diagnosis. – A species of *Ogygiocaris* characterized by a relatively short cranidium with an approximately rectangular outline; parafrontal band highly reduced or absent; preglabellar area very short; preocular area narrow, posterolateral projection extremely short (exsag.). Librigena with narrow border widening

strongly anteriorly; paradoublural line located midway between eye and margin and long genal spines. Axis on thoracic segments with length width ratio of only 25% and an axial width corresponding to slightly more than half the pleural width. Pygidium parabolic to sub-semicircular in outline with no posterior indentation. Axis moderately broad, corresponding to between 74 and 84% of the pleural width; each axial ring with fairly well-developed lateral lobes; pleural ribs in cross-section rounded to slightly flattened. Outer surface covered by moderately dense and fine striation (emended from Henningsmoen 1960).

Remarks. – *Ogygiocaris striolata* appears to be a short-lived and rare species in the Oslo Region, only occurring in a stratigraphically short interval covering the initial change from shale-dominated deposits to more marly deposits dominated by concretionary carbonate nodules.

The species was split into two subspecies by Henningsmoen (1960), mainly based on the density of striation on the pygidium. This splitting seems valid and is retained in this work. Because of a new and better material illustrating parts of the exoskeleton which previously were unknown a full description of each subspecies is given below.

Ogygiocaris striolata striolata Henningsmoen, 1960

Pl. 4, Figs 13–17; Pl. 5, Figs 1–4; Tables 15–16

1960 *Ogygiocaris striolata striolata* n. subsp. Henningsmoen, pp. 221–223, pl. 3, figs 4–10.
1984 *Ogygiocaris striolata striolata* Henningsmoen, 1960 – Wandås, p. 232, pl. 10G–L.

Type stratum and type locality. – Holotype pygidium PMO 72108 and counterpiece PMO 72116 belongs to the top of the Sjøstrand Member. It was collected at Djuptrekkodden just north of Slemmestad, Oslo-Asker district.

Material. – One cephalon (PMO 206.249), two cranidia (PMO 72115, 82450), four librigenae (PMO 72104, 72112, 102.746, 203.223 c/1), one hypostome (PMO 72104), 19 pygidia (PMO 3750, 56558, 56559, 72085 (+72086, 72088), 72104, 72108, 72114, 82335, 82458, 102.746, 203.247/1, 203.248/1, 203.248/2, 203.250, 203.253, 203.254, 206.238, 206.247, 206.250) and one nearly complete specimen (PMO 95829).

Table 15. Cranidial measurements on *Ogygiocaris striolata striolata* Henningsmoen, 1960

PMO number	L	Lg	Lp	Lo	Lt	Lf1	Lf2	Lae	Le	Lb	Lbp	W	Wc	Wo	Wgmin	Wae
82450	~23.8	22.4	~1.4	3	2.3	~3.3	2.8	12	4.7	~7.4	~13.7	~39.1	23.8	14	13.5	15.8
95829	~15.5	14.1	~1.4	1.7	~1.4	~2.1	1.5	8.2		4.6	8.5	~27.6	~15.1	8.9	8.3	10.9

Abbreviations: see Table 7.

Table 16. Librigenal measurements on *Ogygiocaris striolata striolata* Henningsmoen, 1960

PMO number	Le	Lp	Ls	Wbmin	Wp	Wpp	Wl	Wb1	Wb2	Wpm	Terr. doubl.	Terr. surface
72112				4.5	4.7	4.9	9.4		5	14.6	~30	52
72104	5	~14.2	11.1	2.8	2.6	2.8	5.9		3.1	~10.7	~32	
95829	5.6	16.8	13.3	2.5	3.5	3.6	6.4	5.1	2.5	~13.8	20	

Abbreviations as in Table 8.

Diagnosis. – Subspecies of *O. striolata* with distance between one lateral glabellar furrow and the next decreasing evenly from posterior to anterior; preocular area and librigenal border narrow; paradoublural line on cranidium very posteriorly placed, abaxial distance to anterior margin close to one-third the cranidial length. Hypostome with relatively dense terrace-line pattern, anterior end of terrace-lines turned outwards. Pleural furrow on thoracic segments long, corresponding to three-fourths the pleural width. Pygidium with slightly flattened pleural ribs and with 26 to 32 terrace-lines on doublure. Surface of exoskeleton covered by dense and fine striation (emended from Henningsmoen 1960).

Description. – Medium sized *Ogygiocaris*, largest pygidium 41 mm long. Cephalon semicircular in outline, L/W ratio probably around 40%.

Cranidial length approximating 60% of the width. Glabella slightly contracted at middle; length corresponding to more than 90% of the cranidial length. Largest glabellar width located midway between eyes and anterior margin, while minimum glabellar width located between the eyes corresponds to a bit more than half the anterior cranidial width. Occipital ring with distinct posterior furrow. Occipital furrow laterally deep, becoming indistinct mesially, where it turns rearwards into a V-shape around mesial tubercle. Most anterior part of furrow located just above 10% of cranidial length from posterior margin. Mesial tubercle small, situated at about 10% of the cranidial length from posterior margin. Main part of glabella with four pairs of lateral glabellar furrows or pits; the first, S1, only partly connected to axial furrow, terminating slightly more than one-third the glabellar width inside glabella. S2 not or only slightly connected to axial furrow. S3 very shallow, describing a strongly forward convex curve. S4 short, shallow, strongly forwardly directed, starting just inside axial furrow. Posterior lobe, L1, slightly inflated, triangular in outline. L2 is sub-rectangular in outline. L3 and L4 shorter, strongly curved. Frontal glabellar lobe pyriform to subcircular with broadly rounded anterior margin. No anterior pit is observed in axial furrow. Preglabellar area short, less than 10% of cranidial length at sagittal line. Preocular area broader, but still relatively narrow for the genus with a width corresponding to one-third the glabellar width. Preocular area ends just in front of cranidial midline. Paradoublural line curving strongly rearwards from posterolateral margin of preocular field to axial furrow; the initial distance to anterior cephalic margin around 30% of the cranidial length; distance from posterior cranidial margin to abaxial end of paradoublural line corresponding to about 55 to 60% of the cranidial length. Palpebral lobes very small, exsagittal length equivalent to 20% of the cranidial length, situated about one-third the cranidial length from the posterior border or outside posterior half of S2. Posterior fixigena moderately broad, maximal exsagittal length of posterolateral projection close to 15% of cranidial length, transected by moderately deep posterior border furrow. Length of inner part of posterolateral projection corresponding to around 70 to 75% of the abaxial part. Cranidial surface covered by dense striation, which is commonly weathered away. Glabella with forward convex pattern of striation, becoming increasingly more pronounced towards glabellar front. Striation turns rearwards on preocular area. Posterior axial furrow on cranidium with forward concave striation, which changes to obliquely anterolaterally directed striation on the posterolateral projection.

Length of eye about 35% of librigenal field length from posterolateral to anterolateral corner; shallow furrow may be present around eye socle. Paradoublural furrow located slightly outside midline between eye socle and lateral margin. Course of paradoublural furrow sub-parallel to border margin, curving inwards and rearwards towards eye in front. Lateral border narrow, widening strongly anteriorly on specimen PMO 95829, continuing rearwards into a long posteriorly directed cylindrical genal spine, 80% as long as genal field. Surface of genal field and border covered by very fine striation running obliquely out- and forwards, on the border turning in a more exsagittal direction. About 52 striae were counted on the border on specimen PMO 72112. Doublure with 20 to 32 terrace-lines on three examined specimens, strongly concave.

Hypostome subrectangular, only slightly longer than wide, largest width across anterior wing less than one-third from anterior margin. Anterior lobe of middle body rounded subrectangular in outline, reaching the curved anterior margin; anterior lobe delimited posteriorly by distinct middle furrow. Posterior lobe of middle body forward concave, width equivalent to 50% of the maximum width. Maculae situated on anterolateral flanks of posterior lobe. Anterior border exsagittally narrow, continuing out into short, evenly rounded wings. Lateral border length corresponding to slightly more than half the length of hypostome. Posterior border rounded subtriangular. Surface of hypostome covered by relatively dense terrace-line pattern more or less following the general contours, on the middle body forming into a sharp V-pattern. Anterior end of terrace-lines turned outwards.

Thorax of nearly equal width throughout, axis slightly widening towards front. Thoracic axis occupying 20% of the thoracic width, length (sag.) corresponding to 25% of the width. Pleurae short (exsag.) and wide, transition between inner and outer pleural portion located around 40% of the pleural width from axial furrow. Pleural area crossed by moderately wide and deep pleural furrow. No terrace-lines have been preserved.

Pygidium parabolic to sub-semicircular in outline with rounded to nearly straight posterior margin; length around 60% of the width. Axis long and slender, terminating around 25% of the pygidial length from the posterior margin. The axial width varies between 19 and 22% of the pygidial width or between 74 and 83% of the pleural field width. There are 10 to 11 axial rings in addition to the terminal piece and articulating half ring. Pleural field carries eight or nine pleural ribs in addition to anterior half rib. Ribs vaulted to slightly flattened in cross-section. Inter-pleural furrow indistinct, situated on the anterior half of the pleural rib. The furrow widens and deepens somewhat on distal half of rib. Border enfolding the posterior half of the terminal axial piece; border width for adult specimens between 16 and 19% of pygidial width at pleural furrow 5 and 8; outer surface in cross-section straight to slightly concave, doublure concave. The width at pleural furrow 8 corresponds to between 90 and 105% of the abaxial distance between pleural furrow 5 and 9.

Outer surface of pygidium, except for the pleural furrows, covered by very fine and dense terrace-lines. Terrace-lines on each axial segment forming one large central and two smaller lateral, forward convex

lines on posterior part, converging to form a single curve anteriorly. Striation on pleural ribs directed perpendicular to axial line, turning more antero-laterally outside paradoublural line. The number of terrace-lines crossed by a fictive line perpendicular to the border counted to between 50 and 75 on three examined specimens. Pygidial doublure covered by moderately dense terrace-lines arranged in a semiparallel pattern to the margin. There are between 26 and 32 between margin and paradoublural line, becoming increasingly more closely spaced adaxially.

Remarks. – *Ogygiocaris striolata striolata* is morphologically closest to *O. dilatata*, from which it differs on the narrow librigenal doublure, the Bertillon-type pattern striation on the cranidium, the less inflated pleural ribs on the pygidium and by the very fine striation found on the exoskeletal surface and doublure. From the other species of *Ogygiocaris* it is easily distinguished by the broad hypostomal outline, the very short preglabellar area and by the fine terrace-lines and Bertillon-type glabellar striation.

Occurrence. – Elnes Formation. Skien-Langesund: Skinvika (?) at Rognstranda. Eiker-Sandsvær: Råen at Fiskum (level 4.46 to 5.97 m in profile 1, corresponding to the upper Heggen Member). Oslo-Asker: Djuptrekkodden and Elnestangen at Slemmestad (lower Engervik Member); Fornebu in Bærum and at Vigelandsfontenen, Oslo (uppermost Sjøstrand or lower Engervik Member). Modum: Vikersund skijump. Ringerike: beach at Gomnes.

Ogygiocaris striolata corrugata Henningsmoen, 1960

Pl. 5, Figs 5–10; Tables 17–18

1960   *Ogygiocaris striolata corrugata* n. subsp. Henningsmoen, p. 223, pl. 3, figs 1–3, 11.

Type stratum and type locality. – Holotype pygidium PMO 72134 belongs to the uppermost part of the Sjøstrand Member at Djuptrekkodden just north of Slemmestad, Oslo-Asker district.

Material. – Eight cranidia (PMO 72131, 203.081b/8, 203.337b-c/3, 203.343a/2, 206.290/5, 207.427/3, 207.429/2, 207.429/5), seven librigenae (PMO 206.279/1, 206.292b/3, 207.429/3, 207.429/4, 207.432/6, 208.505/6, 208.505/7), one hypostome (203.346/1), four thoracic segments (PMO 206.253/1) and 20 pygidia (PMO 3306, 4740, 72120, 72134, 82438 (two specimens), 82635, 82655, 203.081a/1, 203.081b-d/9, 203.184b/2, 203.337a/1, 203.345a-b/1, 206.251, 206.253/1, 206.275 c/3, 206.286/1, 206.288/2, 206.288/3, 207.430).

Table 17. Cranidial measurements on *Ogygiocaris striolata corrugata* Henningsmoen, 1960

PMO number	L	Lg	Lp	Lo	Lt	Lf1	Lf2	Lae	Lb	Lbp	W	Wc	Wo	Wgmin	Wae
72131	37.5	35.1	2.4	4.6	3.3	6	5.2	18.7	~9.9	~22.9	~55.6	34.3	22.2	20.9	24.8

Abbreviations: see Table 7.

Table 18. Librigenal measurements on *Ogygiocaris striolata corrugata* Henningsmoen, 1960

PMO number	Le	Lp	Wbmin	Wp	Wpp	Wl	Wb1	Wb2	Wpm	Terr. surface
208.505/6	7	20	4.7	4.5	4.9	10.2	6.5	5.5	15.3	~36
208.505/7	7.2	~19.5	4.7	4.1	4.4	~9.0	~6.8	5.6	13.9	~35

Abbreviations see Table 8.

Diagnosis. – Subspecies of *O. striolata* with glabellar width at occipital furrow nearly two-thirds the cranidial width at preocular area. Distance between lateral glabellar furrow S2 to S4 largely equal and distinctly shorter than between S1 and S2; distance from abaxial end of cranidial paradoublural line to anterior margin larger than one quarter the cranidial length. Inner portion of pleura only slightly wider (tr.) than one-third the total pleural width. Pygidium with rounded pleural ribs; moderately dense striation on outer surface; doublure with around 20 terrace-lines (emended from Henningsmoen 1960).

Remarks. – *Ogygiocaris striolata striolata* is morphologically very close to *O. striolata corrugata*, but differs by its possibly shorter maximum length of posterolateral fixigenal projection (~14% of cranidial length on two specimens); slightly narrower librigenal border, with the paradoublural line in general passing closer to outer margin than to eye; relatively wide inner portion of thoracic pleura compared to outer (Wt/Wpl ≈ 40% against 34% on *O. s. corrugata*); larger number of terrace-lines on the pygidial doublure (>25); the denser surface striation and by the somewhat flattened pygidial ribs. The differences are small, but with regard to the high taxonomic splitting of recent shrimps and lobsters they are easily large enough to support two separate species. On the other hand observations from this study suggests there is a stratigraphical gap between the two, although Henningsmoen (1960, pp. 215–216) believed he found *O. striolata corrugata* some metres above an *O. striolata striolata* specimen. As he clearly was a little unsure as to its stratigraphical value (Henningsmoen 1960, p. 216: '…, apparently *in situ*, 3.5 m above …') I regard it as having been misplaced stratigraphically. The suggested stratigraphical gap rather suggests that they belong to an evolutionary lineage, where *O. striolata striolata* evolved from *O. striolata corrugata*. For this reason I follow Henningsmoen (1960) in regarding them as subspecies.

Occurrence. – Elnes Formation. Eiker-Sandsvær: Råen at Fiskum (upper Heggen Member). Oslo-Asker: Djuptrekkodden and Elnestangen at Slemmestad (uppermost Sjøstrand to lowermost Engervik Member) and Hjortnæstangen in Oslo. Modum: Vikersund skijump (upper Heggen Member). Mjøsa: E6 road section at Nydal (upper Heggen Member).

Subfamily Asaphinae Burmeister, 1843

Remarks. – The subfamily has in more recent times been divided into four families and a number of subfamilies (e.g. Balashova 1976). However, in the present study several problems with this splitting have been noted, the two most prominent being the separation of *Ogmasaphus* and *Asaphus* into two different families, although their general morphology leaves no doubt of a very close relationship (for further discussion on relationship see under *Asaphus*), and the similar separation of *Volchovites* and *Pseudobasilicus* into different subfamilies. The latter are here regarded as too near each other to be placed in different subfamilies (for details see discussion of *V. perstriatus*). In summary the observed inconsistencies indicate some underlying problems in the subdivision of Asaphinae as presented by Balashova (1976). Until further revisions on the families and subfamilies of Asaphidae have been carried out it seems most prudent to keep to the older nomenclature proposed in Moore's (1959) *Treatise on Invertebrate Paleontology*.

Genus *Volchovites* Balashova, 1976

Type species. – *Volchovites simonkovensis* Balashova, 1976.

Diagnosis. – Exoskeleton moderately convex with distinct axis; mesial tubercle small, situated just in front of moderately shallow occipital furrow; posterior furrow on fixigena fairly deep; eyes large, situated more than half their length from posterior cranidial margin; a nearly straight posterolaterally directed ridge is seen to leave the central portion of the eye, continuing out to the paradoublural line; paradoublural line fairly distinct, marked by the raised border area; posterolateral librigenal corners slightly pointed, not forming genal spines. Fixigenal surface in front of ridge and inner part of librigenal border bearing numerous small exsagittally directed terrace-lines. Thoracic segments with lateral pleural terminations curving rearwards into sharply pointed posterolateral corners, although they are not nearly as spine-like as on *Pseudobasilicus*; panderian openings on pleurae isolated and small. Pygidium subtriangular, moderately convex with indistinct to moderately developed pleural ribs. Doublures extremely wide with fine and densely spaced terrace-lines (modified from Balashova 1976).

Remarks. – *Volchovites* is very closely related to *Pseudobasilicus* described below but differs in its more convex exoskeleton with higher convexity of the borders; the more anterior placement of the eyes, which are situated more than half their length in front of the posterior cranidial margin; the lack of genal spines; the presence of numerous small exsagittally directed terrace-lines on anterior part of librigena, which are normally absent on *Pseudobasilicus*; the slightly less spine-like thoracic pleural terminations and by the more subtriangular pygidium.

Occurrence. – The genus is found in Baltoscandian marls and limestones of middle to upper Kunda age.

Volchovites perstriatus (Bohlin, 1955)

Pl. 5, Figs 11–15, Pl. 6, Figs 1–8; Text-Fig. 50

1955 *Pseudoasaphus perstriatus* nov. sp. Bohlin, pp. 138–140, text-figs 3, 8B, pl. 5, figs 3–5, pl. 6, figs 1–2; table 1.
1976 *Volchovites perstriatus* (Bohlin, 1955) – Balashova, pp. 35, 36 (affinity).
1982 *Pseudoasaphus perstriatus* Bohlin – Jaanusson, 167, 179 (occurrence).
1984 *Ogmasaphus tropidalox* n. sp. Wandås, pp. 221–222, pl. 3C–L, pl. 4A–C.
1984 *Pseudobasilicus perstriatus* (Bohlin, 1955) – Wandås, pp. 221–222, pl. 4D–G, J.
1985 *Ogmasaphus tropidalox* Wandås – Owen, p. 263, fig. 5n–o, table 3.

Type stratum and type locality. – Holotype pygidium UM.Nr.Ar.4240 collected from the uppermost Holen Limestone at Böda Hamn on Öland, Sweden.

Norwegian material. – Ten cranidia (PMO 67575, 67584, 67587, 67598, 83149, 83178, 89760, 90483, 90542, 106.384), 12 librigenae (PMO 67184, 67574, 67575, 67583, 67595, 83092, 83178, 83238, 90388, 90483, 90542, 207.390/5), one hypostome (PMO 93130), 88 pygidia (PMO 36683, 67147, 67184, 67187 (two specimens), 67574 (three pygidia), 67575 (two specimens), 67577, 67581, 67582, 67585, 67587, 67595, 67598, 67600, 67602, 83062, 83090, 83092, 83093, 83103, 83149 (four specimens), 83154, 83177, 83178 (four specimens), 83238 (two specimens), 83240, 83258, 83260, 83263 (two specimens), 89760, 89763, 89767, 90388, 90389, 90391, 90457, 90462, 90482 (two specimens), 90513, 90515, 90542 (three specimens), 90548, 90564, 93130, 101.738, 102.555, 106.001, 106.019 (two specimens), 106.038, 106.098, 106.143 (three specimens), 106.151 (two specimens), 106.374, 106.383 (two specimens), 106.384 (four specimens), 143.474, 143.476 (three specimens), 207.368/2, 207.378, 207.381/1, 207.390/4, 207.399/3, 207.519)

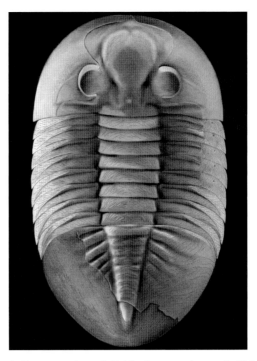

Fig. 50. Reconstruction of *Volchovites perstriatus s.l.* (Bohlin, 1955) from the Elnes Formation.

and 19 partly articulated specimens, including cephalons and specimens composed of pygidium and a few thoracic segments (PMO 33326, 33559, 67596, 67602 (three specimens), 82056, 83246, 93128, 105.899, 105.978, 105.979, 105.998, 105.999, 106.008 (with hypostome), 106.010, 106.090, 207.379/1, 207.380).

Description. – Cephalon with very wide doublure bearing up to about 40 terrace-lines. Preglabellar area short, length corresponding to between 10 and 15% of cranidial length. Anterior cranidial margin comparatively straight with narrow anterior ogive. Eyes of moderate size, length corresponding to 25% of cranidial length and situated between 50 and 75% of their own length in front of the posterior cephalic margin, closest on the smaller specimens. Posterolateral genal corners slightly pointed.

The hypostome has not been described before and is consequently treated in some detail. Hypostome approximately 1.2 times as long as wide, largest width located just behind midline; anterior border short and broadly rounded, laterally ending in moderately short anterior wings; anterior lobe of middle body rounded rectangular in outline, slightly longer than wide, the length corresponding to a little over half the total length; middle furrow short and deep; macula distinct, raised, facet facing laterally to anterolaterally, situated about one-third the hypostomal width from each other; posterior lobe of middle body short and low; anterior part of lateral border located around one-fifth the hypostomal length behind the anterior hypostomal margin; lateral border merges posteriorly with a long posterior border marked by a deep central V-incision, depth of incision corresponding to one-third the hypostomal length, maximal width being just below half the hypostomal width. Hypostome smooth except for some terrace-lines on the wings and anterolateral portion of the anterior lobe of middle body. Terrace-lines are much coarser than those of the cephalic and pygidial doublures.

Pygidium subtriangular, the length, excluding the anterior half ring, corresponding to two-thirds the width; axis moderately narrow, width corresponding to nearly 30% of pygidial width; axial rings numbering 12 to 14; pleural field with six faint to distinct and slightly sigmoidal pleural ribs; border comparetively narrow, width of pleural field at anterior pleural rib corresponding to between 33 and 47% of axial width. Pleural field covered by moderately dense anterolaterally directed pattern of long undulating terrace-lines, terrace-lines becoming sparser near outer margin and may completely dissappear along posterior part of border. Doublure bearing from 40 to nearly 110 very fine and densely packed terrace-lines.

Remarks. – The Norwegian pygidia present quite a large variation with regard to number of doublural terrace-lines, pleural convexity and development of pleural ribs. The number of doublural terrace-lines may be somewhat stratigraphically related with the more dense patterns dominating in the older populations, but a larger sample is needed for a better control on its stratigraphical and potential ecological relations. The pygidial convexity is clearly related to facies type and through this to the stratigraphic level. Pygidia with a large convexity on the abaxial parts of the pleural field (e.g. Plate 6, Figs 3–5) are thus found in the coarser limestone beds, while those with a more flattened outer pleural field and stronger adaxial convexity (e.g. Plate 6, Figs 6–8) are found in the more fine-grained and siliciclastic marls. This change in morphology is probably an evolutionary response to the life on a muddy substrate, where it is an advantage to have a comparatively large surface in order to distribute the weight and thereby prevent the exoskeleton from sinking down into the substrate. This is archieved by *V. perstriatus* by making the outer margins more concave, thereby increasing the sediment touching parts of the animal.

The Norwegian material contains forms, which appear very similar to the closely related but slightly older *V. duplicatus* (Bohlin 1955) of Sweden. Although they are easily differentiated by the slightly narrower pygidial doublure and shorter preglabellar area on *V. perstriatus*, it also seems quite clear that they must belong to a single lineage which developed from this moderately convex species with a dense terrace-line pattern into a generally more flattened species occupying a lower energy, muddy environment.

V. perstriatus is distinguished from the contemporaneous East-Baltic *V. simonkovensis* Balashova, 1976, by its shorter preglabellar area and the less triangular pygidium with its narrower axis. The cranidium may look quite similar to those of *Ogmasaphus*, but is easily distinguished by the marked anterior indention on the glabella.

Ontogeny. – As already noted by Wandås (1984, p. 221) the pleural ribs on the pygidium are generally fainter on the smaller specimens, while they may quite strongly developed on the larger ones. Likewise the relative size of the eyes decreases with increasing size of the animal. The relative width of the pleural field in relation to the axial width may also increase with size, but this may be connected to the overall change in pygidial convexity as specimens with a low pygidial convexity generally also belongs to the small end of the size scale, probably reflecting less favourable conditions in the more muddy environment, which replaced the original calcareous shallow water environment.

Occurrence. – In Norway it is found in the uppermost Helskjer Member and the basal part of the Sjøstrand and Heggen members, Elnes Formation. It is extremely common in the north, while becoming largely absent in the south of the Oslo Region. Eiker-Sandsvær: Stavlum at Krekling (Helskjer Member, 0.4 m above Huk Fm.). Modum: road-cut at Vikersund skijump (Heggen Member, 4.05 m above Huk Fm.). Hadeland: Hovodden (7.9 to 7.95 m above the Huk Fm.), quarry south of road 195 about 1.5 km south of Brandbu (Helskjer Member). Mjøsa: road profile along E6 between Nydal and Furnes (Helskjer Member, level 2.75 to 10.13 m in sub profile 1), Furnes Church (7.75 to perhaps 9.3 m above the Huk Fm.) and beach profile at Helskjer (Helskjer Member, 2.25 to 8.5 m above the Huk Fm.).

Outside Norway it is known from the uppermost Holen Limestone (formerly named the Gigas Limestone) of Öland, Västergötland and Dalarne, Sweden (Bohlin 1955).

Genus *Pseudobasilicus* Reed, 1931

Type species. – *Ptychopyge lawrowi* Schmidt, 1898.

Diagnosis. – Exoskeletal convexity low; cephalon and pygidium with broad, flattened borders; frontal area moderately long; eyes fairly large; eyes generally situated less than half their length in front of the posterior cranidial margin; posterolateral projection on fixigena not increasing in length abaxially; librigenae broad, posteriorly continuing out into genal spines. Hypostome with deeply and broadly indented posterior margin. Thoracic segments with strongly pointed and posteriorly curved lateral pleural terminations; panderian organs developed as small, isolated openings; inner margin of pleural doublure strongly convex. Pygidium generally with a somewhat rounded outline; pleural ribs, if present, distinct and rounded, reaching border; doublure very wide (modified from Moore 1959 and Zhou *et al.* 1998).

Remarks. – The genus contains the two subgenera *P.* (*Pseudobasilicus*) and *P.* (*Pseudobasilicoides*) Balashova, 1971, of which the latter is monospecific. *P.* (*Pseudobasilicoides*) *elegans* Balashova, 1971, was considered by Balashove (1971) to differ from the typical species of *Pseudobasilicus* by the distinct paradoublural line on the cephalon; the strongly outwardly curved anterior branch of the facial suture; a long preglabellar area; a panderian opening close to inner doublural margin and with only three or less terrace-lines in the space adaxially of the panderian opening; a posterior fixigenal area reaching just over halfway out to the genal spine and a fairly narrow doublure, the narrowest central part corresponding to between 40 and 42% of thoracic pleural width. The last two characters are no longer considered diagnostic for the subgenus *P.* (*Pseudobasilicoides*) as *P.* (*Pseudobasilicus*) *truncatus* n. sp. described below has the same characteristics and the forward expansion of the anterior cranidial lobe is only marginally stronger on *P.* (*Pseudobasilicoides*), but apart from this, the diagnostic characters seem to stand.

Subgenus *Pseudobasilicus* (*Pseudobasilicus*) Reed, 1931

Diagnosis. – Asaphid with long preglabellar area; a moderately effaced paradoublural line; a panderian opening on librigena located around two-thirds the librigenal doublure width inside the outer margin and with 6 to 10 doublural terrace-lines between inner margin and panderian opening (modified from Balashova 1971).

Pseudobasilicus (*Pseudobasilicus*) *truncatus* n. sp.

Pl. 6, Figs 9–12, 14

Derivation of name. – Refers to the characteristic deep truncation of the posterior pygidial margin.

1982 *Pseudobasilicus brevicostatus* – Wandås, 138 (occurrence).
1984 *Pseudobasilicus perstriatus* (Bohlin 1955) – Wandås, p. 221, pls 4 K–L, 5A.

Type stratum and type locality. – Holotype PMO 202.077/1 originates from the uppermost Helskjer Member of the Old Quarry at Slemmestad, Oslo-Asker.

Type material. – The holotype, PMO 202.077/1, was collected from 1.24 m up in the Helskjer Member in the Old Quarry at Slemmestad. Paratype PMO 83294 was found 0.15 m above the Helskjer Member at Molleklev, Rognstranda, while paratype PMO 103.717 originates from Vikersund Skijump. Paratype PMO 202.084/1 was collected from 1.26 m above base of the Helskjer Member in the Old Quarry at Slemmestad.

Material. – One librigena (PMO 202.069/1), six pygidia (PMO 102.544, 102.652, 103.674, 103.698, 202.084A/1, 202.096A/1) and seven partly articulated specimens (PMO 83294, 102.556, 103.666, 103.669–70 (one specimen), 103.717, 104.059, 202.069/3, 202.077A/1).

Diagnosis. – Glabellar length corresponding to around 17% of cranidial length; eyes situated more than 95% their own length from posterior cephalic border; cranidial tubercle small. Pygidium fairly wide, strongly truncated posteriorly; axis strongly tapering rearwards, the anterior width corresponding to one quarter the pygidial width; terrace-lines on pygidial doublure very fine and densely spaced.

Description. – Moderately sized for a *Pseudobasilicus*, largest pygidium approximately 19 mm long, while largest cranidium is nearly 25 mm long. Cephalon broadly semicircular in outline, length corresponding to half the width.

Cranidium with a length approximating between two-thirds and three-fourths the width. Glabella slightly contracted at middle, bordered by wide and rather shallow preglabellar and axial furrows. Glabellar length just above 75% of the cranidial length. Smallest glabellar width, corresponding to one-third the cranidial width found at the preocular area, is located between the eyes. Occipital ring with low lobe-like anterolateral corners bordered by shallow to indistinct posterior furrow. Occipital furrow moderately wide and shallow, deepest laterally, becoming more indistinct mesially. Most anterior part of furrow located around 13 to 14% of cranidial length from posterior margin. Mesial tubercle small, situated approximately 15% of cranidial length from posterior margin. Lateral glabellar furrow S1 deep, straight, delimiting fairly well-developed and strongly triangular posterior glabellar lobe; length of lobe corresponding to around one-fourth the cranidial length. Frontal glabellar lobe pyriform in outline, moderately inflated. Parafrontal band present. Preglabellar area moderately wide, around 17% of cranidial length at sagittal line. Preocular area broad, the width corresponding to around two-thirds the glabellar width. Preocular area reaches back to midline between posterior and anterior margin of cranidium.

Paradoublural line meets axial furrow just in front of eye; abaxial end situated 40% of cranidial length inside cephalic margin or nearly 50% of cranidial length from posterior margin. Palpebral lobes small, exsagittal length equivalent to 20% of the cranidial length. Posterior area of fixigena fairly narrow, reaching halfway out to lateral cephalic margin, exsagittal length of posterolateral projection approximating 15% of cranidial length, transected by moderately deep posterior border furrow. Posterolateral projection getting shorter (exsag.) abaxially. Cranidial surface with close to 40 terrace-lines on preocular area. Posterior fixigenal border with closely spaced anterolaterally directed terrace-lines. Anterior branch of facial suture not strongly outwardly curved in front of eyes.

Length of eye between 40 and 45% of librigenal field from posterolateral to anterolateral corner; width of eye (lat.) corresponding to about 40% of the width of the pleural field perpendicular to axis. Eye situated between 95 and 100% its own length in front of the posterior cephalic margin. Paradoublural furrow indistinct but slightly accentuated by faintly raised border located around 5% the distance between eye socle and lateral margin from eye. Course of paradoublural line semiparallel to forwardly diverging from outer margin. A small but distinct panderian opening is situated approximately one-third the doublural width outside the inner margin and is separated from the inner margin by around six to eight terrace-lines. Lateral border moderately broad, the narrowest part corresponding to between 38 and 41% of the thoracic pleural width. Genal spine or angle very short and broad. Surface of genal field with some faint terrace-lines; border densely covered by fine terrace-lines running more or less parallel to exsagittal line. Doublure with up to around 40 terrace-lines running parallel to paradoublural line.

Hypostome unknown.

Thorax widening slightly forwards. Thoracic axis bounded by deep axial furrows, axis comparatively narrow, occupying 25% of the thoracic width, length (sag.) between 25 and 30% of the width. Pleurae ending in short posteriorly pointing flat spines, spines becoming more pronounced on the posterior thoracic segments. Pleural area crossed by moderately deep pleural furrow ending close to posterior margin distally. Panderian openings small and isolated, located approximately one-third the pleural length (exsag.) from its anterior margin and two-thirds the pleural width from the axial furrow. Pleural terrace-lines fairly dense, wavy, directed sub-parallel to exsagittal line. Doublural terrace-lines moderately densely spaced, distally sub-parallel to exsagittal line, adaxially turning abaxially on posterior part.

Pygidium broadly trapezoidal with strong posterior truncation and broadly rounded lateral corners. Pygidial length, excluding the anterior half ring, corresponding to between 50 and 60% of the width. Axis short and wide, strongly tapering rearwards, terminating more than 20% of the pygidial length from the posterior margin (read posterior most margin), delimited by deeply impressed axial furrows. The axial width constitutes 25% of the pygidial width. Axial segmentation clearly defined by distinct furrows on anterior to middle part, becoming shallower on posterior part. There are approximately 12 axial rings in addition to the terminal piece and articulating half ring. Pleural field with $6^1/_2$ to 7 short pleural ribs in addition to anterior half rib. Ribs gently rounded in cross-section, terminating on the inner third of the border. Anterior half rib well defined but narrow (lat.) and very short (exsag.). Interpleural furrow absent. Pleural furrows moderately wide. Paradoublural line located nearly one-third the pleural width from the axis. Largest width of doublure approximately one-third of the pygidial width.

Outer surface of pygidium, except for the pleural furrows, covered by moderately dense terrace-line pattern. Terrace-lines on axial segments forward convex with central bow-shaped connective. Doublural terrace-line pattern generally very fine and dense, the number of terrace-lines at broadest place exceeding 70 on best preserved pygidium (PMO 202.077/1).

Remarks. – *P. (P.) truncatus* n. sp. differs from the typical *Pseudobasilicus* on several accounts, showing large affinities with the genus *Volchovites* treated above. The most conspicuous of these affinities is the anterior librigenal striation, which was said by Balashova (1976) to only occur on species of *Volchovites* and the anterior location of the eyes. Furthermore *P. (P.) truncatus* n. sp. is characterized by very short genal spines transitional to sharply pointed genal corners. The thoracic pleural ends may be slightly less spine-like than those of other species of *Pseudobasilicus*, although still more pointed than on *Volchovites*. These similarities together with a general correspondence of characters suggest a very close relationship between the two genera and may raise questions as to the correctness of the family subdivision proposed by Balashova (1976). For the same reason the changes in subfamily groupings proposed by Balashova (1976) is not adopted here.

Material of *P. (P.) truncatus* n. sp. was due to some similarity originally assigned to *Volchovites perstriatus* (Bohlin 1955) described above, but may be differentiated on its less convex exoskeleton; the small eyes; the presence of genal spines and by a more broadly rounded pygidium with a wide posterior truncation. The latter character, the posterior incision, may become less pronounced in the stratigraphically youngest specimens, but is always present.

P. truncatus n. sp. differs from *P. lawrowi* (Schmidt 1898) of Aseri Stage (C1a), middle Darriwilian, by the generic characters listed above, but also on the deep pygidial truncation and the finer terrace-line pattern on pygidial doublure. With regard to the name *Pseudobasilicus brevicostatus* used by Wandås (1982) in a species list for the basal Elnes Formation at Vikersund Skijump, it was apparently a *nomen nudum*, and was later (Wandås 1984) changed to *P. perstriatus* (Bohlin, 1955), which is described above as *Volchovites perstriatus*.

Occurrence. – Upper part of Helskjer Member to lowermost Heggen Member, Elnes Formation. Skien-Langesund: Molleklev at Rognstranda (0.15 m above Helskjer Member). Modum: Vikersund Skijump (3.1 to 14.9 m above Huk Fm.). Oslo-Asker: The Old Eternite Quarry at Slemmestad (1.18 to 1.31 m up in the Helskjer Member).

Pseudobasilicus (Pseudobasilicus?) sp. A

Pl. 7, Figs 3–7

1984 *Pseudobasilicus perstriatus* (Bohlin, 1955) – Wandås, p. 221, pls 4H–I, 5B.

Material. – Seven pygidia one of wich has posterior four thoracic segments attached (PMO 3392, 33415 (two specimens), 33474, 68334, 105.902, 207.390/3).

Description. – Thoracic axis narrow, occupying just above 25% of the thoracic width; pleural furrows deep and strongly laterally directed.

Pygidium weakly convex, semicircular in outline, the length, excluding the anterior half ring, corresponding to two-thirds the width; axis moderately long and narrow, the width corresponding to just over 25% of the pygidial width, while the length constitutes nearly 80% of the total pygidial length; axis with 12 to 13 axial rings; pleural field bearing $6^1/_2$ to 7 faint to comparetively well developed and moderately short pleural ribs reaching halfway out to lateral margin anteriorly; paradoublural line extending comparatively far out on the pleural region, distance from axial furrow at anterior pleural rib corresponding to nearly 20% of the pygidial width; border slightly concave. Doublural width corresponding to 30% the pygidial width. Pygidial surface covered by fairly dense anterolaterally directed terrace-line pattern, becoming less dense on the borders. Doublure bearing around 50 closely spaced terrace-lines.

Remarks. – *P.* (*P.*?) sp. A differs from *P. lawrowi* (Schmidt 1898) by the less wide and longer pygidial axis; the generally less well-developed pleural ribs and much denser doublural terrace-line pattern. It is differentiated from *P. planus* Balashova, 1971 by the much less posterolaterally directed and shorter (exsag.) anterior half rib; shorter pleural ribs and more laterally directed thoracic pleural furrows. *P.* (*P.*?) sp. A differs from *P. truncatus* n. sp. in the semicircular outline of the pygidium without any posterior truncation; the narrower pygidial axis and by the coarser doublural terrace-lines.

Occurrence. – Lower part of the Elnes Formation. Oslo-Asker: Brocks Lökke in central Oslo (L. G. Koch, personal communication 2006). Hadeland: Hovodden at Randsfjorden (5.80 to 8.35 m above base of formation), Melbostad and at forest road north of Stubstad at Gran. Mjøsa: road profile along E6 between Nydal and Furnes (uppermost Helskjer Member, 6.81 m above the Huk Fm.).

Pseudobasilicus (Pseudobasilicus) sp. B

Pl. 13, Figs 16, 19

Material. – One nearly complete pygidium PMO 83251.

Description. – Pygidium approximately 42 mm long, broadly rounded with a slight posterior truncation and a length, excluding the anterior halfring, corresponding to around 55% of the width. Axis moderately wide with a width corresponding to between 25 and 30% of the pygidial width, while the length corresponds to 80% of the pygidial length; fairly indistinct axial segments numbering 10 excluding the terminal piece and articulating half ring. Pleural ribs nearly effaced. Paradoublural line indistinct, situated very close to the axial furrow, the distance from axial furrow at anterior pleural rib corresponding to 36% of the pleural region width or 17% of the pygidial width. Doublure very wide, the widest part corresponding to one-third of the pygidial width. Pygidial surface covered by moderately dense, anterolaterally directed terrace-lines on the pleural region. Axis bearing one distinct transversely directed terrace-line on each axial segment, just reaching out to the pleural region, although they normally may reach further out. Doublural terrace-lines mostly covered, but where exposed appearing to be fairly coarse.

Remarks. – This taxon is easily differentiated from the other species of *Pseudobasilicus* by its rather indistinct pleural ribs; posterior truncation and fairly coarse doublural terrace-lines. It seems especially closely related to *P. (P.) truncatus* n. sp. from which it differs mainly in the effaced pleural ribs and coarser doublural terrace-lines. It may be questioned whether these characters just reflect local variations, and the pygidium may belong in *P. (P.) truncates*.

Occurrence. – The specimen was collected from a limestone bed at the top of the Helskjer Member, 8.45 m above the Huk Formation at Furnes Church, Mjøsa.

Pseudobasilicus (Pseudobasilicus?) sp. C

Pl. 6, Figs 13, 16

Material. – Two nearly complete pygidia preserved in marly mudstone and concretionary limestone respectively (PMO 33327, 36588).

Description. – Largest pygidium nearly 29 mm long. Pygidium very weakly convex, strongly semicircular in outline, the length corresponding to two-thirds the width; axis clearly delimited by axial furrows; moderately long and narrow, the width corresponding to 20% of the pygidial width and the length approximating 75% of the pygidial length; axis with approximately 12 axial rings. Pleural field bearing $7^1/_2$ long and fairly distinct pleural ribs reaching all the way out to posterior border; paradoublural line running close to axial furrow, distance to furrow at anterior pleural rib corresponding to 15% of pygidial width; border concave with completely flat outer part; doublure fairly wide, paradoublural line at anterior pleural rib located 40% of the pleural region width or 13% the pygidial width from the axial furrow. Pygidial surface bearing moderately dense, transversely directed terrace-line pattern, which turns more anterior on the borders; doublure with coarser pattern.

Remarks. – Unfortunately no diagnostic characters have been observed for a differentiation between the two subgenera established for this genus and a confident assignment is therefore not possibly. *P.* sp. C differs from *P. (Pseudobasilicus) lawrowi* (Schmidt 1898) of the upper Kunda to Aseri Regional Stage, East Baltica area, on the narrow axis; more rounded outline and perhaps also by less rearward directed pleural ribs. *P. (Pseudobasilicus) planus* Balashova, 1971, from upper Kunda to Aseri Regional Stage of the East Baltic area has a less rounded pygidium with a wider axis and distinctly more posteriorly directed anterior half rib. It is easily distinguished from *P. (Pseudobasilicus) truncates* n. sp. and *P. (Pseudobasilicus)* sp. B in the absence of a posterior pygidial truncation and from *P. (Pseudobasilicus?)* sp. A by the much narrower pygidial axis.

Occurrence. – Exact stratigraphic occurrence is unknown, but it probably lies in the Elnes Formation. Hadeland: Lynne farm at Ringstad just east of Gran. Mjøsa: Ålset near Snertingdal.

Pseudobasilicus? sp.

Pl. 10, Fig. 18; Pl. 11, Fig. 1

Material. – One very fragmentary pygidium, PMO 33490, preserved in marly mudstone.

Description. – Pygidium weakly convex with wide and essentially flat concave border and a semicircular outline without posterior truncation; the length, excluding the anterior half ring corresponding to just less than 60% of the width; axis moderately long and narrow, the width corresponding to just less than 25% of the pygidial width, while the length takes up nearly 80% of the total pygidial length; axis with 13 axial rings excluding the anterior halfring and the terminal piece; pleural field bearing eight distinct pleural ribs reaching halfway out to lateral margin anteriorly; paradoublural line not seen. Pygidial surface, except for axis, completely covered by very fine and dense anterolaterally directed terrace-line pattern; axis characterized by more sparse terrace-line pattern with one primary line crossing over each of the axial segments.

Remarks. – The pygidium differs from all the other Norwegian species assigned to this genus by the extremely dense terrace-line pattern; the moderately sharp anterolateral corners; the narrow axis and the eight pleural ribs in the pygidium. It appears quite close to *Pseudobasilicus (Pseudobasilicus?)* sp. A and may turn out to be conspecific with this.

Occurrence. – The specimen was collected from the Elnes Formation, probably the upper part, at Haugslandet, Gran in the Hadeland district.

Genus *Pseudasaphus* Schmidt, 1904

Type species. – *Ptychopyge globifrons* Eichwald, 1857, subsequent designation by Reed (1930).

Diagnosis. – Exoskeleton flat to moderately convex. Cephalon and pygidium with more or less distinct concave border; preglabellar area moderately long; frontal lobe generally slightly pointed anteriorly; eyes normally large; genal corners pointed or with genal spines. Hypostome of *Asaphus* type. Panderian organs developed as notches or separate openings on librigena and as separate openings on thoracic pleurae; inner margin of pleural doublure slightly to moderately convex. Pleural ribs on pygidium weakly developed and limited to inner pleural parts; doublure fairly broad with nearly straight and unbroken, semiparallel terrace-lines; outer surface generally with strong ornamentation, directed anterolaterally (modified from Moore 1959).

Remarks. – Balashova (1976) argued that *Pseudasaphus limatus* Jaanusson, 1953a together with *Ptychopyge aciculatus* Angelin, 1854 and *Valdaites krestcyensis* Balashova, 1976 formed a single well-defined taxonomic group, which she ascribed to a new genus, *Valdaites*. In her diagnosis, Balashova (1976, p. 37) attached special importance to a number of characters including the larger number of exoskeletal terrace-lines; the fine and dense striation found on palpebral lobes and posterior fixigenal areas; the posterior location of the eyes and the narrow (tr.) posterior fixigenal areas. Characters such as the terrace-line density may be discounted because of their large variation within single asaphid taxa. Regarding the location of the eyes, *Pseudasaphus limatus* has smaller and more anterior eyes than outlined in the generic diagnosis, making this character invalid, at least if *P. limatus* should be included. The only feature, which does seem to define and separate the three species from those included in *Pseudasaphus* is the narrow (tr.) posterior fixigenal area, but even here a species like *P. katlinoensis* Balashova, 1973, appears quite close. *Valdaites* is consequently regarded as a junior synonym of *Pseudasaphus*. The same applies to *Mischynogorites* Balashova, 1976, which was erected on strength of a single species, *Asaphus brachyrachis* Törnquist, 1884 from which only the pygidium is known with certainty. As discussed in connection with *Pseudasaphus* sp. B treated below, this species strongly resembles *Pseudasaphus*, and is here regarded as belonging to that genus.

Balashova (1976) divided *Pseudasaphus* into the two subgenera *Pseudoasaphus* (*Pseudoasaphus*) and *Pseudoasaphus* (*Pseudoasaphoides*). *P.* (*Pseudoasaphus*) contains only the type species *P. globifrons* (Eichwald, 1857), while what had originally been termed *P. praecurrens* Schmidt, 1904, *P. katlinoensis* Balashova, 1976, *P. lesnikovae* Balashova, 1976 and *P. janischewskyi* Balashova, 1976 were moved to the other. According to Balashova (1976, p. 42) *P.* (*Pseudoasaphoides*) differs from *P.* (*Pseudoasaphus*) in its relatively flat exoskeleton; shallow pits in the axial furrow beside the anterior glabellar lobe; shallow axial furrows; low eyes located more than half their own length in front of the posterior cranidial margin and by the well-developed panderian opening on the genal corners. A general comparisson of *P. globifrons* with the above mentioned species does not reveal any clear cut differences between the two subgenera, and

the distinction decreases even more when the three species formerly assigned to *Valdaites* (see above) are included. *P. globifrons* may be a little more convex with higher eyes and slightly less developed panderian organs on the librigena, but again I do not find the differences strong enough to support a division into subgenera and therefore regard *Pseudoasaphoides* as a junior synonym of *Pseudasaphus*.

Pseudasaphus limatus Jaanusson, 1953a

Diagnosis. – Cephalon very weakly convex; eyes low, comparatively small, situated more than half their length in front of posterior cranidial margin; posterior facial suture terminating closer to axial furrow than lateral cephalic margin. Pygidium with straight or more frequently truncated posterior end. Exoskeletal surface with dense terrace-line pattern and with fine striation on palpebral lobes (emended from Jaanusson 1953a).

Remarks. – The species has been found to vary somewhat with regard to its terrace-line pattern resulting in some confusion regarding the limits of the taxon (e.g. Henningsmoen 1960, p. 246). The new material collected for this study reveals a strong connection to the stratigraphical level from which they have been sampled with a marked shift in terrace-line pattern in the topmost part of the Elnes Formation. Consequently two subspecies are recognized, *P. limatus limatus* (Jaanusson 1953a), appearing from the very top of the Elnes Formation, and the stratigraphically older *P. limatus longistriatus* n. subsp., occurring in the upper part of the Elnes Formation.

Pseudasaphus limatus limatus Jaanusson, 1953a

Pl. 6, Fig. 15; Pl. 7, Figs 8–15; Pl. 8, Figs 1–3

1884   *Asaphus* cfr. *undulatus* Steinh. – Törnquist, pp. 69–70, pl. 2, fig. 22.
1953a   *Pseudasaphus limatus* n. sp. Jaanusson, pp. 421–422, pl. 3, figs 6–7.
1960   *Pseudoasaphus limatus* Jaanusson, 1953 – Henningsmoen, pp. 245–247; pl. 9, fig. 4, pl. 13, figs 1–5.
1976   *Valdaites limatus* (Jaanusson 1953) – Balashova, pp. 37–39, pl. 3, fig. 11.

Type stratum and type locality. – Holotype pygidium OM.Nr.58931 collected from the Vollen or Arnestad Formation on Fornebu at Oslo, Norway.

Norwegian material. – The Norwegian material includes three cranidia (PMO 40355, 83240, 206.148a/1), three librigenae (PMO 206.132, 206.134/1, 206.137/1), one hypostome (PMO 40400) 14 pygidia (OM.Nr.58931, PMO 3477, 3972, 3973, 20317, 33570, 72067, 206.128, 206.137/2, 206.141, 206.144/1, 206.144/4, 206.148/3, 206.152) and three partly articulated specimens (PMO 3346, 58911, 58917).

Diagnosis. – Subspecies of *P. limatus* characterized by forwardly curving axial terrace-lines just outside axial furrow.

Description. – Largest cranidium 42 mm long, largest measureable pygidium approximately 70 mm long. Cephalon low and broadly semicircular in outline, length corresponding to somewhere around half the width.

Cranidium with a length approximating 80% of the width. Glabella contracted at middle, bordered by wide and very shallow preglabellar and axial furrows. Glabellar length constituting around 85% of the cranidial length. Smallest glabellar width, corresponding to one-third the cranidial width or around 45% the anterior cranidial width, Wc, is located between the eyes. Occipital ring narrow and very weakly differentiated from the rest of glabella. Occipital furrow moderately narrow and extremely shallow, central part located 10% the cranidial length in front of posterior margin. Minute mesial tubercle situated approximately 15% of cranidial length from posterior margin. Lateral glabellar furrows faint, the posterior one, S1, deepest, delimitating moderately sized and markedly triangular posterior glabellar lobe; length of lobe corresponding to around 20% of the cranidial length. Frontal glabellar lobe pyriform in outline, convexity low. Parafrontal band not observed. Preglabellar area moderately wide, around 15% of cranidial length at sagittal line. Preocular area broad, the width corresponding to around two-thirds the glabellar width. Preocular area reaches back to just behind sagittal midline of cranidium. Palpebral lobes small, exsagittal length equivalent to 20% of the cranidial length. Posterior area of fixigena narrow (tr.), width corresponding to 20% of the cranidial width; largest exsagittal length of posterolateral projection corresponding to between 15 and 20% of cranidial length found just behind eye, decreasing rapidly outwards; posterolateral projection transected by moderately deep posterior border furrow. Cranidial surface covered by sparse and evenly spaced pores and some very fine striations located around mesial tubercle, on occipital ring, glabellar front and palpebral lobes.

Librigena and eye only known from fragmentary material, but eyes are low and fairly small, their length corresponding to nearly 25% of the cranidial length and situated more than two-thirds their own length from the posterior margin. Paradoublural furrow generally indistinct, anteriorly lying just up to

the eye socle. Course of paradoublural line semiparallel to outer margin. Lateral border broad. Genal spine fairly short. Doublure generally bearing just over 30 terrace-lines, spacing closest on inner part.

Hypostome only known from very poorly preserved material, but it is of the *Asaphus* type; slightly longer than wide with a broad and squat anterior lobe of middle body and a very deep posterior V-incision into posterior border, corresponding to 40% of the total hypostomal length. Anterior border short and broadly rounded, laterally ending in short anterior wings. Surface covered by fairly large and sparsely distributed pores and a few terrace-lines along the margins.

Thorax broad, flat and widening forward with slightly downward turned outer pleural region. Thoracic axis narrow, constituting just under 25% of the thoracic width, the length corresponding to less than 25% of the width; axis only slightly raised above pleural region, bounded by moderately shallow axial furrows. Pleural region wide, the inner part corresponding to 25% of the thoracic width and transected by fairly deep and only sligthly posterolaterally directed pleural furrows. Panderian openings minute and completely isolated, situated slightly more than 10% the thoracic width from the lateral margin. Posterolateral corners of pleurae extended into short posteriorly directed points. Axial surface covered by dense pattern of pores and fine, forwardly convex terrace-lines, especially concentrated on posterior part. Pleural region characterized by dense pattern of pores and short, anterolaterally directed terrace-lines. Thoracic doublure with slightly convex inward inner margin and somewhere between 25 and 30 curving terrace-lines.

Pygidium broadly trapezoidal with straight or concave posterior margin and sharply rounded lateral corners. Pygidial length, excluding the anterior half ring corresponding to about or slightly more than two-thirds the width. Axis fairly short and narrow, terminating just above 20% of the pygidial length from the most posterior margin, delimited by moderately shallow axial furrows. Axis constitutes 20% of the pygidial width. Axial segmentation quite weakly defined. There are 13 to 15 or perhaps even 16 axial rings in addition to the terminal piece and articulating half ring. Pleural field with about $7\frac{1}{2}$ moderately short pleural ribs in addition to anterior half-rib. Ribs flatly rounded in cross-section and very variable with regard to distictivenes, but generally terminating on the inner half of the pleural region, but they may extend to about one-third from the margin. Anterior half-rib fairly well defined but narrow (lat.) and very short (exsag.). Interpleural furrow absent or very shallow. Pleural furrows moderately narrow. Paradoublural line located just over 40% of the pleural width from the axial furrow at anterior pleural rib. Largest width of doublure approximately one-third of the pygidial width.

Outer surface of pygidium covered by moderately dense terrace-line pattern. Terrace-lines on axial segments convex forward with central bow-shaped connective; each segment bearing one more strongly developed terrace-line, which may continue out onto the pleural field, where it immediately turns forward. Pleural terrace-lines long and anterolaterally directed, bending slightly laterally, where many show a forking pattern, adaxially terminating at the axial furrow. Doublural terrace-line pattern moderately coarse, the number of terrace-lines at broadest place generally lays around 20 to 25, but may reach 30 on a few larger specimens.

Occurrence. – In Norway it is found from the uppermost Håkavik Member and up into the Vollen Formation or perhaps even the overlying Upper Ordovician Arnestad Formation. Skien-Langesund: Gravestranda at Frierfjord (Henningsmoen 1960). Eiker-Sandsvær: Recorded from the Elnes Formation (Henningsmoen 1960). Oslo-Asker: profile at Djuptrekkodden near Slemmestad (level 56.41 to 58.55 m corresponding to the top Elnes Formation to basal Vollen Formation), Håkavik (Håkavik Member and probably lowermost Vollen Fm.), Fornebu, Huk on Bygdøy (Elnes Fm.), Professor Dahls gate 48 and road cut above Lille Frøen in Oslo City (Elnes Fm.). Ringerike: Norderhov Church (Henningsmoen 1960). Hadeland: southern slope on Brandbukampen at Gran.

Outside Norway it has been found in the Folkeslunda Limestone (öfvre grå ortocerkalk) at Kårgärde and in the Dalby Limestone (Ludibundus kalk) at Fjäcka, both in the Siljan district of central Sweden (Jaanusson 1953a) and in a bore core at the town Mukhovtsy in the St. Petersburg district, NW Russia (Balashova 1976).

Pseudasaphus limatus longistriatus n. subsp.

Pl. 8, Figs 4–8

Derivation of name. – Refers to the strongly developed posterolaterally directed pygidial terrace-lines on the inner pleural regions.

1960 *Pseudoasaphus limatus* Jaanusson, 1953 – Henningsmoen, p. 246.

Type stratum and type locality. – Holotype PMO 206.117/1 belongs to the upper half of the Håkavik Member at Djuptrekkodden, Oslo-Asker.

Type material. – The holotype, PMO 206.117/1, was collected from level 56.01 m in the Djuptrekkodden profile, Oslo-Asker, corresponding to the uppermost Håkavik Member. Paratypes PMO 206.111 and PMO 206.115/2 were found at levels 55.83 and 56.00 m in the type profile.

Material. – The material consists of seven fragmentary pygidia (PMO 20325, 206.026, 206.108/3, 206.111, 206.112/1, 206.115/2, 206.117/1).

Diagnosis. – Subspecies of *P. limatus* characterized by an inner pleural region on pygidium with strong, posterolaterally directed terrace-lines reaching far out onto the pleural region.

Description. – Only the pygidium is known from this taxon. Pygidium trapezoidal with straight to strongly truncated posterior margin and moderately sharp anterolateral corners. Pygidial length, excluding the anterior half ring approximating 75% of the width. Axis narrow and of moderate length, terminating 20% of the pygidial length from the posteriormost margin; width of axis constituting 20 to 25% of the pygidial width. Axial segmentation generally fairly distinct, axial rings approximating 14 in number. Pleural field with 7 or 7½ faint and short pleural ribs in addition to anterior half-rib; ribs flatly rounded in cross-section. Anterior half-rib only moderately well defined, narrow (lat.) and short (exsag.). Interpleural furrow absent, at the highest very faintly developed. Pleural furrows moderately narrow (exsag.). Doublure wide, maximal width approximating one-third the pygidial width.

Outer surface of pygidium covered by fairly densely packed, long and slightly undulating terrace-lines. The anterolaterally directed pleural terrace-lines are partly missing on pleural field on three out of four specimens, suggesting this to be the normal appearance for this population. Inner pleural region is characterized by strong, posterolaterally directed terrace-lines reaching from the axis and far out onto the pleural region. Doublure bearing around 19 nearly straight and semiparallel terrace-lines on examined specimens.

Remarks. – The subspecies seems very close to the nominate subspecies and it could therefore be argued whether it really should be ranked as high as the subspecies level. On the other hand, the two taxa do seem to be completely separated stratigraphically although clearly belonging to the same lineage. Furthermore, the new subspecies shows a tendency towards sparse striation on the pygidial pleural field but more specimens are needed before this possible differentiating character may be verified. The material is therefore provisionally distinguished at subspecies level with its stratigraphical position suggesting that it is not simply an ecophenotypic variant.

The reason for the very restricted geographical occurrence of *P. limatus longistriatus* n. subsp. most probably relates to its extremely short stratigraphical range, and hence the likelihood of finding it is very small compared to the typical form occuring through a much longer stratigraphical interval.

Occurrence. – Upper part of the Håkavik Member, Elnes Formation. Oslo-Asker: Djuptrekkodden at Slemmestad (50.23 to 56.01 m in the profile) and Odden at Bekkebukta on Bygdøy.

Pseudasaphus sp. A

Pl. 8, Figs 9–11

Material. – Three fragmentary pygidia preserved in calcareous concretions or silty mudstone (PMO 106.549, 208.643/5, 208.643/7).

Description. – Largest pygidium just under 25 mm long. Pygidium rounded trapezoidal in outline with moderate to slight posterior truncation; length, excluding the anterior half ring, corresponding to approximately 50% of the width although this number probably should be higher due to later deformation. Axis narrow and moderately short, the width corresponding to 17% the pygidial width on specimen PMO 208.643/7, while the length constitutes around 75% of the total pygidial length; number of axial rings approximating 10 in addition to the terminal piece and anterior half ring. Pleural field bearing 7 strongly effaced, short pleural ribs reaching out to paradoublural line, which extends quite far out on the pleural region, doublural width corresponding to just over 25% of the pygidial width; border flat to slightly concave. Pygidial surface covered by fairly dense anterolaterally directed terrace-line pattern, becoming less dense on the border posterorly. Doublure bearing 20 widely spaced terrace-lines.

Remarks. – Although the material is very poorly preserved making an exact identification difficult this species seem fairly close to *P. limatus limatus* described above; it has the same terrace-line pattern on the pygidial surface; a posterior truncation; identical width of doublure and the same number of doublural terrace-lines. It differs in its distinctly less subtriangular pygidial outline and possibly also by a slightly deeper axial furrow and shorter axis.

Occurrence. – Lower to middle part of the Heggen Member, Elnes Formation. Eiker-Sandsvær: Level 1.48 m in profile 6 at Vego near Krekling. Modum: Road-cut at Vikersund skijump (lower Heggen Member, 14.3 m above the Huk Fm.).

Pseudasaphus sp. B

Pl. 7, Figs 1–2

Material. – Two fragmentary pygidia preserved in concretionary limestone (PMO H2707, 61144).

Description. – Largest pygidium (PMO 61144) very large; about 7.5 cm long. Outline strongly rounded with a length corresponding to nearly 75% of the width. Axis moderately long and narrow, the width corresponding to 20% of the pygidial width, while the length occupies around 70% the pygidial length. Axis bearing approximately 16 axial rings. Pleural field with seven strongly effaced pleural ribs, which do not reach beyond paradoublural line; paradoublural line running close to axial furrow, distance to furrow at anterior pleural rib corresponding to 10% of pygidial width; border distinctly concave with nearly horizontal outer part. Doublure wide, the widest part corresponding to just over one-third the pygidial width. Terrace-line pattern moderately dense, consisting of anterolaterally directed, wavy terrace-lines on pleural surface; axis with distinct, rather straight primary terrace-lines with a flat central forward curve; denser anteromedially directed secondary terrace-lines are found between the strong, transversely directed primary ones. Doublure bearing between 30 and 40 moderately coarse and straight terrace-lines, some interspaced with short anterolaterally directed and finer terrace-lines. An irregular furrow on specimen PMO H2707 bounds an area on the anterior right corner characterized by very fine and densely spaced terrace-lines of a more wavy and interrupted nature (Plate 7, Fig. 2).

Remarks. – The taxon shows a strong affinity with *Mischynogorites brachyrachis* (Törnquist, 1884) from contemporaneous beds in the Siljan district of Central Sweden, especially specimen UM nr. Ar. 4126 figured by Jaanusson (1953a; plate 8, fig. 8), but differs in the lack of strong, posteriorly directed terrace-lines that completely dominate the inner pleural region out to the border on *P.? brachyrachis* and by the longer pygidial axis. It is clear that they must belong to the same genus. *Mischynogorites* was erected by Balashova (1976) due to the previous difficulties in assigning the pygidia of *Asaphus brachyrachis* to any known Baltoscandian genus (see Jaanusson 1953a, pp. 446–451). The Norwegian material is like the Swedish species in not being associated with any cephala or thoracic segments that could be assigned unequivocally to the same taxon. Even so it presents a number of diagnostic feature such as the near lack of pleural ribs; the distinctly *Pseudasaphus*-like doublural and axial terrace-line pattern; the very straight-sided axis and the wide doublure, all of which point towards this being a *Pseudasaphus*. The only differences from this genus are found in the more rounded pygidial outline and distinctly concave pygidial border, though the beginnings of a concave outer pygidial margin may be found on especially large specimens of *Pseudasaphus limatus*. Considering the large resemblance between the two species and *Pseudasaphus* they are here assigned to that genus, and *Mischynogorites* Balashova, 1976 is consequently regarded as a junior subjective synonym of *Pseudasaphus* Schmidt, 1904.

The area on the right pygidal corner bearing fine terrace-lines is clearly an abnormality and probably a teratological phenomenon or an old injury which, when healed perhaps over several months, resulted in this abnormal change in terrace-line pattern. Similar cases have been described by Owen (1985).

Occurrence. – Specimen PMO H.2707 originates from the concretionary limestones at Huk on Bygdøy, Oslo-Asker, where it has been registrated as coming from the Elnes Formation. The sample, though, includes fragments of *Reedolithus carinatus*, a trilobite only known from the Vollen Formation and is therefore assigned to this formation. The other specimen, PMO 61144, was found on Fornebu, Oslo-Asker, and has tentatively been assigned to the Late Ordovician Arnestad Formation or younger strata, but could perhaps also come from the underlying Vollen Formation.

Genus *Ogmasaphus* Jaanusson, 1953a

Type species. – *Asaphus praetextus* Törnquist, 1884, by original designation.

Diagnosis. – Cephalon with narrow frontal area; low anterior glabellar inclination; glabella highly pyriform with strongly effaced lateral glabellar lobes and long anterolaterally directed S1 furrows, which nearly separates the posterior part of glabella from the frontal; large eyes; lack of small node behind eyes; more or less concave border and pointed genal angles or spines. Hypostome as in *Asaphus*. Panderian organs moderately small, notch-like on librigena and appearing as separate openings on thoracic pleurae. Thoracic pleurae with straight inner doublural margin. Pygidium characterized by distinct flattened border; moderately faint to strong pleural ribs on inner portion of pleural field; comparatively broad doublure and generally with a moderately dense pattern of anterolaterally directed pygidial terrace-lines. (Modified from Henningsmoen 1960 and Moore 1959).

Remarks. – Henningsmoen (1960, pp. 235–236) noted a high degree of resemblance between *Ogmasaphus* and *Asaphus*. His examination of all the generic characters given by Jaanusson (1953a) revealed that none of the features clearly distinguish the

former from the latter. Thus *Ogmasaphus* merely seemed to differ from *Asaphus* by grades in features, the most important being the slightly wider frontal area on the cephalon and the presence of pleural ribs; broad doublure and a flattened border on the pygidium. Unfortunately all of these characters are also found in species clearly belonging to *Asaphus* as recently redefined by Ivantsov (2003) such as the relatively broad preglabellar field of *Asaphus incertus* Brøgger, 1882 and the beginning of a flat border on *Asaphus ludibundus* Törnquist, 1884. Henningsmoen (1960) concluded that *Ogmasaphus* most probably should be regarded as a subgenus of *Asaphus*.

Balashova (1976, p. 33) did not share this opinion, but separated *Ogmasaphus* from the subfamily that includes *Asaphus* and put it together with genera such as *Volchovites*, *Pseudomegalaspis* and *Pseudasaphus* in the newly established family Pseudoasaphidae Balashova. This was clearly done on strength of an earlier grouping of *Ogmasaphus* with the genus *Ptychopyge* proposed by Jaanusson (1953a, p. 427). As already discussed by Henningsmoen (1960), *Ogmasaphus* differs from *Ptychopyge* in several important features including the lack of a small node behind the eye; the much shorter frontal area and the small or absent genal spines. Balashova's taxonomy is consequently not accepted here.

Based on the Norwegian material examined herein together with the descriptions of other Baltoscandian species of *Ogmasaphus* presented by Jaanusson (1953a) there seem to be a general distinction from the genus *Asaphus* in the typically narrower glabella; the wider preocular cranidial width; the generally longer frontal area; the sharp or spiny genal corners; the generally stronger pleural ribs and paradoublural line on the pygidium; the more developed pygidial border and in the denser and more widespread pleural terrace-line pattern on the pygidium. A complete separation between *Asaphus* and *Ogmasaphus* is provisionally adopted here, although they are clearly closely related and do show some transitional forms.

The Norwegian material described below reveals new taxa within the formerly erected Norwegian species. The material is on the other hand very fragmentary with few articulated specimens and an extremely poor stratigraphical control. As a result the precise taxonomic status and stratigraphical relations of the individual taxa, including the established species, are unclear. It is possible that better material will result in some of the four taxa *O. kiaeri*, *O. jaanussoni*, *O. multistriatus* and *O. sp. A* being reduced to the rank of subspecies or even ecophenotypes.

Ogmasaphus stoermeri Henningsmoen, 1960

Pl. 8, Figs 12–17; Table 19

1884 *Asaphus* confr. *platyurus*, Ang. – Brøgger, p. 259.
1960 *Ogmasaphus stoermeri* n. sp. Henningsmoen, pp. 242–244 (*partim*), pl. 11, figs 1, 2, 5, 6 (pl. 11, figs 3, 4 belong to *Asaphus* and the specimen mentioned from Bø at Gjerpen belongs to *O. kiaeri*).

Type stratum and type locality. – Holotype PMO 4684 from the silty lower to middle part of the Heggen Member exposed in the Teigen Shale Quarry between Skollenborg and Krekling railway station, Eiker-Sandsvær.

Material. – One cranidium (PMO 66657), three librigenae (PMO 60397, 66645, 89770), one poorly preserved hypostome (PMO 66657), 22 pygidia (PMO 4801, 60365, 60379, 60385, 60394, 66610, 66614, 66643, 66648, 66658, 82430, 82688, 83294 (two specimens), 89770, 89772, 102.593, 105.649, 143.552, 208.641/1, 208.642/1, 208.644/1) and eight partly articulated speciemens (PMO 4682, 4684, 58718, 60382, 72069, 89772 (two specimens), 105.922).

Diagnosis. – Species of *Ogmasaphus* characterized by angular to slightly pointed genal corners; a very narrow axis and a narrow pygidial doublure bearing 15 to 20 terrace-lines (emended from Henningsmoen 1960).

Description. – Largest cranidium (PMO 90905) is 28.5 mm long, while largest pygidium (PMO 72069) measures around 49 mm.

Cephalon broadly rounded in outline, the length corresponding to 40% of the width; cranidial width at preocular cheek corresponding to just over 50% of the full cranidial width or approximately 1.6 to 1.8 times the occipital width; glabella well delimited by fairly distinct axial and preglabellar furrows; occipital ring taking up 30% of the cranidial width, anteriorly delimited by moderately shallow occipital ring, becoming more pronounced laterally; posterior glabellar lobes moderately well developed, length corresponding to 20% of the cranidial length; anterior glabellar lobe elongately pyriform; preocular cheek moderately wide, the width corresponding to around 23% of the anterior lobe width; paradoublural line on librigena situated approximately one-third the distance between eye and lateral cephalic margin from eye; genal corners angular to slightly pointed, not extended into spines.

Thorax with narrow axis approximating up to 25% of the thoracic width.

Pygidium broadly rounded semicircular, the length, excluding the anterior half ring, corresponding to generally just over half the width; axis extremely narrow and moderately short, taking up no more than just under 20% of the pygidial width and 70 to 76% of the pygidial length. Pleural field with 6 to $6^{1}/_{2}$ faint ribs reaching out to paradoublural line; paradoublural line subparallel to outer margin; doublure narrow, occupying 20 to 25% of the pygidial width. Pygidial surface covered by moderately dense pattern of anterolateral directed wavy terrace-lines; doublure bearing 15 to 20 terrace-lines diverging markedly from the outer margin in the forward direction.

Remarks. – The pleural ribs on the pygidia become more effaced with increasing size and may completely dissappear on the largest specimens, making them harder to distinguish from pygidia of *Asaphus*. Likewise the pygidial doublure may tend to decrease in width with increasing size, but more material is needed to verify this trend.

The species differs from the other of the genus by the distinctiveness of the posterior lateral glabellar lobes; the extremely narrow and short axis and by the

Table 19. Pygidial measurements on *Ogmasaphus stoermeri* Henningsmoen, 1960

PMO number	Wpl1	Wdoublure	Waxis	Laxis	W	L	Terrace-lines
82688	18	16.5	14.1	26.9	72.7		~20
4684	6.9	6.7	5.6	10.8	31.3	15	
66610		9.2	7.3	15.3	38.6	20.3	
4682	9.4	9.7	7.8	14.9	41.3	21.4	~15
58718		9.8	7.5	17.3	42.9	23.2	16
60394_1		13	11	22.9	57.8	31.1	~20
60379		18.3	15.1	29.7	78.5	41.1	~19
72069	22.4	19.7	17.9	35.4	92.7	47.9	~15
60365		17	13.7	26.1	~68.5	35.9	~20
82430	14.8	12.1	10.8	20.4	57	27.9	17
90560	17.8	13.5	12.6	25.1	69.3	33.1	16

Wpl1, pleural field width at anterior pleural rib; Wdoublure, maximal width of doublure.
Waxis, maximal width of axis; Laxis, length of axis excluding articulating half ring.
W, maximal width of pygidium; L, length of pygidium excluding articulating half ring.
Terrace-lines, maximal number of terrace-lines counted at widest point.

narrow pygidial doublure. The relatively broad frontal glabellar lobe observed on even smaller specimens, together with the narrow pygidial doublure places this early species of *Ogmasaphus* very close to *Asaphus*, suggesting the former was derived from the latter. Even so it clearly does not belong to *Asaphus* as indicated by the relatively distinct pleural ribs and paradoublural line on the pygidium; the relatively long frontal area and in the moderately dense anterolaterally directed pygidial terrace-line pattern so typical for *Ogmasaphus*.

Occurrence. – *O. stoermeri*, which represents one of the oldest species within the subgenus, is mainly found in the more silty parts of middle Heggen Member in the southern part of the Oslo Region, but it does extend down into the more marly basal part of the member. Lower to middle Heggen Member, Elnes Formation. Skien-Langesund: Molleklev at Rognstrand (0.15 m above Helskjer Member), Bø farm at Hoppestad just northwest of Skien. Eiker-Sandsvær: Muggerudkleiva; Teigen shale quarry between Skollenborg and Krekling railway station; Waterfall in the Ravalsjø River near Heistad; Vego at Krekling (level 1.25 to 1.57 m in profile 6, corresponding to the middle Heggen Member). Modum: Road-cut at Vikersund skijump (basal Heggen Member, 5.4 m above Huk Fm.). Hadeland: East profile on point at Brandbu Church (upper Helskjer to basal Heggen Member) and Hovodden (basal Heggen Member, 9.9–16.1 m above the Huk Fm.). Mjøsa: large road-cut some kilometres north of Furnes (Heggen Member).

Ogmasaphus kiaeri Henningsmoen, 1960

Pl. 9, Figs 1–9; Table 20

1909 *Pseudasaphus globifrons*, Eichw., var. – Holtedahl, p. 28 (listed).
1940 *Asaphus platyurus* Ang. – Grorud, pp. 159–160 (occurrence).
1960 *Ogmasaphus kiaeri* n. sp. Henningsmoen, pp. 239–240, pl. 10, figs 1–5, 7–12.
1960 *Ogmasaphus multistriatus* n. sp. Henningsmoen, pp. 241–242 (*partim*), pl. 6, figs 13, 15, 16.
1960 *Ogmasaphus stoermeri* n. sp. Henningsmoen, p. 244 (only).

Type stratum and type locality. – The holotype PMO 19112 comes from the Vollen Formation at Søndre Kojatangen near Vollen, Oslo-Asker.

Material. – The examined material includes 17 pygidia (PMO 3329, 3368, 3566, 3963 (two specimens), 3966, 3986, 4098, 21921, 21922, 33337 (two specimens), 36841–42, 58929 (?) (three specimens), 207.479b/18) and seven fragmentary articulated specimens (PMO 3328, 3330, 3490, 3972, 19112, 58932 and 82666).

Diagnosis. – *Ogmasaphus* species with no genal spine and 27 to 35 terrace-lines on pygidial doublure (emended from Henningsmoen 1960).

Description. – Length of largest cranidium close to 30 mm, while largest pygidium measures 39 mm in length, suggesting the complete exoskeleton could be at least 10 or 11 cm long.

Cranidial length corresponding to 60% of the width. Glabella long and slender on small specimens becoming more squat on the larger cranidia; glabellar length approximating 90% of the cranidial length; occipital ring and furrow effaced; mesial tubercle small but moderately well developed, situated approximately 15% the cranidial length from posterior margin; lateral glabellar furrows, except for S1, completely effaced; S1 shallow, appearing as a direct continuation of the axial furrow; anterior glabellar lobe elongated teardrop-shaped to squat pyriform, bearing a low elongated ridge on central part. Preglabellar area long, around 10% of cranidial length at sagittal line. Preocular area moderately broad, the width corresponding to between 25 and 30% of the anterior glabellar lobe width, while the total preocular cranidial width approximates between 60 and 65% of the cranidial width. Abaxial end of paradoublural line located one-third the cranidial length inside cephalic margin or nearly 60% of the cranidial length in front of posterior margin.

Eye length approximately 30% of the cranidial length. Eye location somewhat variable, generally situated between 50 and 70% of its own length in front of the posterior cephalic margin. Paradoublural

Table 20. Pygidial measurements on *Ogmasaphus kiaeri* Henningsmoen, 1960

PMO number	Wpl1	Wdoublure	Waxis	Laxis	W	L	Terrace-lines
3566		5.1	5.3	10.5		13.2	33
3330	~4,0	6.4			23.5	~12.8	~35
21922	4.5	6.1	6.4	11.1	25.9	13.9	34
3329	5	6.5	5	10.5	21.9	13.4	
4098	6.8	8.8	8.9	14.8	34.8	~18.4	27
3986	3.9	4.9	4.6	7.3	19.2	9.9	28
3963	10.6	11.8	~11.3	19.7	~47.6	~25.0	30
19112	~12.8	14.3	13.7	28.8	57.2	36.3	32
36841_42	15.2	16.3	16	31.5	70.7	39.4	28
58932	~13.1	~13.0	15.3				27

Abbreviations: see Table 19.

line located around 30% of the distance between eye socle and lateral margin from eye. Lateral border moderately wide. Doublure covered by around 27 terrace-lines running parallel to paradoublural line. Panderian opening slit-like. Genal corners sharply rounded with no genal spines.

Thoracic axis occupying one-third the total width; panderian openings moderately small.

Pygidium broadly rounded with flatly convex to slightly concave borders and concave doublure; the length for undeformed specimens, excluding the anterior half ring, corresponds to between 52 and 63% of the width on eight measured specimens; axis relatively broad, the width corresponding to 25% of the pygidial width, while the length takes up 75 to 80% of the pygidial length; axial segments numbering 11 to 12. Pleural field bearing $5^1/_2$ to $6^1/_2$ moderately weak to nearly effaced pleural ribs; pleural ribs flatly rounded in cross-section with shallow central interpleural furrow. Paradoublural line effaced. Pygidial doublure moderately wide, the maximal width corresponding to 25% of the pygidial width; doublural terrace-lines numbering 27 to 35 at widest part. Pygidial surface covered by pattern of short anterolateral directed wavy terrace-lines on central to anterior part, becoming more continous posterolaterally.

Remarks. – *Ogmasaphus kiaeri* differs from *O. multistriatus* in its sparser doublural terrace-lines on the pygidium and possibly by a wider preocular area. The sparse and somewhat fragmentary material makes this differentiation a little uncertain and better material may reduce them to subspecies level. The same may be true for *O. kiaeri* and *O. jaanussoni*, where the latter largely differs by the presence of a librigenal spine and a slightly narrower pygidial doublure and axis.

O. praetextus (Törnquist, 1884) and *O. costatus* Jaanusson, 1953a differ in the shorter frontal area and narrower preocular area and in the case of the former by the more strongly developed pleural ribs on the pygidium.

Occurrence. – The species has been collected from most of the Oslo Region, where it occurs in the upper part of the Elnes Formation and possibly up into the overlying Vollen Formation

or perhaps even the Upper Ordovician Arnestad Formation of the central Oslo Region.

Skien-Langesund: Bø at Hoppestad just north of Skien. Oslo-Asker: Håkavik (Elnes Fm. or perhaps lower Vollen Fm.), southern Kojatange in Asker, Fornebu (?) (probably Arnestad Fm.), Huk on Bygdøy (Engervik/Håkavik Mb.) and Sarpsborggata, Nordraaks gate and the town hall in Oslo City (Vollen Fm.). Hadeland: Skiaker farm just west of Gran (Elnes Fm.). Mjøsa: Helgøya (Elnes Fm.) and road profile along E6 between Nydal and Furnes (18.96 m in subprofile 3 corresponding to the basal 'Cephalopod Shale').

Ogmasaphus jaanussoni Henningsmoen, 1960

Pl. 9, Figs 10–15; Pl. 10, Figs 14; Table 21

1884 *Asaphus platyurus* Ang. – Brøgger, p. 260 (occurrence).
1960 *Ogmasaphus jaanussoni* n. sp. Henningsmoen, pp. 237–238 (*partim*), pl. 9, figs 1, 3 (pl. 9, fig. 2 belongs to *O. multistriatus*).

Type stratum and type locality. – The holotype specimen PMO 61519 was found at Ruseløkkveien in Oslo City. The stratigraphical level is unknown, but most likely it belongs to the Elnes Formation or lowermost Vollen Formation.

Material. – Four articulated but fragmentary speciemens (PMO S.1790–91 (one specimen), 20289, 61519 and 82879), one pygidium (PMO 3327) and possibly one small cranidium (PMO 56267).

Diagnosis. – Species of *Ogmasaphus* with genal spines and narrow pygidial doublure with no more than 24 to 27 terrace-lines (emended from Henningsmoen 1960).

Description. – Largest cranidium 37 mm long, while largest pygidium on same specimen measures nearly 41 mm.

Cranidial length approximating two-thirds the width on holotype, but tectonic deformation may have produced lateral compression of the specimen. Total glabellar length on holotype 93% of cranidial length; occipital ring and furrow effaced; mesial tubercle well developed but moderately small, situated approximately 12% the cranidial length from posterior margin; lateral glabellar furrows completely effaced; anterior glabellar lobe characteristically strongly pyriform, bearing a low elongated ridge on central part; parafrontal band absent. Preglabellar area moderately long, around 7% of cranidial length

Table 21. Pygidial measurements on *Ogmasaphus jaanussoni* Henningsmoen, 1960

PMO number	Wpl1	Wdoublure	Waxis	Laxis	W	L	Terrace-lines
20289	12.2	11.5	14	24.9	57.2	32.3	24
61519	~17.7	14.2	15.7	32.5	~72.9	40.6	~27
3327	7.9	6	5.7	13	~27.6	15.8	22

Abbreviations: see Table 19.

at sagittal line. Preocular area moderately broad, the width corresponding to around 30% of the anterior glabellar lobe width, while the total preocular cranidial width is 63% of the cranidial width on specimen PMO 61519. Abaxial end of paradoublural line located 25% of the cranidial length inside cephalic margin or 60% of the cranidial length in front of posterior margin.

Eye length approximating 30% of the cranidial length. Eye located half its length in front of the posterior cephalic margin. Lateral border narrow with the paradoublural line located 40% of the distance between eye socle and lateral margin from eye. Doublure covered by around 35 terrace-lines on posterior part, decreasing to around 25 on central part. Panderian opening slit-like. Genal corners extended into small genal spines.

Thoracic axis occupies 30% of the total thoracic width.

Pygidium is broadly rounded with concave doublure and weakly convex to slightly concave borders, although this concavity probably reflects later compression. The length, excluding the anterior half ring, corresponds to around 56% of the width on two measured specimens; axis moderately narrow, taking up between 20 and 25% of the pygidial width, while the length corresponds to 80% of the pygidial length; axial segments numbering 12 on holotype. Pleural field bearing around seven nearly effaced pleural ribs cut by central shallow interpleural furrows. Paradoublural line effaced. Pygidial doublure narrow, the maximal width corresponding to 20% of the pygidial width or a little less than the pleural field width at anterior pleural rib; doublural terrace-lines numbering between 24 and 27 on widest part of two examined specimens, semiparallel to outer pygidial margin. Posterolateral pygidial surface covered by fairly dense pattern of anterolaterally directed wavy terrace-lines, becoming sparser anteriorly.

Remarks. – This species is rather close to *O. kiaeri*, but differs on the presence of small genal spines and by the narrow doublure with fewer terrace-lines. From *O. stoermeri* it may be separated on strength of its relatively longer pygidium with a longer and broader axis, which is distinctly wider than the pygidial doublure.

Occurrence. – Oslo-Asker: Slemmestad, Fornebu, Huk (?) on Bygdøy and Oslo City (Håkavik Member and up into the Vollen Fm. and perhaps also down into the basal Elnes Formation).

Ogmasaphus multistriatus Henningsmoen, 1960

Pl. 10, Figs 1–7; Table 22

1890 ny Asaphus-art [new *Asaphus* species] Brøgger, p. 176.
1960 *Ogmasaphus jaanussoni* n. sp. Henningsmoen, pp. 237–238 (*partim*), pl. 9, fig. 2.
1960 *Ogmasaphus multistriatus* n. sp. Henningsmoen, pp. 241–242, pl. 6, figs 13, 16.

Type stratum and type locality. – Holotype pygidium PMO 36822a originates from the Elnes Formation at Storhamar, Mjøsa.

Material. – The examined material comprises 10 pygidia (PMO 33057, 33058 (three specimens), 36822a, 58580, 58928 (three specimens), 82616) and possibly two cranidia (PMO 36822b, 58924) and one cephalon (PMO 21931).

Diagnosis. – *Ogmasaphus* with no or very short genal spines and a broad pygidial doublure bearing 40 to 43 terrace-lines (emended from Henningsmoen 1960).

Description. – Length of largest cranidium possibly belonging to this species 11 mm, while largest pygidium measures close to 12 mm in length.

Cranidial length just over 60% of the width. Glabella long and slender; length slightly over 90% of the cranidial length; occipital ring and furrow effaced; mesial tubercle minute but well developed, situated nearly 15% the cranidial length from posterior margin; lateral glabellar furrows, except for S1, completely effaced; S1 shallow, appearing as a direct continuation of the axial furrow; anterior glabellar lobe elongately teardrop-shaped, bearing a low elongated ridge on central part; parafrontal band absent. Preglabellar area long, nearly 10% of cranidial length at sagittal line. Preocular area moderately broad, the

Table 22. Pygidial measurements on *Ogmasaphus multistriatus* Henningsmoen, 1960

PMO number	Wpl1	Wdoublure	Waxis	Laxis	W	L	Terrace-lines
58928	8	8.9	9.6	15.9	36.9	20.2	42
36822	3.85	5.35	5.2	8.8	20.7	11.1	43
6580	11.7	11.1	12	21.7	48.6	27.5	40
82616	7.6	8.3	8.7	16.1	33	20.7	43

Abbreviations: see Table 19.

width corresponding to between 20 and nearly 25% of the anterior glabellar lobe width, while the total preocular cranidial width approximates 60% of the cranidial width. Abaxial end of paradoublural line located between 25 and 30% of the cranidial length inside cephalic margin or just over 60% of the cranidial length in front of posterior margin.

Eye length around 35% of the cranidial length and situated 37% its own length in front of the posterior cephalic margin on specimen PMO 21931. Paradoublural line located about 30% of the distance between eye socle and lateral margin from eye. Lateral border moderately narrow. Genal corners sharp with no, or at the most very small genal spines.

Pygidium broadly rounded with weakly convex to slightly concave borders and concave doublure; the length, excluding the anterior half ring, corresponds to around 55% of the width; axis moderately broad, the width corresponding to 25% of the pygidial width, while the length takes up 80% the pygidial length; axial segments number 11 to 12. Pleural field bearing 6 to 7 moderately weak to nearly effaced pleural ribs; pleural ribs flatly rounded in cross-section with distinct central interpleural furrow. Paradoublural line moderately effaced. Pygidial doublure fairly wide, the maximal width corresponding to 25% of the pygidial width; doublural terrace-lines numbering 40 to 43 at widest part. Pygidial surface covered by short anterolateral directed wavy terrace-lines, which are not present in pleural furrows.

Remarks. – The pygidium appears intermediate between *Ogmasaphus* sp. A and *O. kiaeri*, but is easily distinguised by the 40 to 43 doublural terrace-lines. A moderately poor librigena, PMO 203.030/1, which most likely belongs to this species, has been found in the uppermost part of the Sjøstrand Member of the Elnes Formation. The librigena has more than 30 closely arranged doublural terrace-lines and lacks genal spines.

Occurrence. – Oslo-Asker: Ildjernet (Vollen Fm.?), ?beach profile at Djuptrekkodden (21.44 m in the profile corresponding to the top Sjøstrand Member), Fornebu (Vollen Fm.?) and the town hall in Oslo City (Vollen Fm.). Hadeland: road-cut between Grinaker and Tingelstad (Elnes Fm.). Mjøsa: Storhamar (Elnes Fm.), Helgøya (?) (Elnes Fm.).

Ogmasaphus furnensis n. sp.

Pl. 10, Figs 15–17; Pl. 29, Figs 5–7, 9; Table 23

Derivation of name. – Refers to the type locality at Furnes Church in the Mjøsa district.

1984 *Asaphus* (*Asaphus*?) sp. Wandås, p. 219, pl. 2 M.

Type stratum and type locality. – Top Helskjer Member at the Furnes Church profile just north of Hamar in the Mjøsa district, Norway.

Type material. – The holotype is the nearly complete, articulated specimen PMO 67181 figured on Plate 29, figures 5–7. Paratype pygidium PMO 67589a is from an unspecified level in the Helskjer Member at the type locality, while paratype pygidium PMO 90498 and PMO 90499a and cranidium PMO 90499b are collected from the 8.3 m level in the type profile, corresponding to the top of the Helskjer Member.

Material. – There are eight pygidia (PMO 36849, 67176, 67180, 67181, 67589a, 90498 (two specimens), 90499a), three fragmentary cranidia (PMO 90498, 90499 (two specimens)) and a nearly complete specimen (PMO 67181).

Diagnosis. – Species of *Ogmasaphus* characterized by 20 or less terrace-lines on the pygidial doublure; a very *Asaphus*-like tail and a pygidial doublure just exceeding one quarter the pygidial width.

Description. – Largest cranidium reaches a length of approximately 23.5 mm, while largest pygidium, PMO 33513, is nearly 27 mm long.

Glabellar length 90 to 93% of cranidial length (two specimens); occipital ring and furrow effaced but present; mesial tubercle well developed but moderately small, situated 15 to 19% the cranidial length from posterior margin; lateral glabellar furrows effaced, but fairly distinct on exfoliated specimens, turned strongly rearwards adaxially; anterior glabellar lobe strongly pyriform, bearing a low elongated ridge on central part; parafrontal band absent. Preglabellar area long, approximating to slightly less than 10% of

Table 23. Pygidial measurements on *Ogmasaphus furnensis* n. sp.

PMO number	Wpl1	Wdoublure	Waxis	Laxis	W	L	Terrace-lines
90499a	8.1	9.4	8.6	13.6	36.7	~17.7	~20
90498		10.9	~9.2	~16.9	39.9	~21.6	19
67586	8.5	11.3	8.75	15.4	39	19.75	
67589a	11.5	11.5	10.5	19.1	45.6	24.5	
67181	6.8	9.4	~7,2	16.3	33	20.6	14

Abbreviations: see Table 19.

cranidial length at sagittal line. Preocular area fairly broad, the width corresponding to between 14 and 17% of the preocular cranidial width. Abaxial end of paradoublural line located around 20% of the cranidial length inside cephalic margin. Cranidial surface largely smooth except for some short terrace-lines around mesial tubercle and, in some specimens also on occipital ring and palpebral lobes. Eye length corresponding to 31% of total cranidial length on PMO 67181. Genal corners sharp, but not ending in genal spines. Inner margin of librigenal doublure bearing around 15 terrace-lines.

Thoracic axis corresponding to 25% of total thoracic width.

Pygidium broadly rounded with no or slight posterior truncation, flat to weakly convex borders and concave doublure; the length, excluding the anterior half ring, corresponds to just over half the width; axis relatively narrow, the width corresponding to 25% of the pygidial width or just below, while the length takes up 80% of the pygidial length; axial segments numbering 11 to 12. Pleural field bears $5^1/_2$ to 7 moderately weak to nearly effaced pleural ribs, which are gently rounded in cross-section with weak central interpleural furrow. Paradoublural line effaced. Pygidial doublure fairly narrow, the maximal width corresponding to 25% of the pygidial width; doublural terrace-lines numbering 15 to 20 at widest part. Pygidial surface covered by moderately sparse pattern of short anterolaterally directed terrace-lines.

Remarks. – The taxon is morphologically close to both *O. jaanussoni* and *O. kiaeri*, but is differentiated from both in the lower number of doublural terrace-lines and, from the former, by a larger maximal doublure width than pleural field width at anterior pleural rib. The pygidium is very similar to those of *Asaphus*, but may be differentiated by its more pronounced pleural ribs.

Occurrence. – The species occurs in the lower part of the Elnes Formation. Mjøsa: Furnes (topmost Helskjer Member around 3.30 m above the Huk Fm.), Flagstadelva north of Hamar (Elnes Fm.).

Ogmasaphus sp. A

Pl. 10, Figs 8–13; Table 24

1960 *Ogmasaphus kiaeri* n. sp. Henningsmoen, p. 240 (*partim*), pl. 10, fig. 6 (only).
1960 *Ogmasaphus multistriatus* n. sp. Henningsmoen, pp. 241–242 (*partim*), pl. 6, figs 14, 17.

Material. – The examined material contains 10 pygidia (PMO 3455, 3466, 4098, 6007, 16853, 33417, 40345, 56581, 56582, 72058) and one partly articulated specimen consisting of thorax and tail (PMO 3466).

Description. – Thoracic pleural furrows terminating aproximately three-fifths the pleural width outside the axial furrow; panderian openings moderately small.

Pygidium broadly rounded with weakly convex to slightly concave borders and concave doublure; the length, excluding the anterior half ring, corresponds to between 53 and 58% of the width on six measured specimens; axis relatively broad, the width corresponding to just over 25% of the pygidial width, while the length takes up 80% of the pygidial length; axial rings numbering 11 to 12. Pleural field bearing $5^1/_2$ to $6^1/_2$ moderately weak to nearly effaced pleural ribs reaching nearly two-thirds the pleural field out towards the margin; pleural ribs weakly rounded in cross-section with central interpleural furrow of varying distinctness. Paradoublural line effaced. Pygidial doublure wide, the maximal width corresponding to between 25 and 30% of pygidial width; doublural terrace-lines generally numbering 50 to 55 at widest part, but may have as many as 58. Pygidial surface covered by short anterolaterally directed wavy terrace-lines.

Remarks. – This material was formerly placed in *Ogmasaphus multistriatus*, but it is readily differentiated by its denser terrace-line pattern on a wider doublure. The same characteristics clearly separate it from all other species within this genus.

The pygidium may look very similar to those of *Pseudobasilicus* sp. A, but is usually easily distinguised

Table 24. Pygidial measurements on *Ogmasaphus* sp. A

PMO number	Wpl1	Wdoublure	Waxis	Laxis	W	L	Terrace-lines
6007	7.4	10.3	10.5	18.8	40.1	22.7	58
40345	7.4	12.1	11.9	20.5	43.9	25.2	51
3455		14.4	12.6	19.3	~54.5		54
56581	9.2	19.6	16.7	31.3	66.9	39	54
72058	5.5	9.7	9.9	16	36.7	19.5	~55
21921	9.3	17.9	15.7	26.1	62.8	34.3	~50
33417			6.8	11.5	26.6	14.5	~50

Abbreviation: see Table 19.

by the relatively larger width, the lower number of pleural ribs, a slightly narrower border and by the anterior pleural furrow, which on *O.* sp. A is wider or of approximately the same width (exsag.) as the anteriormost axial segment, while distinctly shorter on *P.* sp. A.

Occurrence. – Oslo-Asker: Holmen at Vollen (Vollen Fm.); Nye Drammensvei at Gyssestad (Vollen Fm.); road cut at Kilen on Fornebu (Vollen Fm.?); road cut above Lille Frøen and the town hall, both in Oslo City (Vollen Fm. or uppermost Elnes Fm.). Hadeland: Melbostad (Elnes Fm.).

Genus *Asaphus* Brongniart *in* Brongniart & Desmarest, 1822

Type species. – *Entomostracites expansus* Wahlenberg, 1821.

Diagnosis. – Cephalon typically broadly rounded to slightly subtriangular in outline, centrally with a relatively wide and distinctly hourglass-shaped glabella reaching nearly all the way out to the anterior cephalic margin; basal lobes present; eyes typically highly elevated; genal corners rounded without spines; vincular apparatus, if present, represented by pits of various depth; hypostome with deep posterior notch; pygidium with no, or at the most strongly effaced pleural ribs; cephalic and pygidial borders absent or indistinct, convex. (Modified from Jaanusson 1953a; Moore 1959; Ivantsov 2003).

Remarks. – The genus was split into the two sub-genera *Asaphus* (*Asaphus*) Brongniart and *Asaphus* (*Neoasaphus*) Jaanusson when Jaanusson (1953a) revised the asaphids from the south-central Swedish Middle Ordovician. The two subgenera were thought to differ in the width of the pygidial doublure, the size and location of the eyes and in the terrace-line pattern on the pygidial doublure. Later Balashova (1976) erected three further subgenera, although these have largely been overlooked by following authors. Wandås (1984) examined the *Asaphus* material from the lowermost Elnes Formation and summarized the diagnostic differences between the two subgenera *A.* (*Asaphus*) and *A.* (*Neoasaphus*), with that the latter having isolated panderian openings on the thoracic doublure; the eye clearly closer than its own exsagittal length to the posterior cephalic margin and the pygidial doublure narrower with a larger number of terrace-lines. When comparing with the different species of *Asaphus* found within the Elnes Formation this subdivision is sorely tested as the eye location; the pygidial doublural width and terrace-line density show a mix between the two

subgenera in some instances even within the same species. The final character, the panderian opening, appears more or less identical between all species. The subdivision of *Asaphus* into the two subgenera *A.* (*Asaphus*) and *A.* (*Neoasaphus*) thereby does not seem to hold up, at least with the present definition.

Recently the *Asaphus* group underwent a major revision by Ivantsov (2003), who found that most of the groupings were not supported in a phylogenetic view as the proposed diagnostic characters such as the panderian openings and the presence or absence of a stalked eye socle appeared to develop and diss-appear in more than one lineage. Other characters showed gradual transitions from one form to another, often making the assignment of new species to a subgenus an exercise in futility. Consequently he found it best to ignore the subgeneric category and move the only justified 'subgenera' *A.* (*Subasaphus*) Balashova, 1976 and *A.* (*Onchometopus*) Schmidt, 1898 up to generic level. Based on the present material from the Elnes Formation of the Oslo Region I fully agree with Ivantsov (2003); the previous subgeneric divisions among the *Asaphus* species clearly do not represent any natural groupings and furthermore have very vague morphological boundaries leading to more confusion than they help.

Asaphus sarsi Brøgger, 1882

Pl. 11, Figs 7–14; Table 25

1882 *Asaphus striatus*, Sars. var. Sarsi – Brøgger, pp. 94–95, pl. 8, figs 1–3b.
1984 *Asaphus* (*Asaphus*) *striatus* (Boeck, 1838) – Wandås, pp. 218–219 (*partim*), pl. 1J, pl. 2A–D, F, H, K (description partly based on material from true *A. striatus*).
1995 *Asaphus* (*Asaphus*) *striatus* (Boeck, 1838) – Nielsen, pp. 91–95, figs 70–73.
2003 *Asaphus striatus sarsi* Brøgger, 1882 – Ivantsov, pp. 281–284.
2004 *Asaphus striatus sarsi* Broegger, 1882 – Ivantsov, p. 28, fig. 31, diagram 5, 7, 14, 21.

Type stratum and type locality. – The holotype PMO H2618 derives from the Svartodden Member of the Huk Formation in Oslo City, Norway.

Norwegian material. – A large amount of material is stored at the Natural History Museum, University of Oslo, of which the main part comes from the Huk Formation. The examined material from the Huk and Elnes formations includes two cranidia (PMO 201.980B/2, 201.982), seven pygidia (PMO 82163, 201.977/1, 201.977/2, 201.981A/1, 201.992/1, 201.994A/1, 201.995) and seven partly articulated specimens (PMO 1643, 1889, H2616, H2618, 56257, 102.668, 106.046).

Table 25. Pygidial measurements on *Asaphus sarsi* Brøgger, 1882

PMO number	Wpl1	Wdoublure	Waxis	Laxis	W	L	Terrace-lines
H2616	8	16.3	13.5	26.2	~40.4	32.5	21
208531		9.7	11.9	17.2	37.4	22.5	
1643	7.7	9.7	12.6	20.2	~39.2	24.1	~17
106.046	8.7	10.7	13.6	20.4	40.2	23.7	47
1889	~9.9	11.9	12.2	19.6	40.4	23.1	22
56257	7.5	11.7	11.9	18.5	~38.0	22.7	23
201.994/1	10	11	11.8	19.9	~42.2	23.5	
H2618	11	~15.6	18.2	~30.4	51.9	36.5	

Wpl1, pleural field width at anterior pleural rib; Wdoublure, maximal width of doublure.
Waxis, maximal width of axis; Laxis, length of axis excluding articulating half ring.
W, maximal width of pygidium; L, length of pygidium excluding articulating half ring.
Terrace-lines, maximal number of terrace-lines counted at widest point.

Diagnosis. – Cephalon subtriangular, length corresponding to between 50 and 60% of the width; cranidial front moderately pointed; glabellar convexity moderately low; frontal area fairly short, the length corresponding to between 6 and 9% of cephalic length; eyes moderately large and raised well above glabella; hypostome with deep posterior notch corresponding to around 35% of hypostomal length and with gently swollen inner notch margin; pygidium moderately broad with a the length/width ratio around to just over 60%; axis broad and long, the width corresponding to between 28 and 35% of the pygidial width, while the length corresponds to 80% of the pygidial length or sligthly more; axis bearing 10 to 12 fairly distinct axial rings and doublure containing 17? or 18 to 26 coarse terrace-lines (emended from Ivantsov 2003).

Description. – Largest Norwegian cranidium (PMO H2618) 30 mm long. Largest pygidium belonging to same specimen is 39 mm long.

Cephalon subtriangular in outline with parabolic frontal margin. Cephalic length corresponding to between 50 and 60% of the width. Cranidial length approximating 70% of the width. Glabellar convexity moderately low; maximal glabellar width located just behind midline between eyes and anterior cephalic margin corresponds to around 160% of the minimum glabellar width found between the eyes. Occipital ring and furrow weakly developed; mesial tubercle small but fairly distinct, situated between 16 and 19% of cranidial length in front of posterior margin; lateral glabellar furrows, except for S1, largely effaced on outer surface, more pronounced on exfoliated specimens; S1 shallow; anterior glabellar lobe pyriform with slightly pointed front; cranidial front pointed, centrally going out into short ogive; frontal area fairly short, corresponding to between 6 and 9% of cranidial length. Preocular area moderately broad, each side taking up about 12% of the preocular cranidial width.

Eyes fairly large and rather high, their length corresponding to around 27 to 28% of the cranidial length and situated between 85 and 115% of their own length in front of posterior cephalic margin. Paradoublural line on librigena fairly indistinct, located between one-fifth and one-third the distance between eye socle and lateral margin from eye. Librigenal doublure with somewhere between 14 and 17 terrace-lines midway between posterior and anterior end. Genal corners sharply rounded.

Hypostome with deep posterior notch corresponding to around 35% of the total hypostomal length and with a gently swollen anterior notch margin.

Pygidium parabolic in outline with a length, excluding the anterior half ring, corresponding to between 56 and 70% of the width (11 specimens). Pygidium distinctly convex in cross-section with straight to slightly concave outer doublural zone. Axis broad, the width taking up between 28 and 35% of the pygidial width, while the length corresponds to 81 to 86% of the pygidial length (in a single specimen, MGUH 22407 from the Geological Museum of Copenhagen in Denmark, it is as low as 75%); axial segments numbering 10 to 12. Pleural ribs are generally not seen. Paradoublural line indistinct. Pygidial doublure moderately wide, the maximal width corresponding to between 25 and 32% of the pygidial width; doublural terrace-lines numbering 18 to 26 at widest part, although specimens with 17? and 47 terrace-lines have been observed. Pygidial surface generally bearing up to seven primary terrace-lines delineating the pleural segmentation and some lesser and more wavy ones in between. Seemingly completely smooth speciemens do occur, but in general this relates to later exfoliation.

Remarks. – The distinction of this species from *A. striatus* are given below in the discussion on that species.

Occurrence. – The Norwegian members of *Asaphus sarsi* are found in the uppermost Huk Formation and the basal Helskjer Member. Eiker-Sandsvær: Stavlum at Krekling (Helskjer Member, 0.42 m above the Huk Fm.). Modum: Road cut at Vikersund skijump (0.20 m above base of formation). Oslo-Asker: The Old Quarry at Slemmestad (Helskjer Member, 0.18–0.42 m above base), Huk on Bygdøy (Svartodden Member, Huk Fm.), central Oslo (Svartodden Member, Huk Fm.), Tøyen in Oslo, (Svartodden Member, Huk Fm.). Mjøsa: Helskjer on Helgøya (Helskjer Member, 2.8 m above the Huk Fm.).

Outside Norway it is known from the *Asaphus raniceps* trilobite Zone of the Komstad Limestone at Killeröd, Scania, and possibly also from similar beds at Andrarum and Tommarp in southwestern Scania, Sweden (Nielsen 1995). In the St. Petersburg area of Northwestern Russia it has been collected from the upper part of the *Asaphus raniceps* Zone or lower part of the *Asaphus striatus* Zone (Ivantsov 2004).

Asaphus striatus (Sars & Boeck *in* Boeck, 1838)

Pl. 11, Figs 15–16; Pl. 12, Figs 1, 4–8; Table 26

1838	*Trilobites striatus* SS. & Bk. – Boeck, p. 142.	
1882	*Asaphus striatus*, Sars & Boeck, form. *typica* – Brøgger, p. 95, pl. 8, fig. 4.	
1886	*Asaphus striatus*, Boeck – Brøgger, pp. 28, 30, pl. 1, figs 9–9a.	
1940	*Asaphus striatus* (Sars et Boeck MS) – Størmer, pp. 141–142, pl. 3, figs 12–15.	
1963	*Asaphus striatus* – Skjeseth, p. 63 (listed).	
1984	*Asaphus (Asaphus) striatus* (Boeck, 1838) – Wandås, pp. 218–219 (*partim*), pl. 1H–I, ?pl. 2E.	
non 1995	*Asaphus (Asaphus) striatus* (Boeck, 1838) – Nielsen, pp. 91–95, figs 70–73.	
? 2003	*Asaphus striatus striatus* (Boeck, 1838) – Ivantsov, pp. 281–284.	
? 2004	*Asaphus striatus striatus* (Boeck, 1838) – Ivantsov, p. 34, fig. 44, diagram 2, 5, 14.	

Type stratum and type locality. – The lectotype PMO H2634 originates from the Svartodden Member, uppermost Huk Formation, at Eiker, Eiker-Sandsvær.

Norwegian material. – The NHM collections contain a large number of specimens belonging to this species of which the following has been examined: four cranidia (PMO 201.989/5, 202.025/1, 202.074, 206.950/2), one librigena (PMO 202.071/1), two hypostomes (PMO H2634, 106.551), 95 pygidia (PMO H2634 (two specimens), 4798 (four specimens), 4800, 4827, 33513, 60373 (four specimens), 60374 (four specimens), 66613, 66643, 68424, 82212 (two specimens), 83010, 83012, 83017, 83253, 90500, 101.770 (two specimens), 102.570, 103.667, 104.079, 105.927, 106.432, 201.984, 201.987/1, 201.989/2, 201.989/3, 201.989/4, 201.990/1, 201.990/2, 201.992/2, 201.993/2, 201.994/5, 201.998, 202.000/2, 202.001, 202.003/1, 202.004, 202.009/1, 202.011, 202.013, 202.017, 202.030/9, 202.039, 202.041/1, 202.041/3, 202.042/1, 202.042/2, 202.042/7, 202.048/1, 202.049, 202.050/1, 202.051/1, 202.054, 202.055, 202.056/2, 202.057/1, 202.061/1, 202.075, 202.076/2, 202.077/5, 202.084/4, 202.091/2, 202.095/1, 202.104, 202.111, 202.119/1, 202.119/2, 202.119/3,

202.120/1, 202.136/1, 206.946, 206.947/2, 206.947/3, 206.948, 206.949/1, 206.949/2, 206.949/3, 206.951/1, 206.952/3, 206.954/1, 206.954/2, 206.955, 206.957/1) and 23 partly articulated or nearly complete specimens (PMO 4799, 4812, 60374, 60383, 60394, 68413, 82211 (two specimens), 82212, 82315–16 (one specimen), 82666, 90630, 102.471, 102.812, 103.715, 106.520, 106.521, 201.989/1, 202.082/1, 202.123, 202.012/1, 202.082/2, 202.076/4).

Diagnosis. – Cephalon broadly rounded in front, length corresponding to about 45% the width; cranidial front weakly pointed; glabellar convexity moderately low; preglabellar area very narrow or absent; eyes large and fairly high, situated approximately 85% of their own length from posterior cephalic margin; hypostome with deep posterior notch corresponding to between 30 and 35% of hypostomal length and with gently swollen inner notch margin; pygidium moderately broad, the length/width ratio only just surpassing 0.50; axis narrow and moderately short, the width corresponding to 25% of the pygidial width, while the length approximates 80% of the pygidial length; axis bearing 9 to 10 fairly distinct axial rings and doublure containing 18 to 24 coarse terrace-lines (emended from Brøgger 1882; Wandås 1984).

Description. – Largest Norwegian cranidium (PMO 82211) nearly 26 mm long. Largest pygidium belonging to same specimen is 30.5 mm long.

Cephalon wide, the length corresponding to approximately 45% of the width. Cranidial length approximating 70% of the width. Glabellar convexity moderately low; maximal glabellar width located just behind midline between eyes and anterior cephalic margin corresponds to nearly 150% of the minimum glabellar width found between the eyes. Occipital ring and furrow weakly developed; mesial tubercle small but fairly distinct, situated around 17% of cranidial length in front of posterior margin; lateral glabellar furrows, except for S1, largely effaced on outer surface, more pronounced on exfoliated specimens; S1 shallow, appearing as a direct continuation of the axial furrow; anterior glabellar lobe pyriform with rounded front; cranidial front gently rounded with short central ogive; frontal area very short, on some specimens nearly absent. Preocular area moderately broad, each side taking up about 11% of the preocular cranidial width. Cranidial surface with faint forwardly convex terrace-lines on anterior half of frontal lobe.

Eyes large and rather high, their length corresponding to 30% of the cranidial length and situated 90 to 100% of their own length in front of posterior cephalic margin. Paradoublural line on librigena indistinct, located approximately 25% of the distance between eye socle and lateral margin from eye.

Table 26. Pygidial measurements on *Asaphus striatus* (Boeck, 1838)

PMO number	Wpl1	Wdoublure	Waxis	Laxis	W	L	Terrace-lines
66613	~5.3	6.1	~5.9	10.4	~22.5	13	
4827	8.4	11	9.8	14	41.8	18.9	
82666	7.6	11.6	10.3	18.9	45.4	24	~18
H2634	8.9	13.2	~12.5	20.4	~50.5	26.1	23
82211_1	11.4	13.2	13.1	22.9	56	29.3	~18
4798	8.2	9.9	9.1	15.1	~40.2	20.6	24
103.667	9.7	13.8	11.7	21.3	49.5	26.6	20
68424	~6.5	9.6	8.9	13.5	33.1	17.6	18
68413	~8.3	10.9	11.3	18.2	41	22.7	23
83010	~5.2	7.4	5.8	11.5	25.3	14.2	21
83253	~5.9	8.3	7.5	13.2	31.2	16.3	19
90500	9.5	11	9.9	16.8	~39.8	22.1	19
102.812	~5.5	6.8	6.9	~10.7	26.2	13.6	~19
60394_2	~7.0	8.8	8.6	13.4	~34.6	16.2	
33513	~11.5	13.2	12	21.1	52	26.6	~19
66643	9.5	10.9	9.9	18.5	47	24.3	18
60376	5	7.3	6.3	10.5	27	14	
83012	~7.9	~11.4	9.8	~17.2	~40.2	~22.3	
60383	5.3	7.1	~6.5	~10.9	27.6	~13.6	
106.432	10.3	11.7	11.8	18.2	48.5	22.5	20
106.521	11.9	13.9	13.6		55.4	~26.0	
201.989/2			8.4	13.5	32	16.7	
202.082/1	9.5	11.2	9	~15.9	41.7	~21.5	~19

Abbreviation: see Table 25.

Librigenal doublure with approximately 16 terrace-lines midway between posterior and anterior end. Genal corners sharply rounded.

Hypostome with deep posterior notch corresponding to between 30 and 35% of the total hypostomal length and with a gently swollen anterior notch margin. Surface, except for median body, covered by dense terrace-line pattern.

Pygidium semiparabolic in outline with rounded posterior margin and a length, excluding the anterior half ring, corresponding to between 51 and 56% of the width (12 specimens). Pygidium markedly convex in cross-section with slightly concave outer doublural margin. Axis narrow, the width taking up between 23 and 28% of the pygidial width, while the length corresponds to 73 to 81% of the pygidial length; axial segments numbering 10 to 11. Pleural field bearing 5 to 6 nearly completely effaced pleural ribs. Paradoublural line indistinct. Pygidial doublure moderately wide, the maximal width corresponding to between 24 and 29% of the pygidial width; doublural terrace-lines numbering 18 to 24 at widest part. Pygidial surface generally smooth without any terrace-lines, but this may in part relate to a generally poor preservational state of the test surface. Several specimens are characterized by a sparse to moderately dense pattern of anterolaterally directed secondary terrace-lines on the main pleural field, becoming less dense towards the axis from which strong primary terrace-lines reach out before terminating on inner pleural field.

Remarks. – *Asaphus sarsi* treated above was regarded as a junior synonym of *A. striatus* by Wandås (1984) and Nielsen (1995), but is easily differentiated by features such as its more triangular cephalon and wider pygidial axis, which is somewhat broader than the doublure. There is seemingly no difference in the number of doublural terrace-lines on the pygidium as earlier thought by Brøgger (1882), although *A. sarsi* in general tends to lie in the high end of the range. The two taxa were regarded as subspecies by both Brøgger (1882) and Ivantsov (2003) due to their close relationship and a seemingly lack of stratigraphical overlap. The Norwegian material does show a clear overlap in the stratigraphical occurrence, and hence they are here raised to species level.

The terrace-line number on the pygidial doublure of *A. striatus* appears to decrease for the stratigraphically youngest specimens, possibly in connection with the change in lithofacies from limestones to dark siliciclastic mudstones.

The Russian specimens assigned to *A. striatus* by Ivantsov (2003) are characterized by quite well-developed segmental terrace-lines on the pygidium lacking in at least the main part of the Norwegian material. They furthermore appear to have a slightly wider axis, although this to a large degree may relate to their less compressed preservational state. In all cases I do not find these possible differences important enough for a separation into two distinct species, but it may suggest the presence of different subspecies or ecophenotypes; possibly the last as the Norwegian material includes a few specimens featuring strongly developed terrace-lines (see below).

Occurrence. – The species is a characteristic component of the fauna found within the Helskjer Member to lower Heggen Member, where it completely dominates the fauna in the southern and central part of the Oslo Region. Skien-Langesund: Skinnviktangen at Rognstrand (0–0.20 m up in the Heggen Member), railway cut east of the bridge across Bøelven at Skiensdalen (Heggen Member), Bø at Gjerpen (Elnes Formation). Eiker-Sandsvær: Eiker (Helskjer Member), Muggerudkleiva (Heggen Member), Waterfall at Ravalsjøelva near Flata at Heistad (Heggen Member), Stavlum at Krekling (Helskjer Member, 0.06 to 1.16 m above the Huk Fm.). Modum: Road-cut at Vikersundbakken (Upper Helskjer to basal Heggen Member 2.05 to 15.9 m above the Huk Formation). Oslo-Asker: The Old Quarry at Slemmestad (lower Helskjer Member 0.31 to 1.58 m above the Huk Formation), road to skijump at Slemmestad (Helskjer Member). Ringerike: Amundrud farm west of Hønefoss (Helskjer Member). Hadeland: Hovodden (top Helskjer to basal Heggen members, 3.45 to 8.3 m above the Huk Formation), north of Rikoltjernet at Gran (Elnes Fm). Mjøsa: Furnes (top Helskjer Member, 7.75 m above the Huk Formation).

   Outside Norway it is apparently known from the St. Petersburg area of western Russia (Ivantsov 2003).

Asaphus cf. *striatus* (Sars & Boeck *in* Boeck, 1838)

Pl. 14, Figs 1–3

2003 *Asaphus striatus striatus* (Boeck, 1838) – Ivantsov, pp. 281–284.

2004 *Asaphus striatus striatus* (Boeck, 1838) – Ivantsov, p. 34, fig. 44, diagram 2, 5, 14.

Norwegian material. – Four somewhat fragmentary pygidia (PMO 100007, 101.728, 101.737, 101.780).

Remarks. – A few pygidia with strong segmental terrace-lines and short anterolateral directed pleural terrace-lines have been observed within the *Asaphus striatus* zone of the lowermost Elnes Formation. The terrace-line patterns seem largely in agreement with

those observed within the Russian material described by Ivantsov (2003, 2004) and as the pygidia in all other aspects seem well in agreement with the typical smooth Norwegian form of this species, they are here tentatively assigned to *A. striatus*. A full verification must wait until better material has been collected.

Occurrence. – The pygidia all belongs to the Helskjer Member or basalmost Heggen Member, Elnes Formation. Eiker-Sandsvær: Stavlum (0.40–1.16 m above the Huk Formation). Ringerike: Åsen east of Alm (top Helskjer or basal Heggen Member). Outside Norway it is known from the St. Petersburg area of north-western Russia (Ivantsov 2004).

Asaphus narinosus n. sp.

Pl. 12, Figs 11, 17; Pl. 13, Figs 18, 20–21; Table 27

Derivation of name. – The name: 'broad-nosed' refer to the prominent and strongly forward widening glabella.

1984 *Asaphus* (*Asaphus*) *striatus* (Boeck, 1838) – Wandås, pp. 218–219 (*partim*), pl. 2F.

Type stratum and type locality. – The nearly complete holotype exoskeleton PMO 208.580 was collected from the upper Heggen Member (2.68 m below datum in subprofile 4) at Råen near Fiskum in the Eiker-Sandsvær district, Norway.

Type material. – The holotype, PMO 208.580, was collected from the upper Heggen Member, 2.68 m below top of subprofile 4 at Råen near Fiskum, Eiker-Sandsvær. Paratype cephalon PMO 208.566/1 was found in the upper Heggen Member, 0.93 m below the top of subprofile 4. Paratype pygidium PMO 102.650 is from the base of the Heggen Member, Elnes Formation, at a road-cut at Vikersundbakken, Modum. Holotypes and paratypes are all part of the NHM collections.

Material. – The Norwegian material includes one cephalon (PMO 208.566a/1), three cranidia (PMO 208.575/3, 208.607/2,

Table 27. Pygidial measurements on *Asaphus narinosus* n. sp.

PMO number	Wpl1	Wdoublure	Waxis	Laxis	W	L	Terrace-lines
102.650	16.2	14.1	17.6	26.3	63.7	31.9	19
208.636/3	5.7	6.7	7.9	10.6	~28.1	13.9	
208.576	~7.4	~8.9	9.7	13.6	34	17.9	
208.577	~8.2	10.3	12.2	16.7	~41.2	21	20
208.609/3	~8,7	10.3	12.3	17	~41.4	21.6	~21
208580	6.9	~9.1	11.4	15.5	38.7	19.2	~18
208.592/1	~7.2	9.2	10.4	15.3	~36.1	19.9	
208.590	6.5	~8.6	10	12.7	31.9	16.4	21
208.620/4	~5.1	~6.7	7.8	9.5	~26.6	12.7	
208.572			6.8	9.6	23	11.5	
208.591/1			6.6	9	22.9	11.4	
208.578			10.2	15.7	~32.0	19.1	

Abbreviation: see Table 25.

208.634/5), one librigena (PMO 208.618/3), 17 pygidia (PMO 4818, 102.521, 102.650, 105.927, 202.079/1, 202.083/1, 202.085, 202.092, 202.110, 202.132, 208.566/3, 208.569, 208.576, 208.578/1, 208.586/4, 208.607/1, 208.634/6) and one nearly complete specimen (PMO 208.580).

Diagnosis. – Frontal glabellar lobe wide with few anterior terrace-lines; genal corners sharply rounded; thoracic axis moderately wide, clearly expanding towards mid-section; pygidial doublure distinctly narrower than axis; axis moderately broad, corresponding to around 30% of the pygidial width and pygidial doublure with 16 to 21 terrace-lines.

Description. – Largest cranidium (PMO 208.580) just over 19 mm long. Largest pygidium (PMO 102.650) is 32 mm long, although it has been somewhat foreshortened by later deformation.

Cephalon wide, the length corresponding to just over 40% of the width. Cranidial length corresponding to 2/3 of the width. Glabellar convexity fairly low, although this in part may relate to later compression; maximal glabellar width, which is located just behind midline between eyes and anterior cephalic margin, corresponds to between 250 and 270% of the minimum glabellar width found between the eyes. Occipital ring slightly inflated and delimited by moderately weak occipital furrow frontally; occipital ring takes up around 35% of the cranidial width; mesial tubercle small but distinct, situated around 16% of cranidial length in front of posterior margin; lateral glabellar furrows, except for S1, strongly effaced; S1 shallow; posterior lateral glabellar lobe nearly completely effaced; frontal glabellar lobe broadly pyriform to slightly rhombic with relatively broad and flat anterior margin, wide; cranidial front gently rounded with a short ogive; frontal area nearly absent. Preocular area moderately broad, each side taking 10 to 12% of the preocular cranidial width. Posterolateral fixigenal projection slightly shortened (exsag.) on inner part, resulting in a sigmoid anterior margin. Posterior furrow deep, located on anterior part of the posterolateral projection.

Eyes moderately large and low, their length corresponding to around 27 to 28% of the cranidial length and situated approximately 60% of their own length in front of posterior cephalic margin. Paradoublural line on librigena fairly indistinct, marked by a slight furrow located about one-third the distance between eye socle and lateral margin from eye. Librigenal doublure with approximately 30 terrace-lines. Genal corners sharply rounded. Cephalic surface smooth but for about 7 to 10 transversely directed, forward convex terrace-lines located on the anterior half of the frontal glabellar lobe.

No hypostomes are known.

Thoracic axis takes up one-third of the total width, distinctly widening towards the central segments. Surface terrace-lines are moderately coarse.

Pygidium broadly rounded with a length corresponding to half the width. Pygidium convex with straight to slightly concave outer doublural margin. Axis moderately wide, the width taking up between 28 and 32% of the pygidial width, while the length corresponds to 75 to 83% of the pygidial length (11 specimens); axial rings around 10 in number. Pleural ribs nearly completely effaced. Pleural field width at anterior pleural rib corresponding to or exceeding the maximal doublural width. Paradoublural line indistinct, leaving the axial furrow around twice the distance from axis to posterior pygidial margin in front of posterior margin. Pygidial doublure narrow, the maximal width corresponding to between 24 and 27% of the pygidial width (nine specimens) or clearly narrower than axis; doublural terrace-lines numbering 16 to 21 at widest part. Pygidial surface appear to lack any surface sculpture except for some transversely directed terrace-lines on the facets and three pairs of short radial terrace-lines crossing the axial furrow anteriorly.

Remarks. – The species appears morphologically close to *Asaphus ingrianus* Jaanusson, 1953, from which it may be differentiated by the strongly effaced occipital ring and L1 lobes; the slightly larger eyes and by the general lack of strong axial terrace-lines on the pygidium. It also shows a large resemblance to the contemporaneous *Asaphus knyrkoi* Schmidt, 1901, but is distinguished by the larger glabellar widening frontally; the more sigmoid posterior branch of the facial suture; the limitation of terrace-lines to the anterior half of the frontal glabellar lobe, thorax and the pygidial facets and axial furrow and by the distinctly coarser doublural terrace-lines on the pygidium. The species resembles to some degree the slightly younger *Asaphus heckeri* Ivantsov, 2000 from which it is separated by the sharply rounded genal corners and the extremely short frontal area (Ivantsov 2003). It is differentiated from the slightly younger *Asaphus pachyophthalmus* Schmidt, 1901 by the near lack of posterior lateral glabellar lobes; the stronger frontal widening of glabella and by the lack of distinct radial terrace-lines on the pygidium.

Occurrence. – It is found from the Helskjer Member and up into the Sjøstrand and Heggen members of the Elnes Formation. Skien-Langesund: Bø north of Skien (Heggen Member). Eiker-Sandsvær: Råen at Fiskum (upper Heggen Member, 3.12 to 0.93 m below datum in profile 4 and 1.11 m below datum in profile 5). Modum: Vikersundbakken (3.35 to 7.15 m up in the Elnes Fm. or just above the Helskjer Member). Oslo-Asker: The Old Quarry at Slemmestad (1.24 to 1.54 m corresponding to the transition between the Helskjer and Sjøstrand Members). Hadeland: Hovodden (8.3 m above the Huk Formation).

Fig. 51. Reconstruction of *Asaphus raaenensis* n. sp. from the Elnes Formation.

Asaphus raaenensis n. sp.

Pl. 12, Figs 9–10, 12–16; Pl. 13, Fig. 17; Text-Fig. 51; Table 28

Derivation of name. – Referring to the locality Råen at which it was found.

1960 *Ogmasaphus stoermeri* n. sp. Henningsmoen, pp. 242–244 (*partim*), pl. 11, figures 3–4.

Type stratum and type locality. – The holotype PMO 208.524/2 was found at level 16.82 m in sub-profile 1 at Råen near Fiskum, Eiker-Sandsvær, corresponding to the *Pterograptus elegans – Pseudamplexograptus distichus* graptolite Zone boundary in the upper Heggen Member.

Type material. – The holotype, PMO 208.524/2, was collected from the upper Heggen Member, 16.82 m below top of subprofile 1 at Råen near Fiskum, Eiker-Sandsvær. Paratypes PMO 60375a, 60390, 60396 and 60510 all come from the Heggen Member at Muggerudkleiva, Eiker-Sandsvær. Paratype pygidium PMO 203.289/1 was collected from level 36.87 m in the beach profile at Djuptrekkodden north of Slemmestad, Oslo-Asker, corresponding to the middle Engervik Member. Paratype pygidium PMO 203.455 was found in the middle part of the Heggen Member corresponding to level 38.29 m in the beach profile at Djuptrekkodden north of Slemmestad, Oslo-Asker. Paratype cranidium PMO 206.963/1 was found in the upper middle Heggen Member or 16.81 m down in sub-profile 1 at Råen near Fiskum, Eiker-Sandsvær. Holotypes and paratypes are all part of the NHM collections.

Material. – cranidia (PMO S.1783, 60375, 203.261/1, 203.296, 203.403/2, 203.436, 203.437, 205.011/2, 205.015, 206.024, 206.047/4, 206.078, 206.230, 206.234/1, 206.241/1, 208.501/1, 208.511, 208.514, 208.516/2, 208.519/1, 208.538, 208.586/6), five librigenae (PMO 60375, 205.975, 206.066/7, 208.559, 208.591/2), one hypostome (PMO 60396), 101 pygidia (PMO 4633, 4634, 4635, 4825, 60366, 60375 (two specimens), 60385 (two specimens), 60396, 203.289/1, 203.294/1, 203.311, 203.408, 203.443, 203.448/1, 203.448/2, 203.454/1, 203.455, 203.460/2, 203.472, 203.480/3, 203.480/6, 203.483/3, 203.487/3, 205.002, 205.014, 205.064/2, 205.066, 205.078/1, 205.970, 205.980, 206.055/5, 206.068/1, 206.069, 206.108/3, 206.227, 206.243/1, 206.245/6, 206.249/1, 206.269/3, 206.274/4, 206.280/3, 206.286/2, 206.290/1, 208.459/2, 208.460/5, 208.468/4, 208.489/1, 208.500, 208.503/2, 208.503/3, 208.503/4, 208.503/9, 208.503/10, 208.504/5, 208.512, 208.513/1, 208.513/2, 208.516/1, 208.517/1, 208.518, 208.519/2, 208.520/2, 208.520/3, 208.520/4, 208.523/3, 208.526, 208.527/1, 208.528/1, 208.531, 208.533/2, 208.534, 208.535/6, 208.538, 208.539, 208.550/1, 208.577, 208.581/1, 208.586/5, 208.587/1, 208.587/2, 208.588/4, 208.590, 208.591/1, 208.592/1, 208.592/2, 208.597/1, 208.598, 208.602, 208.603/1, 208.605/1, 208.608/1, 208.609/2, 208.609/3, 208.609/4, 208.616/2, 208.620/4, 208.633/1, 208.635, 208.636/3) and 14 partly articulated or nearly complete specimens (PMO 4816, 60366, 60375, 60376, 60385, 60396, 60510, 208.517/1, 208.521, 208.479/1, 208.506/3, 208.523/2, 208.523/4, 208.524/2).

Diagnosis. – Cephalon broadly rounded anteriorly with slightly triangular outline, length corresponding to about 45% the width; cranidial front moderately pointed; glabellar convexity low; glabella strongly widening anteriorly, the maximal width corresponding to more than twice the narrowest part found between the eyes; frontal area present though narrow; eyes moderately large and fairly high, length approximating 30% of the cephalic length and situated approximately 50% of

Table 28. Pygidial measurements on *Asaphus raaenensis* n. sp.

PMO number	Wpl1	Wdoublure	Waxis	Laxis	W	L	Terrace-lines
3972			4	6.4	15.7	8.5	
60396	6.3	8.3	8.6	11.5	~34.0	15.4	
203455	6.5	7.9	9.5	13.8	~33.7	18.1	~17
203311	~5,0	6.9	7.9	11.7	~28.8	14.7	~17
203454	5.4	7.7	9.2	12.9	~33.7	17.2	19
203483/3	7	10.5	10.9	16.2	40	20.8	16
208523	5.3	7.3	8	11.5	28.2	14.6	~16
208524	3.9	5.05	6.2	8.7	~21.9	11.5	~16
4816	6	7.7	9.4	13	33.6	17.2	18
60375	4.9	6.75	7.3	10.4	28	13.8	
60366_2	6.3	5.4	7	10.6	26.8	13.5	
60366_1	~5.6	~7.7	8.4	12.6	31.8	15.5	

Abbreviation: see Table 25.

their own length from posterior cephalic margin and genal corners sharply rounded to slightly pointed.

Pygidium broad, the length corresponding to half the width; axis moderately wide and short, the width corresponding to between 25 and 30% of the pygidial width, while the length approximates 75 to 80% of the pygidial length; axis bearing 9 to 10 generally indistinct axial rings and doublure containing 16 to 19 coarse terrace-lines.

Description. – Largest cranidium (PMO 206.963/1) 21 mm long. Largest pygidium PMO 206.962 is 22 mm long.

Cephalon wide, the length corresponding to approximately 45% of the width. Cranidial length approximating just over 60% of the width. Glabellar convexity low; maximal glabellar width, which is located just behind midline between eyes and anterior cephalic margin, corresponds in general to between 210 and 240% of the minimum glabellar width found between the eyes. Occipital ring slightly inflated and delimited by weak occipital furrow frontally; occipital ring takes up 35% of the cranidial width; mesial tubercle small but distinct, situated around 16% of cranidial length in front of posterior margin; lateral glabellar furrows, except for S1, strongly effaced; S1 shallow; frontal glabellar lobe pyriform to slightly rhombic with relatively broad and flat anterior margin, wide; cranidial front gently rounded; ogive relatively long and narrow; frontal area moderately short. Preocular area moderately broad, each side taking up about 11% of the preocular cranidial width. Postero-lateral fixigenal projection slightly shortened (exsag.) on inner part, resulting in an S-formed anterior margin. Posterior furrow fairly deep, located on anterior part of the posterolateral projection.

Eyes moderately large and rather high, their length corresponding to 30% of the cranidial length and situated 45 to 53% of their own length in front of posterior cephalic margin. Paradoublural line on librigena weak but present, located around one-third the distance between eye socle and lateral margin from eye. Librigenal doublure with approximately 15 to 18 terrace-lines midway between posterior and anterior end. Genal corners sharply rounded to slightly pointed. Cephalic surface generally smooth except for between 5 and 11 transversely directed, forwardly convex terrace-lines located on the anterior half of the frontal glabellar lobe.

Only one fragmentary and somewhat flattened hypostome may with some certainty be assigned to this species. It is slightly longer than wide with a moderately deep posterior notch corresponding to just below 30% of the total hypostomal length. Anterior margin of posterior notch gently swollen. Surface, except for median body, covered by very dense terrace-line pattern delineating the hypostomal outline. Median body bearing sparse but fine terrace-lines running perpendicular to sagittal line.

Thoracic axis takes up nearly one-third of the total width, typically widening somewhat towards the central segments. Surface terrace-lines are generally fairly coarse.

Pygidium broadly rounded with a length, excluding the anterior half ring, corresponding to between 45 and 55% of the width (12 specimens). Pygidium with straight to slightly convex outer doublural margin. Axis moderately wide, the width taking up between 25 and 28% of the pygidial width, while the length corresponds to 75 to 81% of the pygidial length; axial rings numbering 9 to 10. Pleural ribs nearly completely effaced. Paradoublural line indistinct, leaving the axial furrow around twice the distance from axis to posterior pygidial margin in front of posterior margin. Pygidial doublure narrow, the maximal width generally corresponding to between 23 and 26% of the pygidial width; doublural terrace-lines numbering 16 to 19 at widest part. Pygidial surface smooth with terrace-lines limited to the articulating facets and also, in some cases, some sparsely distributed, anterolaterally directed and wavy terrace-lines along posterolateral margin and up to three short and coarse posterolaterally directed terrace-lines anteriorly at transition between axial furrow and inner pleural field. Furthermore some specimens bear the beginning of segmental terrace-lines on the anterior half of the axis.

Remarks. – The perhaps most closely related species is the only slightly older *Asaphus heckeri* Ivantsov, 2000 from the East-Baltic area, from which it is largely distinguished on the strength of the less pronounced occipital ring and L1-lobes; the rounder lateral bend on the facial suture; the presence of a paradoublural line and furrow on librigena; the less pronounced segmentation on the lateral flanks of the pygidial axis and possibly by the more posteriorly reaching terrace-line pattern on the pleural region of some pygidia. Some of these differences may be artefacts of later compression and removal of the shell-material and it is therefore a question as to whether the two taxa should be regarded as subspecies rather than full species. *Asaphus raaenensis* n. sp. is mor-phologically very closely associated with the slightly older East-Baltic *Asaphus laevissimus* Schmidt, 1901, but differs on the slightly shorter frontal area on the cephalon; the lack of terrace-lines on the pleural region of the pygidium and by the more flattened cranidial front, which appear distinctly angular on *A. laevissimus*. The species differs from *Asaphus sarsi* and *Asaphus striatus* described above in the relatively wider glabella; the less distinct posterior glabellar lobes; the moderately narrow pygidial axis and even narrower pygidial doublure with no more than 16 to 19 terrace-lines. It is distinguished from *Asaphus plautini plautini* Schmidt, 1901 by the relatively wider pygidium with its narrower axis and by the much weaker paradoublural furrow on librigena. There may be some resemblance to the older *Asaphus ingrianus* Jaanusson, 1953, but it differs on the more rounded lateral bend of anterior branch of facial suture and by the somewhat narrower pygidial axis with less strongly developed segmental terrace-lines. From *Asaphus narinosus* n. sp. it is characterized by the narrower axis, the less frontally widening glabella and by the longer frontal area on cephalon. It differs from *A. maximus* Brøgger, 1886 by the relatively wider pygidial axis, which exceeds the doublural width.

Occurrence. – The species is found in the upper part of the Heggen Member in the southern Oslo Region and in the Engervik and Håkavik members in the central Oslo Region. Skien-Langesund: Bø north of Skien (Elnes Formation), Fossum (Elnes Formation). Eiker-Sandsvær: Muggerudkleiva at Hedenstad (Heggen Member), waterfall in Ravalsjøelva at Heistad (Heggen Member), Råen at Fiskum (22.81 to 0.96 m below datum in profile 1; 7.25 to 1.14 m below datum in profile 4 and 1.26 to 1.04 m below datum in profile 5). Oslo-Asker: Beach profile at Djuptrekkodden just north of Slemmestad (Engervik to Håkavik Member, 36.17 to 54.84 m above datum), Odden on Bygdøy (Engervik Member). Modum: Mælå.

Asaphus ludibundus sensu lato Törnquist, 1884

Pl. 11, Figs 2–3; Table 29

1909 *Asaphus* n. sp. – Holtedahl, pp. 9, 29, 40.
1960 *Asaphus* (*Neoasaphus*) cf. *ludibundus* Törnquist,
      1884 – Henningsmoen, pp. 234–235, pl. 8, fig. 6.
1960 *Pseudomegalaspis?* sp. Henningsmoen, p. 251,
      pl. 12, fig. 8.

Type stratum and type locality. – Lectotype specimen P.I.L. no. 618 T selected by Jaanusson (1953a) comes from the Dalby Limestone in Kårgärde in the Siljan area, Sweden.

Material. – Two complete if somewhat worn specimens (PMO 36832, 36941).

Description. – Total length of largest dorsal shield (PMO 36941) just over 80 mm. Approximate length of cranidium 24 mm, while pygidium measures 29 mm.

Dorsal surface fairly smooth with moderately shallow axial furrow. Cephalon wide, the length corresponding to between 40 and 50% of the width. Cranidial length approximating two-thirds the width. Occipital ring and furrow seem to be effaced, but are practically worn away; mesial tubercle weakly developed. Anterolateral flanks on anterior glabellar lobe rather well rounded; no sagittal ridge visible on central part of lobe. Frontal area short to moderate. Preocular areas moderately broad, each corresponding to 20% of the anterior glabellar lobe width; cranidial width at preocular area corresponding to between 65 and 75% of the total cranidial width. Abaxial end of paradoublural line located 25% of the cranidial length inside cephalic margin.

Eye length on two measured specimens between 27 and 30% of cranidial length, while the width corresponds to between 35 and 50% of distance from eye and normal to the axis out to genal corner; eye located nearly 60% of its own length in front of posterior cephalic margin. The lateral border width corresponds to slightly over half the distance between eye and cephalic margin. Genal corners rounded.

Thoracic axis occupies between 25 and 30% of the total thoracic width. Fulcrum located moderately far out on the pleura.

Pygidium is broadly rounded with slightly concave outer doublure and convex borders. The length, excluding the anterior half ring, corresponds to around 55% of the width; axis narrow, taking up between 21 and 23% of the pygidial width, while the length corresponds to around 80 to 90% of the pygidial length. Pleural field with completely effaced pleural ribs and paradoublural line. Pygidial doublure width corresponding to between 16 and 21% of the pygidial width; doublural terrace-line density very variable, numbering from 16 to 26 on widest part, converging on outer margin posteriorly.

Remarks. – As already noted by Henningsmoen (1960, pp. 234–235) there is a large resemblance between these specimens and the type form of *A. ludibundus* from the Upper Ordovician Chasmops Beds of Sweden and Norway. As a result they are referred to this species sensu lato. The Norwegian material from the Elnes Formation mainly differs in its narrow axis; wide librigenal field; the presence of a

Table 29. Pygidial measurements on *Asaphus ludibundus* s.l. Törnquist, 1884

PMO number	Wpl1	Wdoublure	Waxis	Laxis	W	L	Terrace-lines
36832	13.3	~7.0	10.3	23.5	44.6	26.4	~26
36941	14.6	10.8	11	22.2	51.8	28.2	~16

Abbreviation: see Table 25.

frontal area and the distal location of the thoracic fulcrum. The axial width, corresponding to 20% of the pygidial width, lies fairly close to the typical form (Waxis/Wpyg. ≈ ¹/₄) found within the Oslo Region at higher stratigraphical levels, and may in this case relate to ecological variations. The narrowness is not related to later flattening as suggested by Henningsmoen (1960, p. 234). Ecological variations could also be responsible for the width of the librigenal field, which following Jaanusson (1953a) varies somewhat within the Swedish material. The last two characters are not regarded as important enough for establishing a new taxon, especially as the frontal area is rather weakly developed.

Asaphus ludibundus s.l. shows a large degree of similarity with *Pseudomegalaspis patagiata* (Törnquist, 1884), which is why specimen PMO 36832 was assigned to *Pseudomegalaspis* by Henningsmoen (1960). Even so *A. ludibundus s.l.* may be differentiated on its much wider librigenal cheeks with rounded posterolateral corners; the complete lack of pleural ribs on the pygidium; the fairly distinct frontal glabellar lobe margin lacking the moderately distinct anterior pit seen on *P. patagiata* and by the generally shorter and squatter glabella.

Occurrence. – Collected from the 'Cephalopod Shale' (Upper Elnes Formation) in the Mjøsa district.

Asaphus sp. A

Pl. 12, Figs 2–3; Table 30

1984 *Asaphus* (*Asaphus*?) sp. Wandås, p. 219 (*partim*), pl. 2G, J.

Material. – One nearly complete specimen PMO 72127 and possibly two additional pygidia (PMO 102.542).

Description. – The cranidium of specimen PMO 72127 is 15 mm long, while the associated pygidium measures around 16 mm in length.

Cephalon wide, the length corresponding to approximately 45% of the width. Cranidial length approximating 60% of the width. Glabella distinctly delimited by axial suture; the convexity is fairly low; maximal glabellar width, which is located just behind midline between eyes and anterior cephalic margin, corresponds to around 170% of the minimum glabellar width found between the eyes. Occipital ring and furrow largely effaced; occipital ring takes up about 37% of the cranidial width; mesial tubercle small but distinct, situated 21% of cranidial length in front of posterior margin; lateral glabellar furrows, except for S1, strongly effaced; S1 rather deep, practically

separating the posterior lateral glabellar lobes from the rest of glabella; frontal glabellar lobe pyriform with flat anterior margin, fairly narrow; cranidial front rather flat to slightly forward concave on each side of the ogive; frontal area moderately short. Preocular area rather broad, each side taking up about 13% of the preocular cranidial width. Posterolateral fixigenal projection decreasing in length all the way from axial furrow and out to abaxial termination. Posterior furrow deep, exsagittally located centrally on the posterolateral projection.

Eyes moderately large, their length corresponding to 32% of the cranidial length and situated their own length in front of posterior cephalic margin. Paradoublural line on librigena indistinct, located nearly 20% of the distance between eye socle and lateral margin from eye. The genal corners are unknown. Cephalic surface smooth but for some transversely directed, forwardly convex terrace-lines located on the anterior half of the frontal glabellar lobe, although they do not continue to the glabellar margin.

Thoracic axis takes up one-third of the total width, slightly widening somewhat towards the central segments. Terrace-lines are limited to a few posterolaterally directed lines on the lateral axial flanks and moderately dense terrace-lines on the pleural facets.

Pygidium broadly rounded with a length, excluding the anterior half ring, corresponding to around 53% of the width. Axis narrow, the width taking up 23% of the pygidial width, while the length corresponds to 80% of the pygidial length; indistinct axial rings numbering 8 or 9 excluding the terminal piece and articulating half ring. Pleural ribs completely effaced, only marked by seven moderately strong posterolaterally directed pleural terrace-lines. Paradoublural line not seen. Pygidial surface covered by moderately dense, transversely directed terrace-lines on the facet and moderately sparse, anterolaterally directed secondary terrace-lines on the central pleural region between the seven primary, posterolaterally directed terrace-lines reaching out from the axis.

Remarks. – The general outline of *Asaphus* sp. A agrees well with *A. ingrianus* Jaanusson, 1953, but for the less forwardly widening glabella; a less strongly developed occipital furrow; the narrower axis and a pygidial pleural region characterized by seven moderately strong, posterolaterally directed primary terrace-lines intersected by anterolaterally directed secondary ones. It differs from *A. latus* Pander, 1830 in the lack of librigenal terrace-lines; the more forwardly located eyes and in the distinctly rounder pygidium. It resembles *A. sulevi* Jaanusson, 1953 somewhat, but differs in the distinctly narrower glabella and axis; the relatively wider preocular fixigenal area and by the strongly limited occurrence of cephalic terrace-lines. A resemblance is also found to the older *Asaphus sarsi* Brøgger, 1882 on which there is a pygidial terrace-line pattern and depth of occipital furrow much like the one shown here, but again the pygidial axis is too wide. It could be an extreme form of *A. striatus* although it differs in a number of characters such as the maximal width of the cranidium and the dense pygidial terrace-line pattern.

Occurrence. – The taxon was collected 10.6 m above the base of the Elnes Formation at Hovodden in the Randsfjord area, Hadeland, and possibly 6.0 m above the Huk Formation in the road-cut at Vikersund skijump.

Genus *Subasaphus* Balashova, 1976

Type species. – *Asaphus laevissimus* var. *laticauda* Schmidt, 1901.

Diagnosis. – Cephalon typically wide; basal lobes present; eyes low; genal corners with spines, rarely obtuse. Pygidium wide with faintly subdivided or smooth axis: Cephalon and pygidium not bordered (emended from Ivantsov 2003).

Subasaphus platyurus (Angelin, 1854)

Pl. 11, Figs 4–5

1854 *Asaphus platyurus* Ang. Angelin, p. 54, pl. 30, fig. 1.
1878 *Asaphus platyurus* Ang. Angelin, p. 54, pl. 30, fig. 1.
1884 *Asaphus platyurus* Ang. – Törnquist, pp. 57–58, pl. 2, fig. 12.
1901 *Asaphus platyurus* Ang. – Schmidt, pp. 55–57, pl. 3, figs 1–7, pl. 12, fig. 19.
1907 *Asaphus platyurus* Ang. – Schmidt, p. 96 (occurrence).
1907 *Asaphus platyurus* Ang. var. *laticauda* F. S. – Schmidt, p. 96 (occurrence).
1953a *Asaphus platyurus* Angelin, 1854 – Jaanusson, p. 398.
1953b *Asaphus (Neoasaphus) platyurus* Angelin, 1854 – Jaanusson, pp. 467–478 (*cum. syn.*), pls 1–2, pl. 3, figs 1–4.
1960 *Asaphus (Neoasaphus) platyurus* Ang. – Jaanusson, pp. 238, 251, 253, 254, 264, 270, 271, figs 6, 19, 23, table 5, pl. 5, fig. 2.
1964 *Asaphus (Neoasaphus) platyurus* Ang. – Jaanusson, pp. 44–45.
1984 *Asaphus (Neoasaphus) platyurus* Angelin (1854) – Wandås, p. 220, pl. 3A.
2003 *Subasaphus platyurus* (Angelin, 1854) – Ivantsov, p. 321.

Type stratum and type locality. – The original type specimen figured by Angelin (1854, pl. 3, fig. 1) has disappeared and so a neotype was selected by Jaanusson (1953b). The neotype pygidium RM Nr. Ar. 21948 originates from Hulterstad Channel on Öland, SE Sweden and was collected from the Aserian Segerstad Limestone (Platyurus Limestone).

Norwegian material. – Two pygidia, one associated with three fragmentary thorax segments, have been assigned to this species (PMO 82209, 101.730).

Diagnosis. – Cephalon with genal spines. Pygidium broadly parabolic to subtriangular in outline, weakly convex with relatively indistinct axis bordered by shallow axial furrows; axial rings and pleural ribs strongly effaced; pygidial surface lacking terrace-lines or at the most with some at the margin; doublure relatively wide, width corresponding to 20% of the pygidial width; doublure with around 25 terrace-lines (modified from Jaanusson 1953b).

Remarks. – Jaanusson (1953b) recognised two subspecies of *Subasaphus platyurus*, but did not find any diagnostic characters for the pygidium. This is why the present material may only be assigned to species. As already noted by Wandås (1984), the presence of this species strongly suggests that the basal mudstone deposits above the Helskjer Member were deposited sometime in the first part of the Aseri Stage.

Occurrence. – In Norway it is only known from the basal mudstones succeeding the Helskjer Member at Furnes, Mjøsa (8.6 m above the Huk Fm.), and the lower Helskjer Member at Stavlum near Krekling, Eiker-Sandsvær (0.3 m above the Huk Formation).
Outside Norway it is known from the Aseri Stage from most of Baltoscandia. Sweden: Jämtland, the Siljan area (Vikarby Limestone), Åland (upper Aseri Stage ?), Öland (Segerstad Limestone) and Västergötland (Vikarby Limestone, upper Aseri Stage) (Angelin 1878; Jaanusson 1953b; Jaanusson 1964; Törnquist 1884). Germany (erratic builders) (Schmidt 1901). Estonia (Aseri Stage) (Schmidt 1901). North-West Russia (Doboviki Formation of the Aseri Stage) (Schmidt 1901).

Genus *Pseudomegalaspis* Jaanusson, 1953a

Type species. – *Megalaspis formosa* Törnquist, 1884.

Diagnosis. – Cephalon and pygidium lacking concave border; posterior lateral glabellar lobes effaced; eyes large; genal angles with spines; hypostome of *Asaphus* type; thoracic axis narrow, corresponding to less than 30% of total thoracic width; thoracic lateral pleural terminations rounded; panderian openings moderately sized; pygidial axis flattened, inner part of pleural field with moderately faint, furrowed ribs; doublure narrow with inner margin nearly parallel to pygidial margin (modified from Jaanusson 1953a; Moore 1959).

Table 30. Pygidial measurements on *Asaphus* sp. A

PMO number	Wpl1	Wdoublure	Waxis	Laxis	W	L
72127	~6.8	~8.1	7.3	12.5	30.8	16.2

Abbreviations: see Table 25.

Pseudomegalaspis patagiata (Törnquist, 1884)

Pl. 13, Figs 1–12; Pl. 29, Figs 3–4; Text-Fig. 52; Table 31

1884   *Megalaspis patagiata* n. sp. Törnquist, pp. 82–83, pl. 3, figs 15–17.

1909   *Asaphus latus*, Pand. var. – Holtedahl, pp. 9, 29, 40.

1909   *Megalaspis patagiata*, TÖRNQ. – Holtedahl, pp. 7, 28, 40.

1913   *Megalaspis patagiata* Törnq. – Hadding, pp. 69–70, pl. 6, figs 20–21.

1953a *Pseudomegalaspis formosa* (Törnquist, 1884) – Jaanusson, pp. 452–456 (*partim*), pl. 10, figs 2–3 (only the Norwegian material).

1953a *Pseudomegalaspis patagiata* (Törnquist, 1884) – Jaanusson, pp. 456–459, pl. 9, figs 4–7, pl. 10, fig. 1.

1960   *Ogmasaphus* sp. – Henningsmoen, pp. 244–245, pl. 12, fig. 9.

1960   *Pseudomegalaspis patagiata* (Törnquist, 1884) – Henningsmoen, pp. 248–251, pl. 12, figs 1–7.

1963   *Pseudomegalaspis patagiata* (Törnq.) – Jaanusson, p. 9, fig. 2, table 3 (occurrence).

1963   *Pseudomegistaspis patagiata* – Skjeseth, p. 64 (listed).

1966   *Pseudomegalaspis patagiata* – Hamar, p. 30.

1995   *Pseudomegalaspis patagiata* (Törnquist, 1884) – Månsson, pp. 99, 101, 105, 106, fig. 5G–H.

2000b *Pseudomegalaspis patagiata* (Törnquist, 1884) – Månsson, p. 12.

2002   *Pseudomegalaspis patagiata* – Pålsson *et al.*, p. 42, fig. 10.

Type stratum and type locality. – Lectotype pygidium LM Nr. LO 627 T originates from the Folkeslunda Limestone (formerly: 'Schroeteri beds') at Kårgärde in the Swedish Siljan district.

Norwegian material. – A very large material has been available at the Natural History Museum, University of Oslo, of which more than 300 skeletal parts have been examined in greater detail. Of those the following 13 PMO-specimens are complete or partly articulated: PMO S.1734, 2325, 3695, 4672, 56556, 82619, 82928, 203.467/1, 203.231/3, 206.219/1, 205.067/2, 208.469 and 208.515a/2.

Diagnosis. – Thoracic axis narrow, corresponding to 25% of the total width; pygidial axis narrow, taking up less than 25% of the pygidial width. Pygidial

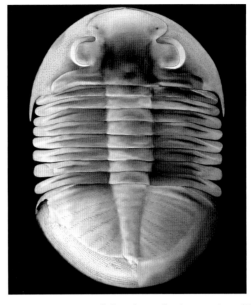

Fig. 52. Reconstruction of *Pseudomegalaspis patagiata* (Törnquist, 1884) from the upper Darriwilian of Scandinavia. Length of specimen close to 5 cm.

doublural terrace-lines generally relatively few, varying between 15 and 23 (emended from Törnquist 1884).

Description. – Medium sized asaphid; the largest cranidium measured being 38 mm long (PMO 207.419/ 1), while largest pygidium measures 51 mm in length (PMO 208.515/1).

Cephalon broadly rounded, the length corresponding to approximately 45% of the width. Cranidial length approximates 60% of the width. Glabella bordered by shallow axial furrow, which deepens slightly posteriorly. Glabella elongated, the width corresponding to between 63 and 68% of the length (five specimens); glabellar convexity moderately low; maximal glabellar width, which is located just behind midline between eyes and anterior cephalic margin, corresponds to between 140 and 160% of the minimum glabellar width found between the eyes. Occipital ring and furrow strongly effaced, flat; occipital ring takes up between 29 and 34% of the cranidial width (five specimens); mesial tubercle minute, situated around 15% of cranidial length in front of posterior margin; lateral glabellar furrows strongly effaced; posterior lateral glabellar lobe effaced but present; anterior glabellar lobe strongly pyriform; anterior part of facial sutures nearly straight, meeting at an angle of approximately 140°, although the anterior-most part of the cranidium is slightly more pointed. Frontal area of cephalon short. Preocular area moderately broad, each side taking up between 13 and

Table 31. Pygidial measurements on *Pseudomegalaspis patagiata* (Törnquist, 1884)

PMO number	Wpl1	Wdoublure	Waxis	Laxis	W	L	Terrace-lines
56556	12.3	9	9.8	24.1	44.1	31.8	
208.515/1	21.3	14.6	18.7	42.4	~93.1	51	18
72074	21.6	~14.2	17.3	35	~77.8	41.8	18
206.242	23.4	15	19.2	38.6	88.9	45.9	~17
36770	~10.7	6.7	9.6	18.5	41.2	24.3	~26
208.469	14.8	8.7	9.5	21.6	51.4	26.4	~18
203.089/1	13.7	~7.8	10.1	~21,1	~50.0	~26.0	~17
203.119/1	~20.7	10.8	16.4	37.7	~70.2	45.5	~16
203.123/5	10.4	~6.3	9.4	17.8	~39.3	21.2	~15
203.338	15.5	~8.5	~11.0	~24.8	51.8	~29.8	~18
203.231/3	7.1	3.3	4.9	9.9	23.9	11.8	~18
203480/2	12	7.2	10	22	45.5	27.9	19
203.289/4	8.1	4.4	6	11.7	25.3	14.8	
203.427/1	10.4	5.5	~6.9	~14.5	34.8	~18.7	22
205.139/4	7.2	2.6	4.1	8.9	21.4	10.4	
205.162/1	7.5	4	5.1	9.4	27.3	12.6	16
205.999	6.4	4.6	5.1	8.9	23.2	~10.5	~16
36769	12.2	7	8.9	18.1	~41.3	24.7	21
72075	6.9	~3.9	5.1	9.9	23.2	12.6	
S.1734	12.8	6.7	9.3	21.2	42.5	25.9	23
3695	10.2	5.3	6.2	13.2	~32.7	17.5	20
208.482	~21.6	~11.2	16.6	~32.6	~71.6	~39.1	
S.3295	14	7	10.2	22.2	44.2	27.1	~22
20193	11.6	5.6	8.7	17	~37.5	20.6	18
208.482	~22.8	~10.5	16.3	32.1	~76.2	38.2	
205.069	6.9	4	?	~10.2	27.8	~13.3	19

Wpl1, pleural field width at anterior pleural rib; Wdoublure, maximal width of doublure.
Waxis, maximal width of axis; Laxis, length of axis excluding articulating half ring.
W, maximal width of pygidium; L, length of pygidium excluding articulating half ring.
Terrace-lines, maximal number of terrace-lines counted at widest point.

17% of the preocular cranidial width. Posterolateral fixigenal projection slightly shortened (exsag.) on inner part, resulting in an S-shaped anterior margin. Posterior furrow fairly shallow, exsagittally located centrally on the posterolateral projection. Surface of cranidium smooth except for some moderately small pores mostly concentrated on fixigenal areas.

Eyes moderately large and high, their length corresponding to around 35 and 40% of the cranidial length and situated approximately 55% of their own length in front of posterior cephalic margin. Paradoublural line on librigena marked by a very faint paradoublural furrow, located around 40% of the distance between eye socle and lateral margin from eye. Librigenal doublure with approximately 16 terrace-lines midway between posterior and anterior end. Genal corners extended into moderately long and strong genal spines. Surface of librigena covered by moderately large pores and a few posterolaterally directed terrace-lines along outer genal spine margin and exsagittally directed ones along inner margin of genal spine.

A single, poorly preserved and possibly somewhat foreshortened hypostome (PMO 203.480b/4) may with some certainty be assigned to this species. The hypostome, which is nearly 9 mm in length, is distinctly wider than long with a very short anterior

border; moderately narrow lateral wings; a compact and somewhat inflated median body and a moderately deep posterior notch corresponding to 30% of the total hypostomal length. The hypostome is strongly exfoliated, which has resulted in the loss of the terrace-lines.

Thoracic axis corresponding to 25% of the total thoracic width.

Pygidium is broadly rounded, weakly convex in cross-section with convex borders. Doublure S-shaped in cross-section with distinctly concave outer part. The pygidial length, excluding the anterior half ring, corresponds to between 50 and 60% of the width; axis long and moderately narrow, taking up between 18 and 24% of the pygidial width, while the length corresponds to approximately 80% of the pygidial length; 12 axial segments present. Pleural field bearing eight to nine somewhat effaced pleural ribs with a very indistinct interpleural furrow. Paradoublural line effaced. Pygidial doublure narrow, the maximal width corresponding to 14 to 17% of the pygidial width; enfolding what corresponds to half the distance from axis to posterior pygidial margin of axis; doublural terrace-lines numbering from 15 to 23 or in a single case even 26 on widest part, converging to outer pygidial margin posteriorly. Terrace-line pattern on pygidial surface quite variable, ranging from pygidia with nearly no terrace-lines inside the paradoublural line to ones where the total pygidial surface is covered by anterolaterally directed, moderately coarse lines. In most cases the terrace-lines are restricted to the pleural ribs and the border outside the paradoublural line.

Remarks. – Henningsmoen (1960, pp. 244–245) assigned the large pygidium PMO 72074 to *Ogmasaphus*, probably because of the wide and broadly rounded outline and the seeming lack of interpleural furrows. The relative width and broadly rounded outline is probably related to a combination of later foreshortening and a slight compression, whereas the apparent lack of interpleural furrows may relate to recent weathering.

The pygidia may resemble those of *Ogmasaphus jaanussoni* described above, but differ in the slightly narrower doublure; the marginally longer axis; the higher number of pleural ribs; the low number of doublural terrace-lines and by the marginal distribution of surface terrace-lines.

The material of *P. patagiata* has been examined with regard to its similarities with *P. formosa* (Törnquist, 1884). As noted by Henningsmoen (1960) the generally more parabolic outline of pygidia and cephala and the more evenly rounded anterior cranidial margin of the type specimen of *P. formosa* all relates to later deformation of the rocks. Likewise the Norwegian material of *P. patagiata* presents a fairly large variation between pygidia with next to no terrace-lines on the pleural field and those nearly completely covered by terrace-lines. The articulating facets are furthermore variable with regard to length and rounding of the anterolateral corners and completely encompass the features seen for the type specimen of *P. formosa*. The only possible differences between the two species are found in the relative widths of the pygidial and thoracic axis and possibly also in the number of doublural terrace-lines on the pygidium, both being distinctly larger on *P. formosa* than on any undoubted specimens of *P. patagiata*. Whether this holds true when a larger sample of *P. formosa* has been investigated only time will show.

Occurrence – Stratigraphically *Pseudomegalaspis patagiata* is found from the middle to upper part of the Elnes Formation and

up into the succeeding Vollen Formation corresponding with the 'Cephalopod Shale' of the uppermost Elnes Formation in the northern part of the Oslo Region. Eiker-Sandsvær: Railway cut near station at Vestfossen (Heggen Member), profile 1 at Råen near Fiskum (upper Heggen Member, 18.92 to 0.64 m below datum), profile 5 at Råen (1.11 m above datum, corresponding to upper Heggen Member), profile 6 at Vego (middle Heggen Member, 1.61 m below datum). Oslo-Asker: Bøveien south of Slemmestad (Sjøstrand to Engervik Member, 0.22 to 13.43 m above datum), beach profile at Djuptrekkodden (middle Sjøstrand to Håkavik Member, 22.13 to 54.25 m above datum), Elnestangen (Engervik Member), Ildjernet (Engervik Member), Odden and Huk on Bygdøy (Engervik Member), the Vigeland fountain in the Vigeland Park of Oslo (Elnes Fm.), Nordråks gate in Oslo, Hagebyen near Ullevål Stadion, Aker (Engervik Member) and at the Geological Institute at Blindern (basal Vollen Fm.). Ringerike: Ringerike (Elnes Fm.). Mjøsa: subprofile 2 along E6 at Nydal (Heggen Member, 0.17 to 1.53 m above datum), subprofile 3 along E6 at Nydal (Heggen Member to basal 'Cephalopod Shale', 2.7 to 20.25 m above datum), Hovindsholm on Helgøya.

Outside Norway it is known from the Jämtland, Siljan and SE Scanian areas in Sweden, where it occurs in the Andersö Shale formation, the Folkeslunda and Furudal limestones and the Killeröd Formation of the Lasnamägi and Uhaku stages and possible in the uppermost Stein Formation (Hadding 1913; Jaanusson 1963; Karis 1982; Månsson 1995, 2000b).

Pseudomegalaspis cf. *formosa* (Törnquist, 1884)

Pl. 13, Figs 13–15; Pl. 29, Figs 2; Table 32

1884	*Megalaspis formosa* n. sp. Törnquist, pp. 80–81, pl. 3, figs 13–14.

1953a	*Pseudomegalaspis formosa* (Törnquist, 1884) – Jaanusson, pp. 452–456 (*partim*), pl. 10, figs 4–7.

1960	*Pseudomegalaspis formosa* (Törnquist, 1884) – Henningsmoen, pp. 249–250.

Material. – A single pygidium (PMO 56347) and a pygidium with the four posterior thorax segments still attached (PMO 112.103) are tentatively assigned to this species.

Description. – Pygidium is broadly rounded, weakly convex in cross-section with convex borders. Doublure weakly sigmoidal in cross-section with concave outer part. The pygidial length, excluding the anterior half ring, corresponds to between 50 and 60% of the width; axis moderately broad, taking up more than 25% of the pygidial width, while the length corresponds to slightly over 80% of the pygidial length; 12 axial rings present. Pleural field bearing eight somewhat effaced pleural ribs with indistinct interpleural

Table 32. Pygidial measurements on *Pseudomegalaspis* cf. *formosa* (Törnquist, 1884)

PMO number	Wpl1	Wdoublure	Waxis	Laxis	W	L	Terrace-lines
56347	14.1	8.8	12.9	22.2	48.9	26.2	31
112.103	16.2	12.15	14.5	~26.3	56.05	~31.7	30

Abbreviations: see Table 31.

furrows. Paradoublural line effaced. Pygidial doublure narrow, the maximal width corresponding to 18–22% of the pygidial width; doublural terrace-lines numbering 30–31 on widest part, converging to outer pygidial margin posteriorly. Pygidial surface covered with moderately sparse pattern of anterolaterally directed, short and wavy terrace-lines, getting denser outside paradoublural line.

Remarks. – A general discussion on the possible differences between *P. formosa* and *P. patagiata* is given above. I have not been able to verify whether the Swedish material of *P. formosa* is characterized by around 30 terrace-lines on the pygidial doublure like the Norwegian specimens tentatively assigned to this species. It may also be that this high number of terrace-lines merely represents an ecological difference from the norm.

Occurrence. – The partly articulated specimen (PMO 112.103) belongs to the middle Fossum Formation at Skinnvika, Skien-Langesund. The exact stratigraphical and geographical location for pygidium PMO 56347 is unknown, but based on the lithology it most probably belongs to the Oslo-Asker district and here either from the upper half of the Elnes Formation or, perhaps more likely, the Vollen Formation. Outside Norway *P. formosa* is known from the Lasnamägian Folkeslunda Limestone of Sweden (Jaanusson 1953a).

Subfamily Isotelinae Angelin, 1854

Genus *Megistaspis* Jaanusson, 1956

Remarks. – The subgeneric relationship of the three species *Megistaspis heros* (Dalman, 1828), *M. heroica* Bohlin, 1960 and *M. lawrowi* (Schmidt, 1906) was regarded with some uncertainty by both Jaanusson (1956) and Bohlin (1960) neither of whom wanted to include them in the subgenus *Megistaspidella* Jaanusson, 1956. Balashova (1976) in her revision of the East-Baltic asaphids raised the former subgenus *Megistaspidella* to genus level and erected a number of new genera and subgenera. *M. lawrowi* was moved over to the newly established genus *Rhinoferus* Balashova, 1976, while *M. heros* and *M. heroica* were assigned to the new subgenus *Megistaspidella* (*Spinopyge*) Balashova, 1976 containing the forms with spiny pygidia and unevenly distributed terrace-lines on the pygidial doublure. Wandås (1984) appears unaware of the work of Balashova (1976) and erected the subgenus *Megistaspis* (*Heraspis*) corresponding very much with Balashova's *Megistaspidella* (*Spinopyge*), although it was better defined and only included *Megistaspis heros*, *M. heroica*, *M. lawrowi* and the new species *M.* (*Heraspis*) *laticauda* Wandås,

1984. In the recent revision of the Scandinavian asaphids of the Middle Ordovician Komstad Limestone Nielsen (1995, pp. 102–104) discussed the higher taxonomy of the megistaspids and rejected the subgeneric classification of Balashova (1976) and returned to the former classification of Jaanusson (1956), adding the four subgenera *Megistaspis* (*Ekeraspis*) Tjernvik, 1956, *M.* (*Paramegistaspis*) Balashova, 1976, *M.* (*Rhinoferus*) Balashova, 1976 and *M.* (*Heraspis*) Wandås, 1984. *M. lawrowi* was removed from *M.* (*Heraspis*) and assigned to *M.* (*Rhinoferus*) due to its faint posterior glabellar hump, which, although typically much better developed, is the hallmark of *M.* (*Rhinoferus*). This low glabellar hump, though, is much closer to what is found among species of both *M.* (*Megistaspidella*) and *M.* (*Heraspis*) (e.g. *M.* (*H.*) *heroica*), the latter also has an identical faint terrace-line pattern on the glabella as earlier noted by Bohlin (1960). The general outline and profile of the cranidium with its fairly smooth and low glabella and the short and narrow anterior cranidial area is furthermore very reminiscent of species such as *M.* (*Megistaspidella*) *triangularis* (Schmidt, 1906) and *M.* (*Megistaspidella*) *lamanskii* (Schmidt, 1906), much more so than of any species belonging to *M.* (*Rhinoferus*), why it is here assigned to *M.* (*Megistaspidella*) and not *M.* (*Rhinoferus*) as proposed by Balashova (1976) and Nielsen (1995).

One of the subgenera rejected by Nielsen (1995) was *M.* (*Spinopyge*) Balashova, 1976. It was originally defined as a subgroup of *Megistaspis* characterized by a relative narrow cranidium; a rounded posterior hypostomal margin; a thorn-like or pointed posterior pygidial margin and by less dense central doublural terrace-lines on the pygidium. As noted by Nielsen (1995, p. 103) the width of the cranidium appear rather dubious as a diagnostic character on the subgeneric level. The same is the case for the doublural terrace-line pattern, which actually is identical to some of the species belonging to *M.* (*Megistaspidella*) such as *Megistaspis* (*Megistaspidella*) *maximus* Wandås, 1984. The presence/absence of a posterior spine on the pygidium is observed in several of the subgenera, which is why it is discarded as a generic character. The final diagnostic feature mentioned for the subgenus is the hypostome. The importance of this feature is uncertain due to lack of well-preserved material, but the differences seem minor and vary somewhat even within the same species. In all cases it seems rather dubious to use this feature as a diagnostic character when we still have no or only a very poor knowledge on the hypostome for many of the species belonging to the different subgenera. For these reasons I fully agree with Nielsen (1995) in the suppression of the name *Spinopyge* Balashova, 1976.

The present work follows the higher megistaspid taxonomy proposed by Nielsen (1995) with the exception of the subgenus *Megistaspis (Heraspis)*, which is here rejected as a junior synonym of *M. (Megistaspidella)*. *M. (Heraspis)* was originally defined as a group of megistaspids characterized by a wide cephalon, which is broader than long; has small eyes and typically a broad but short anteromedian cephalic snout. The thoracic segments have deep articulating furrows and convex half rings posteriorly, while the triangular pygidium has a long terminal spine and a doublure of uniform width (Wandås 1984). The width of the cephalon; the size of the eyes and the length of the cephalic snout are all characters that are easily found within the subgenus *M. (Megistaspidella)* as for example seen on *M. (M.) triangularis*, *M. (M.) lamanskii* and *M. (M.) acuticauda* (Angelin, 1854). The furrow depth and convexity of the posterior thoracic axial half rings do not fall outside the variation seen within *M. (Megistaspidella)* and this character is furthermore largely based on *M. (H.) heroica* as the thorax is poorly known in both *M. (H.) heros* and *M. (H.) laticauda*. Regarding the terminal pygidial spine it is found throughout the subgenera of *Megistaspis*, rendering it invalid for generic taxonomy. The change in width of the pygidial doublure from anterior to posterior is slightly more useful, but even here there is a change of width averaging slightly under 20% within the species assigned to *M. (Heraspis)*, and typically a change of just over 20% in several of the species belonging to *M. (Megistaspidella)* (e.g. *M. (M.) giganteus* Wandås, 1984) The species assigned to *M. (Heraspis)* just represent the outer extremities of the border morpho-space. With regard to the deep cranidial furrows; the presence of cephalic caeca and the median ridge on the preglabellar area these features may also be found on species such as *Megistaspis (Megistaspidella) spinulata* (Bohlin, 1960), *Megistaspis (Megistaspidella) lawrowi*, *Megistaspis (Megistaspidella) triangularis* and *Megistaspis (Megistaspis) elongata* (Schmidt, 1906).

Subgenus *Megistaspis (Megistaspidella)* Jaanusson, 1956

Type species. – *Entomostracites extenuatus* Wahlenberg, 1821.

Diagnosis. – A subgenus of *Megistaspis* with shallow articulating furrows and a triangular cephalon with a median spine or snout. The hypostome is characterized by a subtriangular posterior margin, normally bearing a median terminal point. Thorax with flat axial- and half rings throughout. Pygidium subpara-

bolic to triangular in outline with or without a posterior spine; doublure distinctly widening posteriorly (emended from Jaanusson 1956 and Wandås 1984).

Remarks. – The two species *Megistaspis (Megistaspidella) heroica* Bohlin, 1960 and *M. (M.) laticauda* Wandås, 1984 described below are very closely related and although Wandås (1984) did not find any morphological overlap in the Scandinavian material such features actually do occur among the East-Baltic specimens. Examples of this are illustrated in Schmidt (1906, pl. 7, figs 4–6), where the holotype of *Megistaspis (Megistaspidella) heroica* from NW Russia (Schmidt 1906, pl. 7, figs 4–4a) features a narrow pygidial border with slight traces of distal pleural ribs and a relatively flat and well-defined anterior margin of the frontal glabellar lobe so reminiscent of *M. (M.) laticauda*. Likewise the two East-Baltic cranidia figured by Schmidt (1906, pl. 7, figs 5–6) seem to have alae very weakly developed or lacking, as is the case in *M. (M.) laticauda*. The problem with the East-Baltic material is the lack of detailed stratigraphical knowledge, which could be used to determine whether these specimens represent early forms of *M. (M.) heroica* that were not fully differentiated from the shared ancestor; just represents ecophenotypes of *M. (M.) heroica* or whether the two taxa are best considered as subspecies. The last possibility is in part rejected by the apparent stratigraphical and geographical co-occurrence of the two taxa within the Oslo Region, although better material is needed in order to verify this observation.

The three taxa *M. (M.) giganteus runcinatus* n. ssp., *M. (M.) giganteus giganteus* Wandås and *M. (M.) maximus* Wandås, 1984 revised below represent an unusually clear evolutionary line. The pygidia present the most distinct development, changing from the nearly smooth pygidium of *M. (M.) g. runcinatus* to one with fairly distinct but flattened pleural ribs bearing distinct interpleural furrows in *M. (M.) g. giganteus* to pygidia with high and narrow pleural ribs on which the interpleural furrow has been restricted to the distal half of the pleural ribs (*M. (M.) maximus*). The evolutionary development is somewhat punctualistic in that it takes place within a short stratigraphical interval with long sequences of near stasis both before and after (Fig. 53), though various transitional forms do occur at the stratigraphical boundary at which *M. (M.) maximus* takes over from *M. (M.) giganteus*. The cranidium seem less prone to change than the pygidium with the most marked difference being the development of the effaced L3 and L4 lateral glabellar lobes into small but distinct lateral ridges and the slight increase in eye size occurring between *M. (M.) giganteus* and *M. (M.) maximus*.

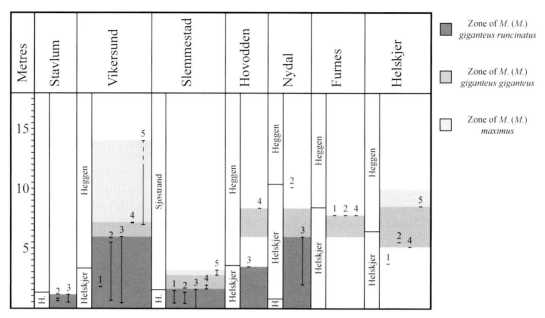

Fig. 53. Range chart showing the stratigraphical distribution of the *Megistaspis* species found in the Elnes Formation of the Oslo Region. The stratigraphical overlaps between *M. (M.) giganteus giganteus* and *M. (M.) maximus* at Vikersund cover a number of transitional forms showing characteristics of both species. 1, *M. (M.) heroica*; 2, *M. (M.) laticauda*; 3, *M. (M.) giganteus runcinatus*; 4, *M. (M.) giganteus giganteus*; 5, *M. (M.) maximus*.

Megistaspis (Megistaspidella) heroica Bohlin, 1960

Pl. 14, Figs 4–5, 7

1898   *Megalaspis heros* Dalm. – Schmidt, pp. 44–45.

1906   *Megalaspis heros* Dalm. sp. – Schmidt, pp. 51–54, pl. 7, figs 4–6.

1949   *Megalaspis heros* (Dalm.) – Bohlin, pp. 539, 547, 555–556, 566.

1951   *Megalaspis* cf. *heros* Ang. – Jaanusson & Mutvei, p. 632 (*partim*), (*non* page 633).

1956   *Megalaspis heros* – Bohlin, p. 129.

1960   *Megistaspis heroica* n. sp. Bohlin, pp. 174–179, text-figs 15–16, pl. 7.

1976   *Megistaspidella (Spinopyge) heroica* (Bohlin 1960) – Balashova, pp. 102–104, pl. 34, fig. 6.

1984   *Megistaspis (Heraspis) heroica* Bohlin, 1960 – Wandås, pp. 223–224, pl. 5C, E, G, pl. 6A, C.

Type stratum and type locality. – Holotype specimen CNI 58/11154 stored at the Central Scientific Research Geological Exploration Museum in St. Petersburg originates from the uppermost Obuchov Fm. (B3b, Kunda Regional Stage, Darriwilian) at Pulkovka, St. Petersburg area, north-western Russia.

Norwegian material. – The species is very rare, but includes three cranidia (PMO 33199, 104.050, 104.051) and 10 or 11 pygidia (PMO 36356 (?), 83202, 83204, 89752, 102.521, 104.048, 106.419, 106.440, 106.519, 201.993/1, 202.118/2).

Diagnosis. – A species of *Megistaspis (Megistaspidella)* characterized by a cephalon, which is broader than

long, triangular with strongly convex sides; a distinct cephalic border posterolaterally; fairly strong glabellar segmentation; preglabellar cranidial width approximately $2^{1}/_{2}$ to 3 times as wide as frontal glabellar lobe and divided by distinct median field ridge; glabellar flanks semiparallel to slightly forwardly narrowing; frontal glabellar lobe not sharply defined anteriorly; alae well developed; posterior border furrows very shallow and broad. The pygidium is broad, triangular with a wide border corresponding to 40% of the pleural field width at anterior pleural rib; pleural ribs 12 to 15, terminating distally at paradoublural line (emended from Bohlin 1960).

Description. – The following generel description is based on the Norwegian material. Largest cranidium PMO 104.050 more than 55 mm long. Largest pygidium, PMO 104.048, 88 mm long.

Cranidium slightly longer than wide. Glabellar margin fairly distinct to slightly obscure, bordered by wide axial furrow. Glabella elongated, the width corresponding to between 60 and 66% of the length (two specimens); glabellar convexity moderately low; maximal glabellar width, which is located at the occipital ring, corresponds to around 115% of the minimum glabellar width found between the eyes. Occipital ring and furrow strongly effaced mesially, becoming stronger laterally; occipital ring takes up between 27 and 30% of the cranidial width (two specimens); mesial tubercle relatively small, situated less than 15% of cranidial length in front of posterior margin or on posterior part of slight glabellar hump; three to four pairs of lateral glabellar furrows, moderately shallow and wide, S1 and S3 most strongly developed, the former widening adaxially; posterior lateral glabellar lobe effaced, laterally continuing out into exsagittally elongated ala; L2 narrow, looking much like L1; L3 strongly curved and wide but short (exsag.); L4 if present narrow (lat.) and effaced; frontal glabellar lobe subtriangular, nearly twice as wide as long, anteriorly continuing out into long and narrow preglabellar ridge the length of which corresponds to 80% of the glabellar length; preglabellar area less than 1.5 and more than 1.2 times as long as glabella. Cranidial width at preocular area corresponding to between 2.2 and 2.5 times the occipital ring width, widest point located just in front of glabella. Paradoublural line marked by moderately distinct paradoublural furrow reaching up to just behind sagittal midline of frontal glabellar lobe. Palpebral area delineated anteriorly by distinct eye ridge. Posterolateral fixigenal projection relatively broad and short. Posterior furrow shallow, accentuated by posterior border ridge. Glabella covered by moderately coarse but faint, exsagitally directed ridges or terrace-lines turning more anterolaterally along the flanks. Preglabellar and preocular areas covered by anterolaterally directed caeca turning slightly outwards anteriorly.

No librigenae, hypostomes or thorax segments are known from Norway, but the Swedish specimens show librigenae characterized by a wide doublure continuing out into a short and broad, posterolaterally directed genal spine.

The eye is small, while the pleural field is covered by very fine caeca (Bohlin 1960; Wandås 1984).

Pygidium is triangular with well-developed posterior spine; pygidium moderately convex with distinctly concave borders from anterior to posterior part. The pygidial length corresponds to between 80 and 90% of the width; axis long and narrow, taking up around 18% of the pygidial width, while the length corresponds to nearly two-thirds the pygidial length; axis with about 20 axial segments. Pleural field bearing around 14 moderately distinct and slightly flattened pleural ribs, the posteriormost becoming more effaced; pleural ribs terminating at paradoublural line; interpleural furrow shallow and mostly restricted to the outer half of the pleural furrows. Paradoublural line distinct, located 40% of the pleural field width inside pygidial margin at anterior pleural rib. Pygidial doublure wide, the maximal width corresponding to 20% of the pygidial width; the width decreasing by nearly 20% from posterior to anterior part. Doublural terrace-lines moderately coarse, probably just over 20 in number, converging to inner margin anteriorly.

Remarks. – As noted by Wandås (1984) the genal spine of *M.* (*M.*) *heroica* does not appear to have been turned inwards posteriorly as originally described by Bohlin (1960). Nielsen (1995, pp. 159–160) described a fragmentary cranidium from the basal part of the *Asaphus raniceps* Zone of the South Swedish Komstad Limestone, which he tentatively assigned to this species. The specimen does resemble *M.* (*M.*) *heroica* except for the very shallow lateral glabellar furrows and a slightly more forwardly located paradoublural line when related to the glabella. The relatively early occurrence of this specimen together with its fragmentary appearence and slightly different features suggests that a 'confer' denotation should be continued until better material has been found.

Occurrence. – In Norway *M.* (*M.*) *heroica* has been found in the Helskjer Member of the basal Elnes Formation. Skien-Langesund: Skinnviktangen at Rognstrand (Helskjer Member). Modum: Road-cut at Vikersund skijump (Helskjer Member, 1.75–7.15 m above the Huk Fm.). Oslo-Asker: The Old Eternite Quarry at Slemmestad (0.39 to 1.42 m above the Huk Formation. Hadeland: Grinaker at Gran (Helskjer Member or lowermost Heggen Member). Mjøsa: Helskjer on Helgøya (3.7 m and possibly also 2 m above the Huk Formation) and road-cut at Furnes Church (7.75 m above base of Helskjer Member).

Outside Norway it is known from the Siljan area and Öland, central to southern Sweden, where it occurs in the *Asaphus raniceps* zone of the uppermost Komstad Limestone and from the uppermost Obuchov Formation (upper Kunda Stage) at Pulkovka in the St. Petersburg area of north-western Russia (Bohlin 1960; Schmidt 1898, 1906).

Megistaspis (*Megistaspidella*) *heroica sensu lato* Bohlin, 1960

Pl. 14, Fig. 6

1984 *Megistaspis* (*Megistaspidella*) *maximus* n. sp. Wandås, pp. 228–229 (*partim*), figs 8–9, pl. 9E.

Material. – One nearly complete if slightly foreshortened specimen PMO 104.047.

Remarks. – This specimen, which has a cranidial length of 110 mm and a pygidial length of 105 to 110 mm, is assigned to *M. heroica* in the broad sense because of the lack of alae; the narrower pygidial doublure and the seemingly lower number of pygidial pleural ribs, although the last character falls well within the variation observed for the Swedish specimens such as that figured by Bohlin (1960, pl. 7, fig. 4). The lack of alae and narrow pygidial border has likewise been noted for the East-Baltic material described by Schmidt (1906, pl. 7, figs 4–6). The latter feature is actually found on the holotype for the species (Schmidt 1906, pl. 7, fig. 4a).

The specimen was earlier assigned to the species *Megistaspis* (*Megistaspidella*) *maximus* by Wandås (1984), but is clearly differentiated on the strongly sigmoid cephalic lateral margins; the more forwardly narrowing glabella with its less inflated lateral glabellar lobes; the near lack of alae and by the much wider cephalic doublure. The pygidium differs in its wider pleural ribs with more pronounced interpleural furrows and by the termination of the pleural ribs at the paradoublural line.

Occurrence. – The specimen most probably belongs to the Helskjer Member of the basal Elnes Formation due to the presence of carbonate concretions in the rock sample. Collected at Øster Øren farm at Vikersund, Modum.

Megistaspis (*Megistaspidella*) *laticauda* Wandås, 1984

Pl. 14, Figs 8–13; Pl. 15, Fig. 1

1984   *Megistaspis* (*Heraspis*) *laticauda* n. sp. Wandås, pp. 224–226, pl. 5D, H, pl. 6B, E, G.

Type stratum and type locality. – Holotype cranidium PMO 102.563 originates from the lowermost Heggen Member, Elnes Fm., at Vikersundbakken in the Modum district, Norway.

Material. – The species is common and the examined material includes 23 cranidia (PMO 101.807 (two specimens), 102.424, 102.563, 102.575 (two specimens), 102.597, 102.618, 102.642, 102.647, 102.654, 106.066, 106.074 (two specimens), 106.537, 106.539, 106.545 (two specimens), 106.547 (two specimens), 106.548 (two specimens), 143.565), 24 librigenae (PMO 90588, 102.550, 102.593, 102.595, 102.598, 102.618, 102.638, 102.639 (two librigenae probably from same individual), 102.652 (three specimens), 106.054, 106.055, 106.107, 106.108, 106.446 + 48 (two specimens), 106.523, 106.526, 106.535, 106.539, 106.544, 106.547) and 90 pygidia (PMO 2390, 2395, 2554, 20218, 33193 (two specimens), 33194, 33195, 83203, 83204, 83245, 89761 (two specimens), 90541, 90564, 101.579, 101.766, 101.789, 101.807, 101.809, 101.810, 102.405 (two specimens), 102.424, 102.453–4, 102.481–2, 102.494, 102.523 (two specimens), 102.542, 102.550, 102.572, 102.579, 102.581, 102.595 (four specimens), 102.597 (two specimens), 102.602 (two specimens), 102.617, 102.618 (four specimens), 102.637, 102.642 (three specimens), 102.652 (five specimens), 102.654, 102.676, 102.677 (two specimens), 104.046 (two specimens), 106.049, 106.054, 106.055 (two specimens), 106.056, 106.066 (three specimens), 106.074 (two specimens), 106.108, 106.109, 106.443 (two specimens), 106.453, 106.526, 106.547 (two specimens), 106.548 (two specimens), 201.987B/4, 202.087A/3, 202.089A/1, 206.952a/1, 206.952/2, 207.520, 211.005).

Diagnosis. – Glabellar segmentation strong; preglabellar cranidial width approximately 2.2 times as wide as frontal glabellar lobe and divided by distinct median field ridge; frontal glabellar lobe with sharply defined and flat anterior margin; alae very weakly developed. Pygidium broad, triangular with a wide border corresponding to just over half the pleural field width at anterior pleural rib; pleural ribs

approximately 17, terminating distally well out onto the border (emended from Wandås 1984).

Remarks. – As already noted above this species is closely related to the contemporaneous *M.* (*M.*) *heroica*, but may be differentiated on the very weakly developed alae; the more pronounced furrows and flattened glabellar front; the coarser cephalic caeca; the shorter preglabellar area; the slightly higher number of pygidial pleural ribs and the more adaxially located paradoublural line on the pygidium, which is crossed by the pleural ribs. A problem with most of these features is that they easily become effaced when the specimens have been exposed to tectonic deformation, thereby erasing the differences between the two species on the less well preserved specimens. The pleural ribs of the pygidium and the presence/absence of alae are therefore often the best characters for a diagnosis. *M.* (*M.*) *heroica* s.l. is even closer to *M.* (*M.*) *laticauda* than the type form, but *M.* (*M.*) *laticauda* may among other things be differentiated on strength of its wide and sharply delineated glabellar front; the relatively wider frontal glabellar lobe and the coarser cephalic caeca.

Occurrence. – Helskjer Member to basal Heggen Member, Elnes Formation. Skien-Langesund: Skinnvikstangen at Rognstrand. Eiker-Sandsvær: Stavlum at Krekling (Helskjer Member, 0.56 to 0.75 m above the Huk Fm.). Modum: Road cut at Vikersundbakken (0.60 to 5.5 m above the Huk Fm.). Oslo-Asker: the Old Eternite Quarry at Slemmestad (Helskjer Member, 0.35 to 1.29 m above Huk Fm.), Meyerløkken and Fredensborg in Oslo. Ringerike: Haug church and Amundrud farm at Hønefoss (Helskjer Member). Hadeland: Stubstad farm at Brandbu (Helskjer Member). Mjøsa: Furnes (top Helskjer Member, 7.75 m above the Huk Fm.), sub-profile 1 at southern road-cut were Fv 84 crosses E6 southwest of Nydal (top Helskjer Member, 10.08 m above the Huk Fm.), Helskjer on Helgøya (Helskjer member, 5.5 m above the Huk Fm.).

Megistaspis (*Megistaspidella*) *giganteus* Wandås, 1984

Diagnosis. – Species of *Megistaspis* (*Megistaspidella*) defined by a strongly triangular cephalon with nearly straight lateral margins and ending frontally in a long snout; a frontal area corresponding to more than the glabellar length and generally bearing a distinct but low inflation just in front of glabella; an only slightly hourglass-shaped and somewhat effaced glabella; a subtriangular pygidium with narrowly rounded posterior margin; a distinctly posteriorly widening pygidial doublure; between 15 and 17 more or less effaced pleural ribs terminating at paradoublural line and interpleural furrows reaching all the way from axial furrow to paradoublural line (emended from Wandås 1984).

Remarks. – The material described as the two sub-species reveals a stratigraphically related change in the pygidium beginning with a highly effaced form, where the axial furrows appear extremely shallow, while the pleural ribs may be completely missing. Upwards the ribs and furrows begin to become better developed although pygidia characterized by moderately deep furrows and well-developed pleural ribs and axis are restricted to a very short interval compared to the total stratigraphical occurrence of the species. Although there are intermediates, the large difference between effaced and ribbed forms together with their clear stratigraphical relationships indicate that they should be regarded as distinct taxa and they are here distinguished as subspecies.

Megistaspis (Megistaspidella) giganteus giganteus Wandås, 1984

Pl. 15, Figs 11–16, 18

1963 *Megistaspis* cf. *centaurus* – Skjeseth, p. 63 (may include *M. laticauda*).
1984 *Megistaspis (Megistaspidella) giganteus* n. sp. Wandås, pp. 226–227, figs 7, 9, pl. 7, pl. 8A–I.

Type stratum and type locality. – Holotype cephalon originates from the top of Helskjer Member, 7.75 m above the Huk Formation in the profile at Furnes Church, Mjøsa, Norway.

Material. – The examined specimens include two cephala (PMO 90493, 90553), four or five cranidia (PMO 67183, 90441–2, 90509, 106.151), two librigenae (PMO 67183, 90388), possibly 40 pygidia (PMO 2392, 67184, 67185, 67571, 67574, 67575, 67577, 67581, 67582, 67590 (two specimens), 67595, 67604, 69254 (?), 83202, 90388, 90390 (two specimens), 90464, 90535, 90538, 90539, 90541 (two specimens), 90546 (three specimens), 90548 (two specimens), 102.443, 102.521, 102.585, 106.016, 106.151, 143.472 (two specimens), 143.473, 202.135A/1, 202.144) and possibly one partly articulated specimen (PMO 106.094 (transitional form to *M. maximus*)).

Diagnosis. – Subspecies of *Megistaspis (Megistaspidella) giganteus* characterized by generally distinct pleural ribs and interpleural furrows and with deeply impressed axial furrow.

Description. – Largest cranidium PMO 90509 broken, but would have reached a length of nearly 150 mm when complete. Largest pygidium, PMO 67574, 91 mm long.

The cephalon is elongated triangular in outline with nearly straight lateral margins and a length corresponding to somewhere between 80 and 120% of the total cephalic width. Cranidium strongly elongated with a snout-like preglabellar area; cranidial length corresponding to somewhere around twice the width. Glabella bordered by wide and shallow axial and preglabellar furrows. Glabella roughly rectangular in outline with slightly narrowed middle, width corresponding to 60% of the length; glabellar convexity low with no glabellar hump; maximal glabellar width located at the occipital ring corresponds to between 115 and 120% of the minimum glabellar width found between the eyes.

Occipital ring and furrow generally strongly effaced; occipital ring takes up just over 30% of the cranidial width; mesial tubercle fairly small, situated approximately 30% of the glabellar length in front of posterior margin; four lateral glabellar furrows very shallow; lateral glabellar lobes effaced, L3 generally slightly more inflated than the others, wide and short (exsag.); frontal glabellar lobe semicircular. Alae fairly distinct, elongated, situated outside L1-lobes slightly more than their own length in front of posterior cephalic margin. Preglabellar area extremely long, weakly concave sagitally while weakly convex laterally, bearing a low and wide, oval high just in front of glabella; an indistinct narrow, sagittal ridge may be traced from the glabellar front and out towards the snout. Maximal cranidial width at preocular area corresponds to twice the occipital ring width, widest point located at glabellar front. Paradoublural line faint, reaching forward to a point outside or slightly behind the anterior margin of glabella. Eye ridge generally strong. Posterolateral fixigenal projection relatively broad and short with fairly deep and well-defined posterior furrow.

Glabella is covered by fine pores and carries short and faint, sagittally directed terrace-lines on posterior frontal glabellar lobe margin. Preglabellar, preocular and posterior fixigenal areas are seemingly free of any sculpture except for the widely spaced doublural terrace-lines on the preocular area; the number of terrace-lines probably exceeds 30.

Eyes relatively large, situated two-thirds their own length in front of posterior cephalic margin. Paradoublural line on librigena marked by shallow paradoublural furrow located between 50 and 60% of the distance between eye socle and lateral margin from eye. Librigenal doublure and border narrow with approximately 15 to 17 terrace-lines. Genal corners extended into long and sharply cylindrical, posterolaterally directed genal spines. Librigenal field smooth, below shell covered by rather fine caeca radiating out from the eye. Border surface smooth but for a few terrace-lines on the proximal part of the genal spine.

No hypostomes are known.

Thorax with narrow axis corresponding to 20% of the thoracic width.

Pygidium triangular in outline with narrowly rounded posterior margin; weakly convex with convex anterolateral and gently concave posterior border; doublure more or less concave throughout. The pygidial length corresponds to between 90 and 95% of the width, on one juvenile specimen decreasing to as much as 80%; axis long and fairly narrow, taking up between 20 and 23% of the pygidial width, while the length corresponds to 80% of the pygidial length; axis with about 23 to 24 axial rings, excluding terminal piece and anterior half ring. Pleural field with 15 to 17 moderately distinct and narrow pleural ribs terminating at paradoublural line. Interpleural furrows fairly deep, reaching from axial furrow to distal termination at paradoublural line, where they widen strongly just before reaching paradoublural line. Paradoublural line generally distinct, located 40% of the pleural field width inside pygidial margin (flattened specimens) at anterior pleural rib. Pygidial doublure narrow, the maximal width corresponding to 17 to 18% of the pygidial width; the width decreasing anteriorly by 20 to 25%. Doublural terrace-lines moderately coarse, 15 to 18 in number, converging to inner and outer margin anteriorly.

Remarks. – M. (M.) *giganteus giganteus* is closely related to M. (M.) *maximus* described below, but may be differentiated on the slightly less hourglass-shaped glabella; the more highly effaced lateral glabellar furrows; the lack of marked lateral glabellar lobes L3 and L4, which appears as distinct transverse ridges on *M. maximus*; a maximal cranidial width at preocular area corresponding to twice the glabellar width; the larger eyes situated no more than two-thirds their own length in front of posterior cephalic margin; the pygidial pleural ribs terminating at the paradoublural line, and which carries interpleural furrows all the way from axial furrow to distal termination. Intermediates between the two species are common at around the 7 m level at Vikersund corresponding to the uppermost occurrence of *M. giganteus* and the lowest appearence of *M. maximus*, clearly indicating a direct phylogenetic relationship between the two.

Fragmentary cranidia may look very similar to flattened specimens of *Megistaspis (Megistaspidella) heroica* treated above, but differs on the relatively longer preglabellar area; the lack of preglabellar caeca; the larger eyes; the more posteriorly located maximal width of the frontal cranidial area; the less pointed anterior glabellar margin and by the less pronounced preglabellar ridge.

Occurrence. – This subspecies is solely found in the Oslo Region, where it occurs from the top Helskjer Member and up into the basal Heggen and Sjøstrand members. Skien-Langesund: Skinnviktangen at Rognstrand, beach north of Omborsnes in Frierfjorden (?) (0.1 to 0.4 m below top of Helskjer Member).

Modum: Road-cut at Vikersund skijump (basal Heggen Member, 7.10–7.15 m above Huk Fm.). Oslo-Asker: The Old Eternite Quarry at Slemmestad (lower Sjøstrand Member, 1.57–1.86 m above Huk Fm.), Tøyen in Oslo (top Helskjer Member). Hadeland: Hovodden at Randsfjorden (basal Heggen Member 8.35 m above the Huk Fm. (these specimens could actually also belong to early forms of *M. maximus*)), Quarry south of road 195 about 1.5 km south of Brandbu towards Tingelstad (Helskjer Member). Mjøsa: Profile at Furnes Church near Hamar (top Helskjer Member, 7.75 m above the Huk Fm.), Helskjer on Helgøya (top Helskjer Member, 5.1 m above the Huk Fm.).

Megistaspis (Megistaspidella) giganteus runcinatus n. ssp.

Pl. 15, Figs 2–10

Derivation of name. – Referring to the effaced pygidial surface.

1984 *Megistaspis (Megistaspidella)* sp. A Wandås, p. 229, pl. 9D.
1984 *Megistaspis (Megistaspidella)* sp. Wandås, pl. 10F

Type stratum and type locality. – Holotype pygidium PMO 106.060 originates from the Helskjer Member, Elnes Formation in the road cut at Vikersund Skijump, Modum, Norway.

Type material. – Holotype pygidium PMO 106.060 comes from the Helskjer Member at Vikersund Skijump. Paratype cranidium PMO 102.642 was collected from the basal Heggen Member at the 5.25 m level in road-cut at Vikersund skijump, Modum. Paratype cranidium PMO 102.425 is from the basal Heggen Member, 5.4 m above the Huk Formation. Paratype cranidium PMO 90584 belongs to the Helskjer Member, Skinnvikstangen at Rognstrand, Skien-Langesund. Paratype librigena PMO 102.598 comes from the basal Heggen Member, 5.4 m above the Huk Formation, road-cut at Vikersund skijump, Modum. Paratype pygidium PMO 201.974/1 originates from 0.09 m below the top of the Huk Formation in the Old Eternite Quarry at Slemmestad, Oslo-Asker. Paratype pygidium PMO 90569 comes from the Helskjer Member, Skinnviktangen at Rognstrand, Skien-Langesund. Paratype pygidium PMO 106.435 is from the basal Heggen Member, 3.70 m above the Huk Formation in road-cut at Vikersund skijump, Modum. Paratype pygidium PMO 102.476 is from the basal Heggen Member, 4.0 to 5.0 m above the Huk Formation in road-cut at Vikersund skijump, Modum.

Material. – The examined material consists of 20 cranidia (PMO 83202, 90562, 90567, 90584, 101.742, 101.810, 102.425, 102.550, 102.575 (two specimens), 102.595, 102.602, 102.642, 102.652 (two specimens), 106.422, 106.533–34 (one frontal area), 202.095B/3, 202.101, 202.112), 12 librigenae (PMO 90559, 90568, 90571, 90584, 101.752, 102.397, 102.598, 106.547, 106.548, 202.006, 202.071A/2, 202.080/1) and 84 pygidia (PMO 2389, 2420, 2628, 59159, 60512, 68424, 69258, 69259, 83202 (two specimens), 83203, 83204, 90478, 90479, 90559 (five specimens), 90562, 90566 (two specimens), 90567, 90568, 90569, 90570, 90572, 90573, 90574, 90576, 90580 (two specimens), 90581 (two specimens), 90583, 90584, 90589, 10010, 101.722, 101.723, 101.726, 101.731, 101.771, 102.425, 102.462, 102.473, 102.476, 102.518, 102.550, 102.579, 102.593, 102.595 (two specimens), 102.597, 102.600, 102.602, 102.616, 102.634, 102.643 (two specimens), 102.652 (two specimens), 102.678, 106.055, 106.060, 106.434, 106.435,

106.464, 106.466, 106.524, 106.525, 106.527, 106.548, 143.536, 201.974/1, 202.010, 202.045A/2, 202.093B/2, 202.094, 202.131A/1, 206.947b/4, 206.953, 207.361/1, 207.384/1).

Diagnosis. – Subspecies of *Megistaspis (Megistaspidella) giganteus* characterized by forwardly located mesial tubercle and a relatively wide pygidium with effaced axis and pleural ribs.

Description. – The description of *M. (M.) giganteus runcinatus* n. ssp. is largely identical to that of the nominal subspecies described above and the following description will therefore focus on the features, which differ.

Largest cranidium PMO 102.595 broken, but more than 56 mm in length. Largest pygidium, PMO 106.434, fragmentary, but has a length of nearly 80 mm. Cranidium with a preglabellar area taking up at least 55% of the cranidial length. Glabella featuring slight glabellar hump with fairly obscure mesial tubercle situated approximately 36% of the glabellar length in front of posterior margin (three specimens). The occipital ring takes up around 35% of the cranidial width. Alae are distinct.

Eyes relatively small, situated at about their own length in front of posterior cephalic margin. Librigenal border with approximately 28 terrace-lines on specimen PMO 102.598.

Pygidium subtriangular in outline, strongly effaced. The pygidial length corresponds to between 70 and 85% of the width (eight specimens); axis narrow, occupying between 19 and 21% of the pygidial width (seven specimens). Pleural field with 14 to 16 or 17 strongly effaced and narrow pleural ribs. Paradoublural line generally effaced, located one-third to two-fifths the pleural field width inside pygidial margin (flattened specimens) at anterior pleural rib. Width of pygidial doublure decreasing by 30 to 50% from posterior to anterior part, most on the oldest specimen from the base of the formation. Doublural terrace-lines 15 to 21 in number.

Remarks. – *M. (M.) giganteus runcinatus* n. ssp. differs from *M. (M.) giganteus giganteus* by the presence of a slight glabellar hump; the anterior location of the mesial tubercle; the slightly wider occipital ring; the distinct alae; smaller eyes; strongly effaced pygidium, which is clearly wider than long and by the relatively narrow and more pronounced posteriorly widening pygidial doublure.

The Svartodden Member of the underlying Huk Formation contains pygidia characterized by an even more effaced surface; a stronger posterior widening of the doublure; by being more elongated and by having a slightly more parabolic outline than those of *M. (M.) giganteus runcinatus* n. ssp.. At present it is not possible to verify whether there may be an abrupt or a more gradual development from this older material to *M. (M.) giganteus runcinatus*, but based on the gradual change in the pygidial smoothness and posterior widening of the doublure observed within the Elnes Formation it seems quite likely that the change may be more gradual than punctualistic. The elongated pygidia of the uppermost Huk Formation form an evolutionary link between *M. (M.) giganteus runcinatus* n. ssp. and the older but morphologically very similar *M. (M.) bombifrons* Bohlin, 1960 from Sweden, which is mainly characterized by the stronger effacement of the dorsal shield; the near lack of alae; the more pronounced posterior widening of the doublure and by the extremely shallow axial furrows. At the moment it is unclear whether it is possible to differentiate between the two or whether they just represent one continuous evolutionary line without any distinct morphological boundaries. In the latter case they should probably not be regarded as separate species.

Occurrence. – This subspecies is solely found in the Oslo Region, where it ranges from the basal Helskjer Member into the lowermost Heggen Member. Skien-Langesund: Skinnviktangen at Rognstrand (Helskjer Member), beach north of Omborsnes in Frierfjorden (0.1–0.4 m below top of Helskjer Member). Eiker-Sandsvær: Mellemåsen at Vestfossen (Helskjer Member), Stavlum (0.45–1.10 m above Huk Fm.). Modum: Road cut at Vikersund Skijump (Helskjer to basal Heggen Member, 0.4 to 5.95 m above the Huk Formation), rivulet at Heggen Church. Oslo-Asker: The Old Eternite Quarry at Slemmestad (Top Huk Fm. to Helskjer Member, -0.09 to 1.52 m above Huk Fm.), Rodelökken, Oslo. Ringerike: Ridge east of Alm farm just south of Randsfjorden (Helskjer Member). Hadeland: Hovodden at Randsfjorden (Helskjer Member, 3.45 m above the Huk Fm.). Mjøsa: Eastern side of point at Nes Church (3.0 m above the Huk Fm.), sub-profile 1 at northern road-cut where Fv 84 crosses E6 southwest of Nydal (Helskjer Member, 1.92–5.94 m above the Huk Fm.).

Megistaspis (Megistaspidella) maximus Wandås, 1984

Pl. 15, Figs 17, 19–22; Pl. 16, Figs 1–2

1984 *Megistaspis (Megistaspidella) giganteus* n. sp. Wandås (*partim*), pl. 8I (cranidium).
1984 Megistaspis (*Megistaspidella*) *maximus* n. sp. Wandås, pp. 228–229 (*partim*), figs 8–9, pls 8J–L, 9A–C, F (*non* pl. 9E: paratype PMO 104.047 = *M. (M.) heroica sensu lato*).
1984 *Megistaspis (Megistaspidella)* sp. B Wandås, p. 229, pl. 10A.

Type stratum and type locality. – The holotype cranidium PMO 102.561 comes from Vikersundbakken in Modum, where it was collected from 12 to 14 m up in the Elnes Formation corresponding to the lower part of the Heggen Member.

Material. – The examined material consists of one cephalon (PMO 102.651), 11 cranidia (PMO 33325, 67592, 83049, 89790, 100.915, 102.435, 102.591, 102.625, 102.731, 105.949, 106.079), two librigenae (PMO 101.803, 103.657), 53 pygidia (PMO 2555, 2556, 3393, 10006, 11443, 11647, 20233, 33325 (two specimens), 67166, 67592, 83048, 83160, 83161, 83205, 90457, 90465, 90468, 101.786–87, 101.793, 101.801, 102.402 (transitional form from *M. giganteus*), 102.403 (two specimens), 102.431, 102.436, 102.442, 102.470, 102.477, 102.504, 102.517, 102.528, 102.603, 102.625, 102.646, 102.663, 102.816, 103.653, 103.656, 103.657, 103.660, 103.661 (two specimens), 105.638, 105.927 (transitional form to *M. giganteus*), 106.042, 106.043, 106.059, 106.062, 143.492, 202.154, 202.156, 202.165) and two partly articulated specimens (PMO 56341–42, 202.163/1).

Diagnosis. – Species of *M.* (*Megistaspidella*) characterized by a strongly triangular cephalon with nearly straight lateral margins and a pointed snout-like front; frontal area flattened, length surpassing the glabellar length; glabella with distinct alae and inflated L3 and L4 lobes; genal spines long; pygidium wider than long, triangular with narrowly rounded posterior end; axis distinctly delineated; approximately 14 pleural ribs high and narrow, reaching outside the paradoublural line; interpleural furrows at the highest developed on the distal half of the pleural ribs (emended from Wandås 1984).

Description. – *M.* (*M.*) *maximus* is morphologically close to *M.* (*M.*) *giganteus giganteus* described above and the following description will therefore concentrate on the features, which differ.

Most complete cranidium PMO 89790 has a length of 76 mm. Largest pygidium PMO 102.402 measures just over 100 mm in length.

The cephalon has slightly sigmoid anterolateral margins. Cranidium with a preglabellar area taking up at least 70% of the cranidial length. Glabella, which is bordered by moderately deep axial and preglabellar furrows, is distinctly hourglass-shaped with a slight glabellar hump. Occipital ring and furrow laterally distinct, becoming more effaced mesially; four lateral glabellar furrows moderately distinct; lateral glabellar lobes L3 and L4 appearing as small transverse ridges on each side of glabella. Alae fairly large, situated slightly less than their own length in front of posterior cephalic margin. Preglabellar area flat. Maximal cranidial width at preocular area corresponds to between 2.1 and 2.4 times the occipital ring width.

Eyes situated 75 to 90% of their own length in front of posterior cephalic margin.

Pygidium moderately wide, the length corresponding to approximately 80% of the width; axis narrow, occupying between 17 and 21% of the pygidial width and bearing no more than about 21 axial rings, excluding terminal piece and anterior half ring. Pleural field bearing 14 to 14$\frac{1}{2}$ distinct and narrow pleural ribs terminating outside the paradoublural line;

interpleural furrows, if present, only occurring on the distal two-thirds of the pleural ribs. Paradoublural line located no more than one-third the pleural field width inside pygidial margin (flattened specimens) at anterior pleural rib. Pygidial doublure narrow, the maximal width corresponding to 15 to 16% of the pygidial width; the width decreasing by 35 to 40% from posterior to anterior. Doublural terrace-lines number 19 to 22, converging to inner margin anteriorly.

Remarks. – The differences found between this species and the closely related *M.* (*M.*) *giganteus* described above are given in the discussion of *M.* (*M.*) *giganteus giganteus*.

Occurrence. – The species has only been described from the Oslo Region, where it occurs in the basal Heggen and Sjøstrand members of the Elnes Formation. Skien-Langesund: Skinnviktangen at Rognstrand. Modum: Road-cut at Vikersund skijump (6.95 to 12 or 14 m above Huk Fm.), road-cut east of Heggen Church (Heggen Member). Oslo-Asker: The Old Eternite Quarry at Slemmestad (basal Sjøstrand Member, 2.73–3.16 m above Huk Fm.), Geitungholmen outside Slemmestad (Helskjer Member), Akersbakken and Frognerhaugen in Oslo City (Helskjer Member). Ringerike: Haug Kirke at Hønefoss, Northern Svartøy in the Tyrifjord, ridge east of Alm farm just south of the Randsfjord. Hadeland: Hofskampen at Gran, road-cut just south-east of Røykenvik, Hovodden at the Randsfjord (basal Heggen Member, 8.30–13.85 m above Huk Fm.), Stubstad farm at Brandbu, Amundsrud (Helskjer Member), quarry south of road 195 about 1.5 km south of Brandbu in the direction of Tingelstad (Helskjer Member). Mjøsa: Helskjer on Helgøya (basal Heggen Member, 8.5 m above the Huk Fm.).

Subfamily Niobinae Jaanusson *in* Moore, 1959

Genus *Niobe* Angelin, 1851

Diagnosis. – Short frontal area; a well-marked and forward widening glabella; generally strongly developed alae; eyes situated flush up against the axial furrow at transverse midline of cranidium; distinct paradoublural lines; rounded genal corners; pygidium with wide axis, distinct flattened border and generally with well-developed pleural ribs (emended from Moore 1959).

Subgenus *Niobe* (*Niobe*) Angelin, 1851

Type species. – *Asaphus frontalis* Dalman, 1827 (by subsequent designation Vogdes 1890).

Diagnosis. – Pygidium characterized by the strongly inflated pleural furrows reaching out onto the border, where their distal terminations are clearly delineated; border distinct, flattened or concave (Moore 1959).

Niobe (*Niobe*) *frontalis* (Dalman, 1827)

Pl. 16, Figs 3–13

1828 *Asaphus frontalis* Dalman, pp. 46–47, 69 (diagnosis).
1953 *Niobe* cf. *frontalis* – Størmer, p. 70.
1955 *Niobe frontalis* (Dalman) – Bohlin, pp. 143–148, pl. 6, figs 5–9.
1963 *Niobe frontalis* – Skjeseth, p. 63.
1984 *Niobe frontalis* (Dalman 1827) – Wandås, pp. 229, 231, pl. 10B–E, H, I.
1995 *Niobe* (*Niobe*) *frontalis* (Dalman 1827) – Nielsen, pp. 166, 172, 173, fig. 127.

Type stratum and type locality. – Lectotype specimen Ar.15970 selected by Bohlin (1955) originates from the lower middle Darriwilian 'Gigas' limestone at Ljung in Östergötland, Sweden.

Norwegian material. – The examined material includes 13 cranidia (PMO 67163, 67184, 67593, 83200, 83259, 90513 (two specimens), 101.702, 101.770, 103.674, 106.064, 143.471, 202.056C/3), four librigenae (PMO 67183, 83178, 106.378, 202.103), three hypostomes (PMO 83146, 102.481, 106.378), 55 pygidia (PMO 67163 (three specimens), 67187 (three specimens), 67576, 67579, 67593, 83149, 83177, 83178 (four specimens), 83180, 83238, 83245, 83260, 90390, 90513 (two specimens), 90516, 90535 (two specimens), 90536, 90539, 90586, 102.430, 102.465, 102.512, 102.572, 102.593, 102.598, 102.618, 103.704, 104.046, 106.012, 106.056, 106.070 (two specimens), 106.098 (two specimens), 106.377, 106.384, 106.443, 106.456, 106.547, 143.471, 202.005, 202.042A/10, 202.047A/1, 202.087A/1, 202.180, 207.399/1) and four partly articulated specimens (PMO 89394, 102.503, 106.053 (two specimens)).

Diagnosis. – Species of *Niobe* (*Niobe*) characterized by fairly deep lateral glabellar furrows or pits; relatively transversely directed terrace-lines on a long frontal area; fine terrace-lines covering much of the glabella; pygidial length corresponding to less than half its width; pygidial border comparatively wide and by six or $5^{1}/_{2}$ pleural ribs on the pygidium (emended from Dalman 1828).

Description of Norwegian material. – Largest cranidium PMO 67593 has a length of nearly 36 mm, while largest pygidium PMO 67576 measures at least 39.5 mm in length.

The Norwegian material is generally somewhat deformed by later compression and tectonic movement, making it difficult to reconstruct the length/width ratios, but the cranidial proportions appear similar to those of the Swedish specimens. Cranidium with relatively short glabella taking up between 83 and 86% (three specimens) of the cranidial length. Glabella distinctly widening forward, the smallest width between the alae corresponding to between 65 and 75% of the maximal width located at posterior part of frontal lobe. Occipital ring and furrow mesially distinct, becoming more effaced laterally; occipital ring turned slightly rearwards mesially in some specimens, resulting in a shortened central part of the occipital ring; occipital ring occupies between 60 and 65% of the cranidial width; mesial tubercle small, situated approximately one-third of the glabellar length in front of posterior margin; three lateral glabellar furrows moderately deep on internal moulds, more effaced on well-preserved specimens with test; frontal glabellar lobe trapezoid with flattened front marked by fairly distinct mesial truncation. Alae moderately well developed. Preglabellar area long, corresponding to between 14 and 17% (three specimens) of the cranidial length, anterior margin characterized by slight ogive. Preocular area moderately broad, the width corresponding to around 10% of the preocular cranidial width. Posterolateral fixigenal projection moderately broad, the width corresponding to around 30% of the cranidial width, while the length corresponds to approximately one-third the glabellar length. Cranidial surface covered by moderately fine, exsagittally to anterolaterally directed striae on anterior and posterior part of glabella and by slightly coarser, transverse to anterolaterally directed terrace-lines on the preglabellar and preocular area. The latter may form a wide 'V' centrally, though mostly they are directed in a nearly straight transverse line, becoming more anterolaterally directed abaxially. Some moderately coarse, posterolaterally directed terrace-lines are found on the posterior fixigenal projection of at least one specimen (PMO 67593).

Eyes of moderate size, their length corresponding to 20% of the cranidial length and situated at glabellar midline. Paradoublural line on librigena marked by deep paradoublural furrow and depressed librigenal field; doublure wide, the width corresponding to between 50 and 60% of the eye length. Genal corners well rounded. Librigenal field appearing smooth.

Hypostome subrectangular in outline with well-rounded front and moderate posterior truncation; widest posteriorly, the width corresponding to around 80% of the length or slightly below. Anterior lobe of middle body longer than wide, length corresponding to around 70% of the total hypostomal length. Posterior lobe of middle body divided into two smaller lobes by slight central depression connecting the posterior pit with deep middle furrow. Anterior border practically absent; anterior wings narrowly rounded; posterolateral border wide, posteriorly nearly split in two by marked truncation.

Thoracic axis taking up around 30% of the total width. Axial rings bearing fine, forwardly convex terrace-lines, changing to a more exsagittal direction on the inner portion of pleura. Distal pleural terminations evenly rounded; doublure bearing moderately dense pattern of strictly exsagittally directed terrace-lines.

Pygidium broadly rounded, flatly convex with flattened to very slightly concave posterior border; length corresponding to just under half the width. Axis broad, occupying between 25 and 28% of the pygidial width (five specimens); axis with eight distinctly W-shaped axial rings, excluding terminal piece and anterior half ring. Pleural field bearing $5^1/_2$ to 6 broad and distinctly inflated pleural ribs with well delineated distal terminations located midway between paradoublural line and outer pygidial margin; interpleural furrows absent. Paradoublural line generally indistinct, but marked by slight posterior bending of pleural ribs and shallow paradoublural furrow located a little over 25% of the pygidial width inside margin at anterior pleural rib. Maximal border width corresponding to between 14 and 17% of the pygidial width (five specimens). Pygidial doublure wide, the maximal width corresponding to 26 to 29% of the pygidial width (five specimens). Doublural terrace-lines moderately coarse, 18 to 19 in number. Surface terrace-lines moderately dense only becoming slightly denser at the transition from border to pleural ribs; terrace-lines transversely to anterolaterally directed. Axis covered by distinct and strongly forward convex terrace-lines, which disappears towards the axial flanks.

Remarks. – The Norwegian material differs from the south Swedish material described by Bohlin (1955) by the longer (exsag.) and wider posterolateral fixigenal projection; the less pronounced anterior cranidial ogive and by the smaller mesial tubercle. Overall the differences appear small and most probably reflect local variations rather than being of taxonomic significance.

Bohlin (1955) found the transverse course of the preglabellar terrace-lines and the absence of terrace-lines on the posterolateral fixigenal projection to be two of the most important diagnostic features for *N. (N.) frontalis*. The first feature is actually quite variable within the Norwegian material encompassing forms with both 'V' shaped and straight terrace-lines. The presence of terrace-lines on the posterolateral fixigenal projection is likewise found to occur within the Norwegian material, suggesting that the terrace-line pattern and distribution should be used with great care when comparing species.

N. (N.) tjernviki Nielsen, 1995 from the Middle Ordovician Komstad Limestone of Sweden is morphologically close to *N. (N.) frontalis*, but Nielsen (1995) listed a number of differentiating characters. Most of these do not hold up to closer examination,

especially when including the Norwegian material, but *N. (N.) tjernviki* may still be separated on the slightly more extensive external terrace-line pattern on the cranidium; the distinctly more elongate pygidium with a length corresponding to slightly more than half the width; the narrower pygidial border (less than 15% of the pygidial width as measured from the illustrations *in* Nielsen 1995, figs 129, 130) and by the denser terrace-line pattern on the pygidium. Furthermore it may perhaps be differentiated by a more developed cranidial ogive as suggested by the reconstruction given by Nielsen (1995, fig. 128). The inferred differences in palpebral lobe size; depth and distinctiveness of the various pits and furrows and in the hypostome are not supported by the Norwegian material.

N. (N.) frontalis differs from *N. (N.) schmidti* Balashova, 1976 from Darriwilian deposits in the East-Baltic area in its wider cephalic doublure and preglabellar area; the slightly wider pygidial border and by the more extensive external terrace-line pattern on the pygidium.

N. (N.) frontalis differs from the contemporaneous Russian *N. karneevae* Balashova, 1976 by the presence of six (cf. 5) pleural ribs on the pygidium, the more flattened glabellar front and by the more deeply depressed lateral glabellar furrows.

Occurrence. – In Norway it is found from the middle Helskjer Member and up into the basal part of the succeeding Sjøstrand and Heggen members, Elnes Formation. Skien-Langesund: Skinnvikstangen at Rognstrand (Helskjer Member). Eiker-Sandsvær: Stavlum at Krekling (Helskjer Member, 1.12 to 1.16 m above Huk Fm.). Modum: Road-cut at Vikersund skijump (basal Heggen Member, 2.5 to 6.67 m above the Huk Fm.). Oslo-Asker: The Old Eternite Quarry at Slemmestad (Helskjer to basal Sjøstrand Member, 0.55 to 3.46 m above Huk Fm.). Hadeland: Hovodden at Randsfjorden (basal Elnes Formation). Mjøsa: Furnes Church (top Helskjer Member, 7.75 m above the Huk Fm.), sub-profile 1 at northern road-cut were Fv 84 crosses E6 southwest of Nydal (top Helskjer Member, 10.10 m above the Huk Fm.), Helskjer on Helgøya (Helskjer Member, 3.7 to 5.5 m above the Huk Fm.).

Outside Norway it is known from the uppermost Holen Limestone of Öland, Östergötland and Västergötland, Sweden (Bohlin 1955; Wandås 1984).

Superfamily Cyclopygoidea Raymond, 1925

Family Nileidae Angelin, 1854

Genus *Nileus* Dalman, 1827

Type species. – *Asaphus (Nileus) armadillo* Dalman, 1827 (by subsequent designation Hawle & Corda 1847).

Diagnosis. – Genus defined by strictly parallel-sided dorsal exoskeleton with evenly rounded front and rear; cephalon subreniform, evenly convex with indistinctly defined, narrow border, occipital furrow effaced; glabella slightly convex, almost parallel-sided, indistinctly defined anteriorly; anterior and posterior areas of fixigenae extremely narrow; eyes large, semicircular; librigenae narrow with rounded genal corners or with small genal spines. Thorax with eight segments characterized by wide, parallel-sided, slightly convex axis and rounded pleural extremeties. Pygidium nearly semicircular with effaced or indistinct axis and smooth pleural fields. Exoskeletal surface minutely pitted (modified from Moore 1959).

Nileus armadillo (Dalman, 1827)

Pl. 16, Figs 14–18; Pl. 17, Figs 1–11; Text-Fig. 54

1827   *Asaphus (Nileus) Armadillo* Dalman, pp. 61–63, pl. 4, fig. 3a–e.
1882   *Nileus armadillo*, Dalm. – Brøgger, pp. 63–65.
1904   *Nileus armadillo*, Dalman – Schmidt, pp. 64–68 (*partim*), pl. 8, fig. 15, *non* pl. 8, figs 12–14, 16–18.
1909   *Nileus armadillo*, DALM. – Holtedahl, pp. 8, 29, 40.
1909   *Nileus armadillo*, DALM., var. *depressa*, S. & B. – Holtedahl, pp. 7–8, 28–29, 40.
1913   *Nileus Armadillo* Dalm. – Hadding, pp. 72–73, pl. 7, figs 11–12.
1953   *Nileus armadillo* – Størmer, p. 83.
1972   *Nileus armadillo* (Dalman, 1827) – Schrank, pp. 365–367, pl. 6, figs 1–3, 5–6.
1972   *Nileus platys stigmatus* sp. et subsp. n. Schrank, pp. 353, 368–369, pl. 6, fig. 4, pl. 7, fig. 3, pl. 8, figs 1–6, pl. 9, fig. 1.
1972   *Nileus platys platys* subsp. n. Schrank, pp. 353, 369–371, pl. 9, figs 2–4, pl. 10, figs 1–2.
1980   *Nileus armadillo* Dalman – Reyment, fig. 3a–e (= illustrates enrolled specimen originally figured *in* Dalman 1827, pl. 4, fig. 3b).
1984   *Nileus armadillo*? (Dalman, 1827) – Wandås, p. 232, pl. 11C.
1995   *Nileus platys* Schrank, 1972 – Månsson, p. 117, fig. 4G–L.
2000b   *Nileus* cf. *armadillo* (Dalman, 1827) – Månsson, pp. 12–13, fig. 8b–c.
2002   *Nileus* cf. *armadillo* (Dalman, 1827) – Pålsson et al., p. 42, figs 8k and 10.

Type stratum and type locality. – Both Wandås (1984) and Nielsen (1995) referred to Reyment (1980) as having selected specimen Ög. 109 figured by Dalman (1827, pl. 4, fig. 3b) as lectotype

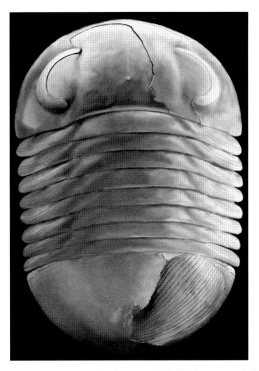

Fig. 54. Reconstruction of *Nileus armadillo* (Dalman, 1827) from the Elnes Formation. Length of specimen 4 cm.

for *Nileus armadillo*. Reyment (1980, p. 2) wrote: 'Figs. 3, a–d, e. *Nilleus armadillo* Dalman. Figured in pl. 4, fig. 3b. One of the specimens on which *N. armadillo*, the type species of *Nileus*, was founded. Husbyfjöl, Östergötland. Collected by J.W. Dalman. Ög 109'. It is clear from this text that no lectotype was selected at the time, although Ög 109 is the most logical choice for a lectotype as the rest of the type material is missing. For this reason specimen Ög 109 is here designated as lectotype for *Nileus armadillo* (Dalman, 1827).

Norwegian material. – The examined material from the Elnes Formation includes more than 300 specimens of which five are complete or partly articulated (PMO 2456, 21915, 82325, 106.514, 203.543/1) and nine represents cephala (PMO 56219, 33520, 72733, 72780, 72782, 203.378/50, 205.100/2, 207.479/8, 207.481/9). The rest consists of disarticulated skeletal parts.

Diagnosis. – Cephalon semielliptical in outline, twice as wide as long and strongly convex. Axial furrows almost straight, gently forward converging; cranidial front gently rounded; glabella relatively long; glabellar tubercle located approximately one-third the cranidial length from posterior margin. Pygidium is characterized by an inner doublural margin with distinct outwards flare below anterior pleural furrow and normally by the restriction of the surface terrace-lines to the articulating facets, anterolateral corners and posterior border. Pygidial doublure typically covered with 15 to 20 terrace lines (emended from Nielsen 1995).

Description of Elnes Formation material. – Largest specimen, PMO 2456, 82 mm long, slightly tectonically elongated, with a 30 mm long cephalon and an approximately 23 mm long pygidium.

Cephalon semielliptical to slightly semicircular in outline, strongly convex transversely as well as sagittally; length corresponding to between 55 and 60% of the width of unflattened specimens. Axial furrows shallow but generally marked by a moderate change in slope between glabella and fixigena, though this may be nearly absent on some of the smaller specimens; axial furrows posteriorly nearly straight, slightly forward converging, turning outwards in front of palpebral lobes. Glabella elongate, rectangular to slightly hourglass-shaped, anteriorly ending in gently curved front, width corresponding to 80% of the length; occipital ring and furrow effaced, only seen on internal moulds; mesial glabellar tubercle small, situated 23 to 31% of the cranidial length from the posterior margin; lateral glabellar furrows and muscle scars generally indistinct, only seen on internal moulds; a sagittal keel reaches from the mesial tubercle and forward to the anterior part of the glabella; no or very indistinctly developed mesial boss at the front. Anterior part of cranidium with small, but broad ogive. Frontal area flattened into short rim, which may include the anterior cranidial margin. Palpebral lobes moderately sized, their length corresponding to between 40 and 50% of the cranidial length, largest on undeformed specimens. Posterior fixigenal projections triangular in outline, appearing rather long in relation to the cranidium. Librigena comparetively broad with wide flattened border on anterior to middle part, which dissappears posteriorly; genal corners somewhat angular, not broadly rounded. External surface of librigena generally free of terrace-lines, but some librigenae with a few terrace-lines on the border are known from the upper part of the formation.

Hypostome typical for the genus; lateral border gradually decreasing in width posteriorly. Terrace-lines moderately fine and dense, transversely directed.

Thoracic axis occupying approximately half the thoracic width and bordered by fairly shallow axial furrows. Pleural area with moderately deep pleural furrow. Terrace-lines sparse on outer surface, seemingly absent on the thoracic axis, while represented by around three transversely directed terrace-lines on the pleural ridge.

Pygidium moderately convex, semielliptical in outline; length corresponding to 55 to 60% of the width; axis nearly effaced, strongly triangular, length corresponding to around 70% of the pygidial length, while the width generally corresponds to less than 40% of the pygidial width. Articulating furrow fairly deep on internal moulds, but absent on outer surface. Axis bearing four to five axial rings in addition to the axial piece and articulating half ring, each axial ring bearing a pair of shallow muscle scars. Pleural field generally smooth, but may bear fine, indistinct pustules radiating out from the posterior half of the axis. Border, except on the stratigraphically youngest specimens, moderately narrow, concave. External terrace-lines sparse, short, restricted to the articulating facets and anterolateral margin. The stratigraphically younger forms may carry long, margin sub-parallel terrace-lines on the border. Doublural terrace-lines numbering 15 to 17.

Remarks. – Although the addition of the taxon from the Elnes Formation results in a relatively long stratigraphical occurrence for *Nileus armadillo*, the morphological resemblance leaves no doubt that it should be regarded as one species. There are some minor differences from the revised description made by Nielsen (1995); the most important being the lack of a frontal cranidial boss and the more posterior position of the glabellar tubercle on the Elnes material. This is a common variation also seen in the older material from the Huk Formation and contemporaneous Komstad Limestone of southern Sweden, and should only be regarded as reflecting the natural variation within a population.

The Norwegian material from the Elnes Formation is contemporaneous with *Nileus platys* Schrank, 1972, described from erratic boulders in Germany. A closer comparison of the two reveals that, although most Norwegian specimens lack the faint radial ridges seen on the pygidium of the latter, some do have it and most of the others may simply lack it for taphonomic reasons. There seem to be no other differentiating character between the late Darriwilian representatives of *Nileus armadillo* and *Nileus platys stigmatus* Schrank, although the flattened rim in front of the glabella generally seems to be slightly better developed. They are consequently regarded as conspecific.

Nileus armadillo differs from the type material of *Nileus platys platys* Schrank, 1972, in the typically narrower and less well-developed pygidial border, although the border becomes wider on the stratigraphically youngest Norwegian specimens from the top of the Heggen Member in the Mjøsa district. As illustrated below such differences fall well within what is to be expected for local variations and subspecies differentiation is considered unnecessary.

Nileus armadillo shows some stratigraphical and geographical variation in the pygidial border width as well as in the presence of terrace-lines on the pygidial border. Thus in the road section at Nydal in the northern Mjøsa districts the border changes from narrow to wide between 21 and 24 m above datum and adds continuous terrace-lines between 26 and 30 m. At Slemmestad in the central Oslo Region only forms characterized by a narrow border without continuous terrace-lines are found, at least up to the base of the Håkavik Member. To the south in Eiker-Sandsvær the only two pygidia on which external terrace-lines may be observed both have long terrace-lines on the narrow border. These pygidia are stratigraphically contemporaneous with the ones from Slemmestad, suggesting that the terrace-line pattern is ecophenetic rather than genetically controlled. This is somewhat surprising as earlier authors (e.g. Schrank 1972; Nielsen 1995) found the pygidial terrace-lines to be one of the better diagnostic characters when differentiating between species and subspecies such as those of *Nileus depressus* (Sars & Boeck *in* Boeck, 1838). On the other hand it corresponds well with observations made on the early Darriwilian representatives of *N. armadillo*, which Nielsen (1995, figs 149–150) found to be quite variable with regard to pygidial terrace-line patterns, which vary from nearly smooth to terrace-lines covering both the anterolateral corners and a large part of the pleural field. The geographical variation in the terrace-line pattern and pygidial border width together with a rather variable width and sharpness of the frontal flattened area on the cephalon strongly supports the synonymy of *N. platys* within *N. armadillo*.

Nileus armadillo is differentiated from the lowermost Upper Ordovician *N. globicephalus* Schrank, 1972 from erratic boulders of Rügen, Germany, by the less rounded genal corners; the wider librigenae with the more strongly flattened border and by the more strongly developed frontal area. A small pygidium and cranidium from the Sandbian Pagoda Formation of southern Shaanxi, China, has a strong resemblance to *Nileus armadillo*, but differs on the seemingly less forwardly narrowing glabella and by the relatively wider pygidium with its apparently lack of terrace-lines (Yin *et al.* 2000).

Occurrence. – In Norway *Nileus armadillo* is found from the uppermost Hukodden Member, basal Huk Formation, and up to the top Heggen, Håkavik and 'Cephalopod Shale' members (Nielsen, 1995; this study). Specimens belonging to the Elnes Formation are known from most of the Oslo Region as shown by the following list. Eiker-Sandsvær: Vego and Råen at Fiskum (top Heggen Member, 3.35 to 3.66 m below the top of the section in profile 1). Oslo-Asker: the beach profile at Djuptrekkodden (middle Sjøstrand Member to basal Håkavik Member, 2.21 to 54.48 m above base of profile), Elnestangen (Engervik Member), Huk on Bygdøy, Hjortnæstangen in Oslo (Engervik Member), Vigeland Park in Oslo. Hadeland: Lynne and Hofstangen at Gran. Mjøsa: road-cut at Furnes Church (Helskjer Member), Helskjer on Helgøya, sub-profiles 2 and 3 at northern road-cut where road Fv 84 crosses highway E6 southwest of Nydal (lower to top Heggen Member, 0.50 to 1.58 m in sub-profile 2 and 0.42 to 21.46 m in sub-profile 3), behind churchyard in Hamar, Flakstadelva in Hedmarken.

Outside Norway it is known from erratic boulders from Germany, which have been transported by ice from Baltoscandia during the Pleistocene (Schrank 1972). The boulders belong to the upper Darriwilian Nileus or Schroeteri Limestone and the Sandbian Ludibundus Limestone (Schrank 1972).

Nileus armadillo is also common in the upper Darriwilian to lower Sandbian Anderö Shale formation in Jämtland, central Sweden, and in the Darriwilian (uppermost Volkhov to Uhaku Regional Stage) Komstad Limestone and Killeröd Formation of Scania, southern Sweden (Månsson 1995; Nielsen 1995; Pålsson *et al.* 2002).

Nileus depressus (Sars & Boeck *in* Boeck, 1838) ssp.

Pl. 17, Figs 12–13

      1984 *Nileus* sp. Wandås, p. 232, pl. 11A–B.
non 1995 *Nileus depressus parvus* n. subsp. Nielsen, pp. 280–282, fig. 204.

Material. – The material consists of two moderately well preserved cephala (PMO 89762, 89795) and a fragment of the latter cranidium (PMO 82322).

Description. – Cranidium 7 mm long. Cephalon semielliptical in outline, strongly convex transversely as well as sagittally; length corresponding to 67% of the width. Axial furrows and glabella completely effaced, leaving no boundary between the glabella and the palpebral lobes. On the internal mould an unusual and deep furrow is seen inside the lateral margin of the glabella extending from the palpebral lobe to the eye-socle furrow below the eye. Glabella elongate with a moderately straight front, width corresponding to 67% of the length; occipital ring and furrow effaced, internal moulds lowered in relation to the main glabella; lateral glabellar furrows practically absent internally as well as externally; mesial glabellar tubercle small, situated just over 40% of the cranidial length from the posterior margin; a slight sagittal keel reaches from the mesial tubercle and forward to the anterior part of the glabella on internal mould; no mesial boss is seen at the front. Cranidial front without ogive. Frontal rim absent. Palpebral lobes fairly large, their length corresponding to 56% of the cranidial length. Posterior fixigenal projections triangular in outline, relatively short. Librigena narrow with rounded angular genal corner and strongly vaulted flanks.

Remarks. – This specimen was tentatively assigned to his new subspecies *Nileus depressus parvus* by Nielsen (1995, pp. 280–282), although the norwegian specimen is somewhat younger. He noted a difference in the eye socle furrow impinging upon the palpebral lobes, which is a unique character for the subspecies and in the more evenly sagittal vaulting of the cranidium. To this is added the relatively narrowness of the frontal glabellar area, which together with the other characters suggests the Norwegian specimen probably does not belong to this subspecies.

It differs from *Nileus depressus depressus* (Boeck, 1838) from the Darriwilian Komstad Limestone of Sweden by the less forwardly diverging lateral sides of the frontal glabellar area; the indistinct axial furrows and by the stronger convexity of the cranidium.

Occurrence. – The specimen was collected from the top of the Helskjer Member, 7.95 m above the Huk Fm. at the road-cut near Furnes Church in the Mjøsa district.

Superfamily Trinucleoidea Swinnerton, 1915

Family Trinucleidae Hawle & Corda, 1847

Subfamily Trinucleinae Hawle & Corda, 1847

Genus *Botrioides* Stetson, 1927

Type species. – *Trinucleus coscinorinus* Angelin, 1854 (= *Botrioides bronni* (Boeck, 1838); see Owen 1987, p. 80).

Diagnosis. – Fringe narrow, declined. Up to four I arcs and E arcs present. Pits on upper lamella in deep radial sulci; on lower lamella sulci absent or restricted to E arcs. Genal prolongation absent. Lateral eye tubercles present (Owen 1987).

Remarks. – The genus was thoroughly revised by Owen (1987), who reassessed the Middle Ordovician trinucleids from Scandinavia. A problem, though, was the poor correlation available between some formations and the limited stratigraphical information attached to some of the old museum collections (A.W. Owen, personal communication, 2008), leading to difficulties in establishing the stratigraphical relationships between species. This is perhaps best illustrated by the two species *B. efflorescens* and *B. foveolatus* treated below. The latter was thought to occur before the former, whereas *B. efflorescens* is here shown to be the older species and probably ancestral to the other as they occur in a continuous succession but seemingly without stratigraphical overlap in both the Elnes Formation and the contemporaneous Killeröd Formation in southern Scania. It is less clear how the closely associated *B. margo* Owen, 1987 fits into the evolutionary line, but most probably it forms an off-shoot from a shared ancestor, which again composes a side branch of all the other *Botrioides* species described so far. The only possible exception

is *B.* sp. A described below, which in many ways resembles the *B. foveolatus* species group, although the pygidium assigned to it clearly belongs to the *B. bronnii* species group of Owen (1987). The problem may only be solved when a specimen containing both pygidium and cephalon has been found.

Eighteen of the more important cephalic characters and character states defining the species belonging to *Botrioides* and some closely allied genera were run through a parsimony analysis with branch-and-bound algorithm and Wagner optimization (Fig. 55; Table 33). The result, which yielded 12 shortest trees of 38 character changes, presents a consensus tree where *B. simplex*, *B. impostor*, *B. bronnii* and *B. broeggeri* are grouped together and *B.* sp. A forms the link to the *B. foveolatus* species group of Owen (1987). The consensus tree is interesting in that the first four species occur in the same order in which they appear stratigraphically, suggesting they may have developed from one another. The *B. foveolatus* species group *sensu* Owen (1987) did not split out into a seperate branch on the cephalic characters alone, as it undoubtedly should be due to the shared synapomorphies

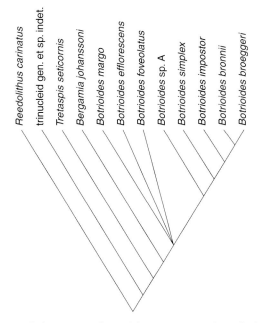

Fig. 55. Phylogenetic analysis of the genus *Botrioides* and related genera. The parsimony analysis presents 12 shortest trees of 38 character changes with branch-and-bound algorithm and Wagner optimization. The cephalic characters and character states used for the analysis are listed in Table 33, where the characters with question marks were excluded from the analysis, while the character states for E2, I1, fringe width and presence/location of lateral eye tubercles, respectively, were consolidated into multi-state characters to prevent them from being given extra weighting. *Reedolithus carinatus* is used as an out-group, while *Tretaspis seticornis* is included to re-examine the proposed placement of the unidentified trinucleid as a phylogenetical link between *Botrioides* and *Tretaspis* (see Hughes *et al.* (1975)).

such as the presence of three or more I arcs, the high number of sulci in each arc and probably by the high number of axial rings on the pygidium. The poor result on this part is probably due to, that less important/unstable characters useful only for species level differentiation gets too much weight and characters from the pygidium would probably be needed before the *B. foveolatus* group truly branches out from the rest of the genus *Botrioides* and not just ends up forming the link between the *B. bronnii* group and the *Bergamia*.

Finally, the analysis supports Owens (1987) association of *Botrioides* with *Bergamia*, while rejecting the hypothesis of Hughes *et al.* (1975, pp. 562–563) that the unidentified trinucleid described below should form the link between *Botrioides* and *Tretaspis*.

Botrioides simplex Owen, 1987

Pl. 18, Figs 10–15

1913　*Trinucleus coscino(r)rhinus* Ang. [*sic*] – Hadding, pp. 74–75, pl. 7, fig. 19.

1984　*Botrioides? bucculentus* (Angelin, 1854) – Wandås, p. 234, pl. 12G, J, pl. 13F.

1987　*Botrioides simplex* sp. nov. Owen, pp. 87–88, pl. 12, figs 7–11.

2000b　*Botrioides simplex* Owen, 1987 – Månsson, p. 14.

2002　*Botrioides simplex* – Pålsson *et al.*, fig. 10.

Type stratum and type locality. – The holotype cranidium PMO 83021 comes from the upper part of the Helskjer Member. The sample most probably originates from around 7.75 m above the Huk Formation in the road-cut near Furnes Church, Mjøsa.

Norwegian material. – The examined material consists of one cephalon (PMO 202.294), five cranidia (PMO 83021, 90565, 102.680 (two specimens), 202.292/2), five fringes (PMO 104.061, 104.064 (two specimens), 202.273/3, 202.293/4) and five complete if poorly preserved specimens (PMO 104.055 (three specimens), 104.064 (two specimens)).

Diagnosis. – Pseudofrontal lobe subspherical, overhanging inner part of narrow fringe which comprises 14 to $16^{1}/_{2}$ short sulci containing arcs E_1 and I_n ($n = 3$). E_1 pits on lower lamella decrease in size abaxially, divided by ridges, which are strongly developed mesially. Lateral eye tubercle generally well developed, located centrally on fixigena opposite S3. Pygidium characterized by broad border; three or four axial rings and two pleural ribs (modified from Owen 1987).

Description. – Except in the presence of $2^{1}/_{2}$ to 3 pleural ribs on the pygidium instead of 2, the present material adds nothing new to the description given by Owen (1987).

Remarks. – The species resembles the slightly younger *B. bronnii* (Boeck, 1838) in the general cephalic outline with the subspherical pseudofrontal lobe and narrow fringe, but is differentiated by the generally slightly less overhanging pseudofrontal lobe; the more anteriorly located lateral eye tubercle and by the absence of an E_2 arc. The lower side of the fringe show a large similarity to that of the younger *B. broeggeri* (Størmer, 1930), but is characterized by a typically slightly higher number of sulci and a more pointed anterior margin due to the more forward reaching glabella.

B. simplex is indicated to occur in the Andersö Shale formation in Jämtland and more precisely in the middle marly interval corresponding to the top of the Elnes Formation (Månsson 2000b; Pålsson *et al.* 2002). This is an extremely late occurrence when compared to the Oslo Region and requires further taxonomic investigation.

Occurrence. – In Norway the species has only been found in the basal part of the Elnes Formation. Modum: road-cut at Vikersund skijump (7.90 to 12.50 m above the Huk Formation). Oslo-Asker: the Old Eternite Quarry at Slemmestad (basal Sjøstrand Member, 5.60 to 5.68 m above the Huk Formation). Mjøsa: road-cut at Furnes Church (top Helskjer Member, most probably 7.75 m above the Huk Formation).

Outside Norway it is known from the Stein Formation and possibly the basal Andersö Shale formation at Andersön in Jämtland, central Sweden (Owen 1987).

Botrioides impostor Owen, 1987

Pl. 18, Figs 21–24; Pl. 19, Fig. 1; Text-Fig. 56

1887　*Trinucleus bucculentus,* Ang. – Brøgger, p. 17.

1930　*Trinucleus bronni* (Sars et Boeck) – Størmer, pp. 19–21 (*partim*), pl. 2, figs 1–7, text-figs 5, 16c, 43 (*non* text-fig. 6).

1953　*Trinucleus bronni* – Størmer, pp. 61, 83.

1975　*Botrioides? bronnii* (Sars & Boeck *in* Boeck, 1838) – Hughes, Ingham & Addison (*partim*), pl. 4, fig. 44 only.

1987　*Botrioides impostor* sp. nov. Owen, pp. 85–86, pl. 11, figs 18–20.

Type stratum and type locality. – The holotype cranidium PMO H0566 comes from the middle to upper Elnes Formation at Djuptrekkodden north of Slemmestad, Oslo-Asker.

Material. – The examined material consists of two cephala (PMO H417, 206.072/1), 86 cranidia (PMO H330 + H580 (11 specimens), H418–19 (five specimens), H0566, 4086, 33269 (three specimens), 33480, 101.701, 202.931, 202.988 (two specimens), 203.037/2, 203.054/2, 203.113/7, 203.116/7, 203.118/2 (two specimens), 203.118/4, 203.118/8 (eight specimens), 203.118/13, 203.119/5, 203.119/7 (five specimens), 203.119/11 (two specimens), 203.126/6, 203.130/27, 203.130//28, 203.130/33 (two specimens),

Table 33. Diagnostic features and character states defining the different species of *Botrioides* and some selected trinucleid genera

	Reedolithus carinatus	*Trinucelid gen. et sp. indet.*	*Tretaspis seticornis*	*Bergamia johanssoni*	*Botrioides margo*	*Botrioides efflorescens*	*Botrioides foveolatus*	*Botrioides* sp. A	*Botrioides simplex*	*Botrioides impostor*	*Botrioides bronnii*	*Botrioides broeggeri*
E2 present anteriorly	–	–	+	+	+	+	–	–	–	–	+	+
E2 complete	–	–	–	–	+	–	–	–	–	–	+	+
E3 and E4 present	–	–	–	–	+	–	–	–	–	–	–	–
I1 present	+	+	+	+	+	+	+	+	+	+	+	–
I1 complete	+	+	+	+	+	+	+	–	–	+	–	–
I2 present	+	+	+	+	+	+	+	+	–	–	–	–
I3 present	+	+	+	+	–	+	+	–	–	–	–	–
I4 present	+	+	–	–	–	+	–	–	–	–	–	–
Pits of arcs I2 – In arranged in sulci	–	–	–	+	+	+	+	+	+	+	+	+
Generally more than 17 sulci in half fringe	+	+	–	–	+	+	+	–	–	–	–	–
Sulci slit-like	–	+	–	+	+	+	+	+	–	–	–	–
Radii-breakdown posteriorly	+	+	+	–	–	–	–	–	–	–	–	–
Concentric ridges between arcs	+	+	+	–	–	–	–	–	–	–	–	–
Strong radial ridges between pits on lower lamella	–	(–)	–	?	–	–	–	?	+	–	–	+
Fringe narrow	–	–	–	–	–	–	–	–	+	+	+	+
Fringe wide	+	–	–	–	–	–	–	–	–	–	–	–
Fringe strongly widening laterally	+	+	+	–	–	–	–	–	–	–	–	–
Fringe border wide	–	–	–	–	+	–	–	–	–	+	–	+
Posterior prolongation of gena slight	–	–	–	–	+	+	+	+	+	+	+	+
Lateral eye tubercles present	+	+	+	–	+	+	+	+	+	+	+	+
Lateral eye tubercles posterior	+	–	–	–	–	+	–	+	–	+	+	+
Occipital spine present	+	–	–	–	–	–	–	–	–	–	–	–
Cranidium strongly reticulate	+	+	–	–	–	–	–	(–)	–	+	+	+
Axial rings on pygidium > 6	–	?	+	+	+	+	+	(–)	–	–	–	–
Pleural ribs on pygidium strong	–	?	–	–	?	+	–	(+)	+	+	+	?

Brackets around some of the character state symbols mean that the indicated state is the typical for that species, although some specimens may differ from the norm.

203.185/3, 203.214/2, 205.248, 205.978, 205.985, 205.986/1, 205.992, 205.998/2, 206.005/5, 206.012/2, 206.013, 206.014/1, 206.016/5, 206.023/1, 206.023/2, 206.026/3, 206.040/2, 206.043/2, 206.044/2, 206.051/5, 206.056/2, 206.066/4, 206.081/1, 206.093/2, 206.098/1, 206.098/2, 206.098/3, 206.119/1, 206.157/7, 208.582), 57 fringes (PMO H330 + H580 (28 specimens), H401, H418–19 (six specimens), 40304, 84698, 102.492, 104.061, 203.118/6, 203.118/11, 203.127/2, 203.198/2, 206.005/4, 206.005/6, 206.011/3, 206.012/3, 206.016/4, 206.025/2, 206.049/2, 206.066/13, 206.068/5, 206.068/6, 206.083/5, 206.092/6, 206.119/1), three pygidia (PMO 205.988/1, 206.006/2, 206.043/3) and two nearly complete specimens (PMO H0574, 202.988).

Diagnosis. – Pseudofrontal lobe oval to subspherical, only slightly overhanging narrow fringe which comprises complete arcs E$_1$ and I$_n$ and in some cases also I$_1$; fringe border broad. Posterior part of fringe widening, allowing for the addition of a few adventitious I pits. Pits, except posteriorly, set in short sulci; pits numbering 13^1/$_2$ to 17^1/$_2$ on a half fringe. Lateral eye tubercle subdued, situated opposite L2. External surface of glabella and genal lobes reticulate. Median tubercle present only on anterior thoracic axial ring. Pygidium transverse, bearing three axial rings, two pleural ribs and a broad border (modified from Owen 1987).

Description. – The present material adds nothing new to the description given by Owen (1987).

Remarks. – Newly prepared material indicates that the species in general has a slightly more subspherical glabella than thought by Owen (1987), who probably based most of his description on the slightly forelengthened holotype cranidium. This means that the cranidium may be confused with that of *B. broeggeri* (Størmer, 1930) but it differs on the posteriorly widening fringe with adventitious pits; the absence of an E2 arc; the precense of an I1 arc; the slightly higher number of pits; the absence of or faintly developed ridges between pits on lower lamella and by the still relatively longer and slightly overhanging glabella. It is distinguished from *B. bronnii* (Boeck, 1838) by the much smaller inflation of the pseudofrontal glabellar lobe; the lack of a frontal E2 arc and by the presence of a wide fringe border. *B. simplex* Owen, 1987 is likewise differentiated on a more bulbous frontal glabella; the lack of a wide fringe border and by the distinctly forwardly located lateral eye tubercles.

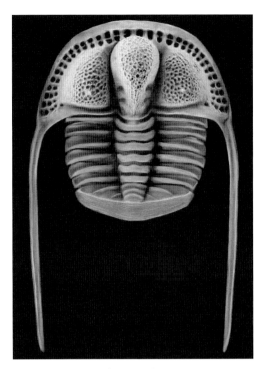

Fig. 56. Reconstruction of *Botrioides impostor* Owen, 1987. Length of specimen, excluding genal spines, 1 cm.

Occurrence. – Eiker-Sandsvær: Stavlum (Helskjer Member, 0.76 m above Huk Formation); profile 1 at Vegum (top Heggen Member); profile 1 (upper Heggen Member, 6.55 to 9.11 m below datum) and profile 4 at Råen near Fiskum (Heggen Member, 2.70 to 4.25 m below datum). Modum: road-cut at Vikersund skijump (8.0 to 11.25 m above the Huk Fm.) and close to Fure. Oslo-Asker: Bøveien south of Slemmestad (lower Engervik Member, 13.33 m above datum), the Old Eternite Quarry at Slemmestad (38.64 m above the Huk Fm.), the beach profile at Djuptrekkodden (middle Sjøstrand to Håkavik Member, 6.56 to 56.03 m above datum), north-west side of Ildjernet (Håkavik Member), Killingen at Oslo (Håkavik Member) and at the Geological Institute at Blindern (basal Vollen Fm.). Ringerike: beach at Gomnes (upper Elnes Fm.). Hadeland: south of Skiaker (Elnes Formation) and Hovstangen (Elnes Formation).

Botrioides bronnii (Boeck, 1838)

Pl. 19, Figs 2–6

1838	*Trilobites bronnii* Sars and Boeck – Boeck, p. 144 (diagnosis).

1987	*Botrioides bronnii* (Boeck, 1838) – Owen, pp. 81–85 (*cum. syn.*), pl. 2, 1–17.

1995	*Botrioides bronnii* (Boeck, 1838) – Månsson, pp. 101–102, figs 6A–G, 7.

Type stratum and type locality. – The lectotype cephalon PMO 61752 selected by Størmer (1940) was collected at Wraatz's Løkke in Oslo (probably located between Drammensveien and Thomas Heftyes gate in Gimle). Exact stratigraphical level is

unknown, but most likely upper Elnes Formation or Vollen Formation.

Norwegian material. – Of the material examined six cephala (PMO 202.952/4, 202.960/4, 202.961/1, 202.962/1, 202.963/1, 202.965/3), 30 cranidia (PMO H480, H481, H489, H0572, 4087, 36623, 40277, 61752, 82342 (two specimens), 105.964 (two specimens), 113.224, 113.225, 202.953/3 (two specimens), 202.987, 203.081/16, 203.081/17, 203.088/2, 203.109/21, 203.121/2, 203.130/6, 203.130/16, 203.130/21, 203.130/37, 203.162/4, 206.026/2, 206.110/5, 206.210/14, 206.296/2), 13 fringes (PMO H0572, 82310, 82661 (three specimens), 104.079, 203.130/25, 203.340/3, 206.042/2, 206.245/1, 208.496, 208.501/2), two pygidia (PMO 202.355/8, 202.960/5) and four complete or partly articulated specimens (PMO H482, 202.932/1, 202.960/1, 206.194) belonged to this species.

Diagnosis. – Pseudofrontal lobe subspherical, typically strongly overhanging narrow fringe comprising complete arcs E_1 and I_n and in some instances also E_2. Arc I_1 may be present anterolaterally. Pits set in short sulci; number of pits in half-fringe $12^1/_2$ to 16. Fringe border narrow. Lateral eye tubercle very subdued, situated opposite L2 or posterior part of S2. External surface of pseudofrontal lobe of large specimens with concentric ridges superimposed on coarse reticulation. Median tubercle on thoracic axial rings. Pygidium transverse, bearing up to four axial rings, four pleural ribs and a broad border (modified from Owen 1987).

Description. – The cephalon, thorax and pygidium of the species were described in detail by Størmer (1930, description for *B. bucculentus*) and Owen (1987). It should however be noted that E_2 does not always form a complete arc but may be restricted to the anterior part as seen on Plate 19, figures 2–3. Furthermore there seem to be more variation in the inflation of the pseudofrontal lobe than first thought by Owen (1987).

Remarks. – The species is generally easily identifiable through the generally strongly inflated and forwardly overhanging pseudofrontal lobe; the two E arcs frontally and by the slightly forward widening fringe without a wide border. The fringe typically is also more pointed anteriorly because of the forward reaching glabella. The ridges between the pits on the lower lamella are less strongly developed than in *B. simplex* Owen, 1987 and *B. broeggeri* (Størmer, 1930).

Occurrence. – In Norway *B. bronnii* is known from the lower Heggen and middle Sjøstrand members and up into the Fossum and Vollen formations. Skien-Langesund: Skinnviktangen (lower Fossum Fm., 27.15 m above phosphoritic layer at the base of the Elnes Formation). Eiker-Sandsvær: Profile 1 at Råen near Fiskum (upper Heggen Member, 5.30–15.04 m below datum). Modum: Road-cut at Vikersund skijump (lower Heggen Member, 15.9 m above the Huk Formation) and close to Fure. Oslo-Asker: Bøveien south of Slemmestad (top Sjøstrand to lower Engervik Member, lower 3.5 m to 13.33 m above datum), the south-west profile in the Old Eternite Quarry at Slemmestad (Sjøstrand Member, 7.36 m above the Huk Formation), the north-east profile in the Old Eternite Quarry at Slemmestad (Sjøstrand Member, 38.67 m above the Huk Formation), beach profile just north of Djuptrek-kodden (Sjøstrand to top Håkavik Member, 2.19–55.18 m above datum), Engervik at Slemmestad, Blindern (basal Vollen Fm.) and Wraatz´s Løkke in Oslo. Ringerike: unspecified locality. Mjøsa: beach at Hovindsholm on Helgøya (Heggen Member, 18.9 m above the Huk Fm.).

Outside Norway it is known from the Middle Ordovician Killeröd Formation and surrounding deposits in Scania, south-eastern Sweden (Månsson 1995; Owen 1987).

Botrioides broeggeri (Størmer, 1930)

Pl. 19, Figs 7–11

1930   *Trinucleus hibernicus* Reed var. *bröggeri* n. var. Størmer, pp. 24–27, text-figs 12, 13, 16d, pl. 3, figs 1–14.

1953   *Trinucleus hibernicus bröggeri* – Størmer, pp. 62, 82.

1953   *Trinucleus* cf. *hibernicus bröggeri* – Størmer, pp. 72, 73.

1953   *Trinucleus* aff. *hibernicus* – Størmer, p. 80 (partim) (listed).

1975   *B.?* *hibernicus broeggeri* (Størmer, 1930) – Hughes, Ingham & Addison, p. 562.

1987   *Botrioides broeggeri* (Størmer, 1930) – Owen, pp. 86–87, pl. 12, figs 1–6.

Type stratum and type locality. – The type cephalon PMO H0553 belongs to the Vollen Formation at Gullerud near Norderhov, Ringerike.

Material. – The examined material consists of eight cephala (PMO H0553, H0567, 66633 (three specimens), 81903 (two specimens), 206.081/3), six cranidia (PMO H0545, H0560, H0563 (two specimens), 205.157/3, 206.100) and 18 fringes (PMO H0563 (three specimens), H0567, H579, 66633 (six specimens), 82661 (three specimens), 103.978, 205.105/1, 205.107/3, 205.120).

Diagnosis. – Pseudofrontal lobe subcircular in outline, weakly differentiated from the rest of the glabella, not overhanging fringe; arcs E1 and In complete, only other arc, E2, commonly incomplete posteriorly; pits 12 to 15 in number, on lower lamella situated in deep sulci divided by strong ridges; fringe border wide on upper lamella; lateral eye tubercle prominent, situated posteriorly opposite L2. External surface of genae and glabella coarsely reticulate (Owen 1987).

Description. – A detailed description was given by Størmer (1930) with some amendments by Owen (1987). The new material adds nothing new.

Remarks. – The species strongly resembles *B. bronnii* (Boeck, 1930) from which it was most probably

derived. It is differentiated on the less strongly inflated pseudofrontal lobe, which generally does not overhang the fringe anteriorly; the moderately wide border on the fringe and the typically more strongly developed sulci on the lower fringe lamella.

It is differentiated from *B. impostor* Owen, 1987 by the more compact glabella; the two E arcs; the deep sulci and by the lack of additional pits posteriorly on the fringe.

Occurrence. – Skien-Langesund: Skinnviktangen (top Elnes Fm.). Eiker-Sandsvær: Rønningsfossen at Flata (Elnes Fm., 46.6–50.5 m above datum), Krekling (Fossum or upper Elnes Fm.) and profile 1 at Råen (upper Heggen Member, 3.41–9.13 m below datum). Oslo-Asker: beach-profile at Djuptrekkodden (top Engervik to Håkavik Member, 40.42–54.37 m above datum). Ringerike: Gullerud near Norderhov (Vollen Fm.).

Botrioides efflorescens (Hadding, 1913)

Pl. 19, Figs 12–15, 20

1913   *Trinucleus coscino(r)rhinus* Ang. – Hadding, pp. 74–75 (*partim*), pl. 7, figs 18, 20 (*non* pl. 7, fig. 19 = *B. simplex*).

1913   *Trinucleus efflorescens* n. sp. Hadding, p. 75, pl. 7, fig. 21a–c.

1927   *Botrioides coscinorrhinus* (Angelin) – Stetson, pp. 90, 97, pl. 1, fig. 12.

1927   *Trinucleus efflorescens* (Hadding) [*sic.*] – Stetson, pp. 90, 96, pl. 1, fig. 4.

1935   *Trinucleus efflorescens* Hadding – Thorslund *in* Thorslund & Asklund (1935), pp. 20, 60, pl. 2, figs 11–12.

1987   *Botrioides efflorescens* (Hadding, 1913) – Owen, p. 91, pl. 13, figs 5–13.

1995   *Botrioides efflorescens* (Hadding, 1913) – Månsson, p. 103 (*partim*), figs 6H, J–K, 7 (*non* fig. 6I = *B. foveolatus*).

2000b *Botrioides efflorescens* (Hadding, 1913a) – Månsson, p. 14, fig. 9l.

2002   *Botrioides efflorescens* (Hadding, 1913) – Pålsson *et al.*, figs 9f, 10.

Type stratum and type locality. – The holotype (by monotypy) cranidium LO 2543T is from the lower Andersö Shale at Andersön in Jämtland, central Sweden.

Norwegian material. – The examined material consists of three cephala (PMO 3181, 3184, 203.198/1), 17 to 18 cranidia (PMO 82236, 82246, 82249, 82342, 82620 (?), 203.102/2, 203.130/13, 203.153/1, 203.161, 203.183/1, 203.185/4, 203.196/2, 203.201/4, 203.207/1, 203.220/11, 203.340/4, 206.296/3, 208.599/1), one fringe (PMO 203.337/4), two pygidia (PMO 82342, 203.176/14) and one complete specimen (PMO H0570 – H0570a).

Diagnosis. – Pear-shaped glabella with moderately swollen pseudofrontal lobe, moderately overhanging fringe. Arcs E_1, I_1 and I_n complete; E_2 and I_2 present anteriorly; I_2 present posteriorly and I_3 restricted to anterolateral part. Number of pits in half-fringe 16 to 21 1/2. Fringe distinctly narrowing posteriorly. Lateral eye tubercle small, situated opposite anterior half of L2. External surface of genae smooth or finely granulate. Pygidium with narrow border, seven to nine axial rings and up to three weak pleural ribs (modified from Owen 1987).

Description. – For description see Hadding (1913). The complete specimen is too poorly preserved to yield any detailed information on the thoracic morphology.

Remarks. – *B. efflorescens* is morphologically close to the younger *B. foveolatus* (Angelin, 1854) discussed below, but differs in the presence of small E_2 pits anteriorly; the posteriorly clearly narrowing fringe as the number of arcs decreases to four and by the generally slightly more posterior location of the lateral eye tubercles. The pygidium is readily differentiated on the sparse and weak pleural ribbing and the on average slightly higher number of axial rings (7–9 instead of $7 \pm 1(?)$).

As noted by Owen (1987) the high number of axial rings on the pygidium together with features such as the extremely large number of I arcs clearly places *B. foveolatus*, *B. efflorescens* and *B. margo* Owen, 1987 from the contemporaneous Gullhögen Formation of Sweden in a separate group from the rest of the genus. The stratigraphical separation of the two Norwegian species *B. efflorescens* and *B. foveolatus* furthermore indicates that they may form an evolutionary lineage. *B. foveolatus* was originally thought to be the older species (Owen 1987, text-fig. 3), but this was solely due to the poorer knowledge of the stratigraphical correlation between the different formations at that time.

Occurrence. – In Norway *B. efflorescens* is found from the top of the Sjøstrand Member and up into the lower part of Engervik Member and in the upper part of the Heggen Member. Eiker-Sandsvær: profile 1 at Råen (upper Heggen Member, 9.06 m below datum), profile 4 at Råen (upper Heggen Member, 4.25 m below datum). Oslo-Asker: road-cut along Bøveien south of Slemmestad (upper Sjøstrand to lower Engervik Member, 3.13–13.43 m above datum), beach profile just north of Djuptrekkodden (upper Sjøstrand Member to lower Engervik Member, 22.62–34.10 m above datum) and Mariedalsveien 13 in Oslo (Sjøstrand Member). Mjøsa: Hovindsholm on Helgøya (Elnes Formation).

Outside Norway it is known from the lower Anderö Shale member at Östre Ottsjö and Andersön in Jämtland and the middle Killeröd Formation of south-eastern Scania, Sweden (Hadding 1913; Thorslund & Asklund 1935; Størmer 1930; Nilsson 1952; Bergström 1982; Karis 1982; Owen 1987; Månsson 1995, 2000b; Pålsson *et al.* 2002).

Botrioides foveolatus (Angelin, 1854)

Pl. 19, Figs 16–19

1857   *Trinucleus foveolatus* Ang. – Kjerulf, p. 286.
1878   *Trinucleus foveolatus*. n. sp. Angelin, p. 85, pl. 41, figs 2–2b.
1927   *Trinucleus foveolatus* (Angelin) – Stetson, pp. 90, 97, pl. 1, fig. 10.
1930   *Trinucleus foveolatus*. Ang. – Størmer, pp. 16–18 (*partim*), pl. 1, figs 4–10, 12–13 (*non* pl. 1, fig. 11 = *Lonchodomas rostratus*).
1930   *Trinucleus foveolatus* Ang. var. *intermedius* n. var. Størmer, pp. 18–19, pl. 1, figs 1–3.
1963   *Trinucleus faveolatus* – Skjeseth, p. 64.
1975   *B.? foveolatus* (Angelin, 1854) – Hughes *et al.*, p. 562.
1975   *B.? foveolatus intermedius* (Størmer, 1930) – Hughes *et al.*, p. 562.
1987   *Botrioides foveolatus* (Angelin, 1854) – Owen, pp. 90–91, pl. 12, figs 15–16, pl. 13, figs 1–4.
1995   *Botrioides efflorescens* (Hadding, 1913) – Månsson, p. 103 (*partim*), fig. 6I only.

Type stratum and type locality. – Neotype cranidium RM Ar2310 selected by Størmer (1930, p. 16) from Angelin's material in the Stockholm Riksmuseum is from the Elnes Formation at Huk on Bygdøy, Oslo-Asker.

Norwegian material. – The material examined includes four cephala (PMO 203.378/28, 205.070, 205.113/3, 205.116/4), 75 cranidia (PMO H391, H392, H492, H493, H494, H0533, H0538, 33277, 82317, 82623, 82624, 82923 (three specimens), 82925 (three specimens), 203.244/2, 203.268/1, 203.281/2, 203.293/11, 203.300/1, 203.301/2, 203.302/2, 203.306/4, 203.307/1, 203.316/7, 203.321/1, 203.321/4, 203.321/5, 203.322, 203.323/3, 203.325/2, 203.325/4, 203.325/8, 203.327/9, 203.331, 203.388/2, 203.391/4, 203.409/1, 203.415/2, 203.426/4, 203.469/2, 205.041, 205.073, 205.079/3, 205.080/2, 205.081/1, 205.081/3, 205.081/6, 205.081/7, 205.083/5, 205.083/8, 205.083/11, 205.095/4, 205.101/1, 205.107/2, 205.117/5, 205.122/1, 205.122/2, 205.125/2, 205.151/1, 205.156/2, 205.158/1, 205.159/2, 205.161/2, 205.162/2, 205.173/6, 205.178/7, 205.197/7, 205.199/1, 205.207/6, 205.207/7, 205.211/2, 206.945/5), 19 fringes (PMO 82622, 82623, 82923, 82624, 82925 (four specimens), 203.281/8, 203.292/2, 203.321/8, 203.391/5, 205.078/2, 205.091/3, 205.116/5, 205.116/6, 205.118/4, 205.173/2, 205.967/3), 23 pygidia (PMO H488, H0538, 82923 (two specimens), 82925, 202.355/8, 203.268/7, 203.302/3, 203.316/6, 203.326/3, 203.349/3, 203.351/1, 203.368/1, 203.425/1, 203.461/6, 205.023/2, 205.075, 205.081/4, 205.081/5, 205.083/7, 205.083/12, 205.105/5, 205.177/1) and two partly articulated specimens (PMO 203.543/3, 205.081/10).

Diagnosis. – Pear-shaped glabella increasing evenly in convexity to the mid-part of pseudofrontal lobe, overhanging fringe only slightly. Arcs E_1, $I_{1–3}$ and I_n situated in long sulci over whole fringe. I_4 present posteriorly on some specimens. Pits in each half arc numbering 17 to $22^1/_2$. Fringe of even width to slightly narrowing posteriorly. Small but distinct lateral eye tubercles situated just in front of S2. External surface of glabella and genae finely reticulate, fine concentric ridges present on mesial part of glabella. Pygidium with narrow border; about seven axial rings and five pleural ribs (Owen 1987).

Description. – A detailed description of pygidium and cephalon was presented by Størmer (1930) and Owen (1987).

A single, worn specimen, PMO 203.543/3, showing a partly articulated thorax has been found recently. The thorax seems essentially identical to that of *B. bronnii*.

Remarks. – For comparison and comments on phylogenetic relationship with *B. efflorescens*, see discussion of that species.

Occurrence. – The species is found from the middle Engervik to basal Vollen Formation (Hamar 1966, p. 30), the uppermost part of the Heggen Member and the basal Fossum Formation to the south. Skien-Langesund: Rognstranda (basal (?) Fossum Fm.). Eiker-Sandsvær: profile 1 at Vego (top Heggen Member, upper 1.5 m of the profile). Oslo-Asker: beach profile just north of Djuptrekkodden (middle Sjøstrand Member to basal Håkavik Member, 34.92–45.62 m above datum), Elnestangen north of Slemmestad (Engervik Member), Huk on Bygdøy (Engervik Member), Blindern (basal Vollen Fm.) and Stensberggata in Oslo (Engervik Member?), Hadeland: unspecified locality. Mjøsa: Toten (Elnes Fm.).

Outside Norway *B. foveolatus* is known from bed 'n' at the top of the Killeröd Formation in southern Scania, Sweden (Månsson 1995, fig. 6I). Karis (1982, p. 82) listed the species from the upper Andersö member.

Botrioides sp. A

Pl. 18, Figs 16–20

1987 *Botrioides* sp. A Owen, p. 88, pl. 12, figs 13–14.

Material. – The material, which in general is very poorly preserved, consists of four cranidia (PMO 82168, 103.701, 103.702, 202.292/9), possibly one fringe (PMO 202.295) and one pygidium (PMO 81265).

Description. – Cranidium semicircular to subtrapezoidal in outline. Glabella increasing in convexity (tr.) forwards towards the pseudofrontal lobe which has little independent convexity in relation to the rest of the glabella. Occipital ring and furrow, occiput, S1 and S2 resemble those of *B. impostor*. Distal part of L2 joined by weakly developed bridge to pseudofrontal lobe. Pseudofrontal lobe subspherical, though not much wider than posterior part of glabella, only slightly overhanging fringe. Axial furrow broad and

shallow, bearing deep fossula just in front of maximal glabellar width. Quadrant-shaped genal lobe moderately convex exsagitally as well as transversely, evenly rounded in transverse cross-section; eye tubercle fairly prominent, situated posteriorly outside S1 or posterior part of L2.

Fringe relatively broad, widening laterally and with a narrow border. Arcs E_1 and I_n complete, containing 16 pits in a half fringe. Arc I_1 present along the whole genal lobe, but missing in front of the glabella. Arc I_2 restricted to the posterior part of the fringe. Pits situated in deep and narrow sulci reaching all the way across the fringe; sulci becoming less developed posterolaterally on fringe.

Test surface, lower lamella of fringe, hypostome and thorax unknown.

A pygidium has tentatively been assigned to this species based on its stratigraphical relation to cranidium PMO 82168. Pygidium transverse, length approximating one-third the width. Axis weakly convex, tapering evenly posteriorly, extending slightly onto the broad posterior border. The axis carries three distinct axial rings in excess of the terminal piece and anterior half ring. Axial furrow shallow. Pleural field with two distinct pleural ribs reaching all the way out to the posterior border. Border steep with sharp inner margin. Pygidial surface smooth.

Remarks. – The species closely resembles *B. impostor* Owen, 1987, but is differentiated on its narrower fringe border and the slightly more inflated pseudofrontal lobe. The long sulci reaching all the way out to the outer fringe is strongly reminiscent of *B. foveolatus* (Angelin, 1854) and *B. efflorescens* (Hadding, 1913), but *B.* sp. A can be distinguished from these by having only one complete I arc and, if the pygidium is correctly assigned, in the fewer axial rings and pleural ribs. Even so *B.* sp. A could easily form the transition between the two morphological subgroups of Owen (1987, p. 80, table 1), the *B. bronni* species group and the *B. foveolatus* species group, which in main are differentiated on the number of I arcs and axial rings. However, because of lack of complete specimens and the slightly doubtful association of the pygidium with the cranidia it seems most prudent to follow the lead of Owen (1987) in keeping the species unnamed.

Occurrence. – The species occur in the lower to middle part of the Elnes Formation. Modum: road-cut at Vikersund skijump (Heggen Member, 12.2 m above the Huk Formation). Oslo-Asker: the Old Eternite Quarry just north of Slemmestad (basal Sjøstrand Member, 5.68 m above the Huk Formation) and in exposure at the old skijump at Slemmestad (uppermost Sjøstrand Member).

Trinucleid gen. et sp. indet.

Pl. 19, Figs 21–23; Text-Fig. 57

1975 *Botrioides*? sp. Hughes *et al.*, pp. 562–563, pl. 4, figs 48–51.
1984 *Botrioides*? sp. Wandås, pp. 234–235 (*partim*), pl. 12H, K–L.
1987 Trinucleid gen. et sp. indet. Owen, p. 100, pl. 14, fig. 3.

Material. – The material is restricted to one cephalon (PMO 87252) and a poorly preserved and strongly flattened cranidium (PMO 208642/2).

Description. – Cephalon semicircular in outline. Glabella pear-shaped to slightly ovoid, becoming increasingly more convex (tr.) anteriorly towards pseudofrontal lobe. Occipital ring and furrow, occiput and deep and narrow S1 and S2 furrows closely resemble those of *B. foveolatus*, especially in the fairly narrow and posteriorly directed abaxial ends of the furrows. S3 very shallow, located just behind maximal glabellar width. Distal part of L2 joined by weakly developed bridge to pseudofrontal lobe. Pseudofrontal lobe subspherical, distinctly wider than posterior part of glabella, clearly overhanging fringe. Axial furrow very broad and shallow posteriorly, strongly narrowing along pseudofrontal lobe, bearing deep but small fossulae slightly in front of widest part of glabella. Quadrant-shaped genal lobe moderately convex exsagitally as well as transversely; alae weak, located opposite slightly inflated occiput; eye tubercle prominent, situated outside anterior part of S2, connected to axial furrow by weak eye ridge. Genal prolongation marked.

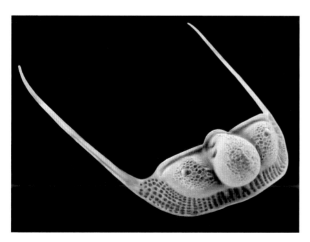

Fig. 57. Reconstruction of cephalon for trinucleid gen. et sp. indet. In oblique anterolateral view. Length of glabella approximately 3 mm.

Fringe narrow mesially, steep, widening posteriorly; outer border narrow. Arcs E_1, I_1, I_2 and I_n complete, the E_1 arc containing 22 and I_n 19 pits in a half arc. Arc E_2 present posterolaterally, melting together with E_1 at anterolateral corners. Arcs I_3 and I_4 present posterolaterally, the former dying out against I_2, while the latter fuse with I_n. The arcs break slightly up at posterior border where a few additional pits may appear, but no sixth I arc described by Owen (1987, p. 100) is observed. Pits situated in deep and rather narrow sulci frontally, reaching all the way across the fringe; sulci gradually breaking up posterolaterally, where slight concentric ridges form between I_1 and I_2, I_2 and I_3 and to some degree between I_3 and I_4.

Test surface covered by coarse reticulation centrally on glabella and genae, becoming finer and then dissappearing along anterior, posterior and lateral flanks.

Lower lamella of fringe, hypostome, thorax and pygidium unknown.

Remarks. – As noted by Hughes *et al.* (1975) and Owen (1987) this species has many similarities with the genus *Botrioides* in the glabellar and genal outline. On the other hand, it is clearly differentiated on the posterolaterally strongly widening fringe lacking major sulci; the break down of the regular radii posterolaterally; the distinct posterior prolongation of the genae; the concentric ridges between the pit arcs posteriorly on the fringe; the wide and shallow posterior part of the axial furrow with weak anterolaterally directed alae and by the restriction of E_2 to the posterior part of the fringe.

Several of the above-mentioned differences are on a generic level, clearly indicating that it should not be grouped together with *Botrioides* but rather with genera such as *Tretaspis* as earlier suggested by Hughes *et al.* (1975). Owen (1987) suggested an affinity with *Reedolithus* on strength of some similarities in the cranidial outline. This, though, do not seem likely due to the lack of an occipital spine; the presence of I arc sulci; the short anterior fringe; the moderately strong glabellar inflation and the forward location of the lateral eye tubercles, all of which favor an affinity with the subfamily Trinucleinae Hawle & Corda, 1847 (cf. Reedolithinae Hughes *et al.*, 1975). At the moment the taxon does not seem to fit into any of the established genera.

Occurrence. – Lower part of the Heggen Member. Eiker-Sandsvær: profile 6 at Vego near Fiskum (lower middle Heggen Member, 1.25 m below datum). Hadeland: Hovodden at Randsfjorden (lower Heggen Member, ~16 m above the Huk Formation).

Family Raphiophoridae Angelin, 1854

Genus *Ampyx* Dalman, 1827

Type species. – Ampyx nasutus Dalman, 1827.

Diagnosis. – Cranidium bearing long and slender frontal spine with cylindrical cross-section; glabella with three pairs of muscle scars on posterior flanks, a fourth pair situated distally on the occipital furrow; bacculae usually well developed; facial sutures more or less sigmoid. Librigenae continuing out into long and slender posteriorly directed spines with cylindrical cross-section. Thorax consisting of six segments; widening forwards to between segment two and three at which point the lateral thoracic margin turns sharply adaxially; posterior thoracic segments with pleural furrows concave forwards. Pygidial axis with two paired rows of muscle scars; pleural fields generally bearing a single pair of forwardly concave pleural furrows (from Fortey 1975; Moore 1959 and Whittington 1950).

Ampyx nasutus Dalman, 1827

Pl. 20, Figs 1–5

> 1827 *Asaphus (Ampyx) nasutus* Dalman, pp. 253–254, 279, pl. 5, fig. 3a–c.
>
> 1852 & 1854 *Ampyx nasutus* Dalman, 1827 – Angelin, pp. 19–20, 81, pl. 17, fig. 1–1 c, pl. 40, figs 4–4a.
>
> 1882 *Ampyx nasutus* Dalm. – Brøgger, p. 58, pl. 5, fig. 15a, b.
>
> 1894 *Ampyx nasutus* Dalm. – Schmidt, pp. 77–80, pl. 6, figs 1–10.
>
> 1950 *Ampyx nasutus* Dalman, 1827 – Whittington, pp. 554–556, text-fig. 6A–B, pl. 74, figs 3–9.
>
> 1973 *Ampyx nasutus* Dalman, 1827 – Modliński, p. 57, pl. 5, fig. 1.
>
> 1973 *Lonchodomas volborthi* Schmidt, 1894 – Modliński, pl. 5, fig. 2.
>
> 1984 *Ampyx nasutus* Dalman, 1827 – Wandås, p. 235, pl. 12D–F, I.
>
> v 1995 *Ampyx nasutus* (Dalman, 1827) – Nielsen, pp. 341–345 (*cum. syn.*), figs 245A–B, 246A–L, 247A–B.

Type stratum and type locality. – Neotype RS Ar. 13394 selected by Whittington (1950) from the collections at the Swedish State Museum. The neotype comes from the lower Darriwilian Komstad Limestone at Västanå (referred to as Husbyfjöl by Dalman, 1827), Östergötland, Sweden.

Norwegian material. – The species is fairly common in the Huk Formation underlying the Elnes Formation, but from the Elnes Formation only the few specimens listed below are known. Three partly articulated specimens (PMO 67587, 68304, 104.057), 14 cranidia (PMO 33326, 33463, 67187 (three specimens), 67575, 68447, 90388, 102.647, 102.659, 106.055, 160.911, 202.056A/1, 202.066A/2) and 22 pygidia (PMO 33317, 33318, 33323, 33326, 33472, 67157, 67180, 67575, 67576, 90535, 90553, 102.410, 102.424, 102.621, 102.622, 102.659, 104.057, 106.548, 202.045A/1, 202.050B/14, 202.077C/6, 206.964).

Description. – Cephalon subtriangular in outline; the cranidium with distinctly inflated and relatively wide glabella occupying between one-third and two-fifths of the cranidial width; glabella clearly overhanging anterior margin; bacculae indistinct to weakly developed; posterior border short, only widening (exsag.) slightly laterally; facial suture forming distinct angle with cephalic margin anterolaterally, in dorsal view continuing posterolaterally in a nearly straight to slightly forward concave line to just inside genal spine at posterolateral corner.

Genal spines about twice as long as cranidium, in cross-section rounded with distinct groove on upper side. Librigenal field approximately four times as wide as border.

Pygidium rounded triangular, width approximately twice the length; axis broad, occupying about 30% of the pygidial width; Anterior pleural furrow concave forwards, fairly deep and narrow V-shaped laterally; greatest border width corresponding to one-third the pygidial length, narrowing to about half of that at posterior end; transition between border and pleural field with no more than an indistinct rim. Terrace-lines on border curving downwards posteriorly.

Remarks. – Nielsen (1995, p. 342) noted that Schmidt (1907) indicated a first appearance of *Ampyx nasutus* in the deposits below the *Asaphus (A.) expansus* trilobite zone or below the Kunda Regional Stage. However, Schmidt (1894, p. 80; 1907, p. 64) placed the first appearance within the *Asaphus (A.) expansus* zone, corresponding to his B_{2b} zone. *Ampyx nasutus* thereby seems to be the diagnostic for the Kunda Regional Stage.

Occurrence. – *Ampyx nasutus* occurs in the East Baltic area and south-eastern Scandinavia from the base of the *Asaphus expansus* Zone (basal Kunda Regional Stage) and up into the *Asaphus raniceps* Zone (upper Kunda Regional Stage) (Schmidt 1907; Nielsen 1995). In southern Norway it appears at the base of the *Asaphus expansus* Zone and continues at least up into the basal part of the Heggen Member at Modum and top Helskjer Member at Helskjer, Mjøsa (Wandås 1984). Wandås (1984) indicates that it can be found as high up as 19.75 m above the Huk Formation at Helskjer and 11.85 m at Modum, but examination of his collection reveals these to be *Ampyx mammilatus*. The highest verified occurrence is 9.33 m above the Huk Formation at Vikersund, Modum, corresponding to the lower part of the Heggen Meber (Wandås 1984, fig. 2). Modum: Vikersund Skijump. Oslo-Asker: The Old Quarry near Slemmestad (Helskjer

Member), Frøen in Oslo. Hadeland: Hovstangen and Røisumgård at Gran and Hovodden (Helskjer Member, 6.0 m above the Huk Fm.). Mjøsa: Furnes Church (top Helskjer Member, 7.75 m above the Huk Fm.) and along E6 between Nydal and Furnes.

Ampyx mammilatus Sars, 1835

Pl. 20, Figs 6–12; Table 34

1835   *Ampyx mammillatus* spec. nov.? Sars, pp. 335–336 (*partim*), pl. 8, figs 4 c–d.

1854   *Ampyx mammillatus.* Sars. – Angelin, p. 80, pl. 40, fig. 3–3 c.

1857   *Ampyx mamillatus* Sars – Kjerulf, p. 285.

1878   *Ampyx mammillatus.* Sars. – Angelin, p. 80, pl. 40, figs. 3–3 c.

1909   *Ampyx mammillatus,* Sars – Holtedahl, p. 28.

1913   *Ampyx clavifrons* n. sp. – Hadding, pp. 73–74, pl. 7, figs 15a–b, 16.

1940   *Ampyx mammilatus* Sars – Størmer, pp. 130–132, text-fig. 3, fig. 4 c–d, pl. 2, figs 5–12.

1963   *Ampyx mammilatus* – Skjeseth, p. 64.

1984   *Ampyx nasutus* Dalman, 1827 – Wandås, p. 235 (*partim*), (only occurence).

2000a   *Ampyx clavifrons* Hadding, 1913 – Månsson, p. 322, fig. 2, fig. 6a–k.

2000b   *Ampyx clavifrons* Hadding, 1913 – Månsson, p. 14, fig. 9e–g.

2002   *Ampyx clavifrons* Hadding, 1913 – Pålsson et al., p. 42, figs 9a–b, 10.

Norwegian material. – Three partly articulated specimens (PMO 33339, 33572, 206.964), 56 cranidia (PMO S.1811, 3691a, 3968, 11687, 11688, 33390, 33391, 33578 (three specimens), 36515, 36640 (two specimens), 36741, 36833, 36835, 36848, 56389, 56390, 56390b, 56396, 56401 (two specimens), 56403 (seven specimens), 61749, 67181 (two specimens), 67589, 72706a, 72707, 74540, 82404, 82579, 82640, 82866, 82867, 82912, 83002, 83003, 83004, 91120, 141.903, 203.267/3, 203.475b/2, 205.025b/3, 207.419/4, 207.419/5, 207.440, 207.459/2, 207.461/1) and 44 pygidia (PMO H0541 (two specimens), 3691, 3691a, 33119, 33341, 33511, 33530, 36661, 36778, 36779, 36851, 56389a, 56389b (two specimens), 56392, 56392a (two specimens), 56393, 56395, 56396, 56401, 67191, 67281, 72100 (two specimens), 74539, 82405, 82866, 83005, 105.964, 202.045A/1, 203.318/2, 203.378 c/56, 203.543/4, 205.047, 205.053, 205.063, 205.092, 205.095 c/8, 205.103/3, 205.231/1, 207.439/1, 207.443/17).

Table 34. Pygidial measurements on *Ampyx mammilatus* Sars, 1835.

PMO number	L	W	Wa	Wb1	Wb2
56396	~7.9	17.7	4.3	~0.7	~0.8
56393	8.2	15.6	3.8	0.8	~0.7
67281	6.3	12.3	3	~0.7	~0.5
36851	5.8	13.5	3.1	0.9	0.7
33341	4.4	~10.4	2.5		0.5

L, length of pygidium; W, width of pygidium; Wa, width of axis. Wb1, border width at widest point; Wb2, border width at posterior end.

Diagnosis. – Cranidium distinctly trapezoidal in outline with almost straight anterior and posterior margins; width of anterior margin corresponding to one-third the cranidial width; glabella not overhanging frontal border, somewhat inflated in front, fairly broad and low posteriorly; cylindrical frontal glabellar spine directed forwards and upwards at an angle of approximately 40°; bacculae well developed; posterior border and furrow wide.

Pygidium triangular; width of axis generally only about 20% of pygidial width but may vary a little; anterior pleural furrow forwardly concave, other furrows effaced or absent; pygidial border nearly vertical, narrow, bearing around five to eight terrace-lines parallel to margin (emended from Månsson 2000a).

Remarks. – The specimens found in the contemporaneous Andersö Shale formation of central Sweden (Månsson 2000a,b; Pålsson et al. 2002) were assigned to *Ampyx clavifrons* Hadding, 1913. A comparison show a high degree of similarity between the two taxa and the only differences are found on the cranidia, where the specimens are characterized by a generally slightly narrower posterior furrow and a more deeply impressed posterolateral pit. The differences are insufficient for specific separation and consequently *A. clavifrons* is regarded as a junior synonym of *A. mammilatus*. Moreover, intermediates are present in the Mjøsa districts indicating that the differences should be seen as typical ecological/geographical variations rather than subspecies variations.

Størmer (1940, p. 131) suggested that the species may belong to *Raphiophorus* Angelin, 1854, rather than *Ampyx*, but lacked specimens showing the number of thorax segments. The recent investigation of the collections housed in the Natural History Museum of Oslo has uncovered a partly articulated specimen (PMO 33572; Pl. 20, fig. 6) including a pygidium and six thoracic segments. This together with a distinctly non ovate glabella and a comparatively long pygidium with a short vertical border clearly separates *A. mammilatus* from the genus *Raphiophorus*.

Both Nielsen (1995, p. 352) and Månsson (2000a, p. 326) assigned the species to the genus *Cnemidopyge* Whittard, 1955 based on its strong resemblance to the contemporaneous *C. costata* (Angelin, 1854) from Baltica. It differs in its strongly trapezoidal cranidium with a nearly straight anterior margin; the straight pleural furrow on the anterior thoracic segment; the pygidial border, which is clearly distinguishable from the pleural field, and by the lack of distinct pleural ribs on pygidium. The characters are more in agreement with *Ampyx* than with *Cnemidopyge*.

Occurrence. – Middle to upper Elnes Formation. Skien-Langesund: Rognstranda. Oslo-Asker: Håkavik, Djuptrekkodden at Slemmestad (upper half of the Engervik Member), Flakstadelven, Paradisbukta and Hukodden on Bygdøy, Hjortnæstangen at Oslo. Hadeland: Hofstangen, Haugslandet, south of Skiaker and road at Grinaker, Gran. Mjøsa: Flagstadelva north of Hamar, beach at Hovinsholm on Helgøya, Furnes Church and along E6 between Nydal and Funes.

Outside Norway it has been recorded from Jämtland in central Sweden (= listed as *A. clavifrons*; Månsson 2000a,b; Pålsson *et al.* 2004).

Ampyx sp. A

Pl. 20, Figs 15, 17

Material. – A single librigena (PMO 208.542b/2) and two small pygidia (PMO 208.542b/5, 208.552b/2).

Remarks. – The librigena is very poorly preserved, but appears narrow, posteriorly continuing out into a prismatic genal spine. The pygidium closely resembles that of *Ampyx nasutus sensu stricto*, but differs in the inner margin of the posterior border describing a slightly more forwardly convex curve between the axis and anterolateral corners. In addition the terrace-lines on the posterior border are convergent posteriorly rather than turning downwards as in the type form. The specimens were assigned to the genus *Ampyx* on the basis of the pygidial morphology, but larger specimens may indicate that it belongs to *Lonchodomas*.

Occurrence. – The specimens were collected from the upper part of the Heggen Member, 9.13 to 19.63 m below datum in profile 1 at Råen near Fiskum, Eiker-Sandsvær.

Genus *Cnemidopyge* Whittard, 1955

Type species. – *Trinucleus nudus* Murchison, 1839.

Diagnosis. – Cephalon sub-semicircular with broadly rounded anterior cranidial border; glabella pyriform, ending in long and cylindrical or triangular frontal spine, bacculae usually distinct; facial suture marginal on anterior cephalic border, forming sigmoid curve on posterolateral part of genae.

Thorax nearly rectangular, composed of six segments, the first macropleural.

Pygidium slightly smaller to larger than cephalon, fairly long, the length corresponding to around 30 to 50% of width; axis narrow, occupying approximately one-fifth of pygidial width; pleural field bearing around seven to 10 more or less distinct pleural ribs; border narrow, transition to pleural field generally indistinct (emended from Moore 1959).

Cnemidopyge costata (Angelin, 1854)

Pl. 21, Figs 1–7

1835	*Ampyx mammillatus* spec. nov.? Sars, pp. 335–336 (*partim*), pl. 8, fig. 4a–b.
1854	*Ampyx costatus.* Boeck. – Angelin, p. 80, pl. 40, fig. 1–1e.
1857	*Ampyx costatus* Boeck – Kjerulf, p. 286.
1878	*Ampyx costatus.* Boeck. – Angelin, p. 80, pl. 40, fig. 1–1e.
1890	*Ampyx costatus*, S. & B. – Brøgger, p. 175.
1909	*Ampyx* n. sp., aff. *A. costatus*, Boeck – Holtedahl, p. 28.
1940	*Ampyx costatus* Boeck MS. – Størmer, p. 132, text-figs 3, 4a–b, pl. 2, figs 13–18.
1964	*Ampyx (Cnemidopyge) costatus* (Boeck) – Jaanusson, table 3.
1995	*Cnemidopyge costatus costatus* – Nielsen, p. 352.
non 1995	*Cnemidopyge costatus* n. subsp. A – Nielsen, pp. 350–352, fig. 253a–i (possibly = *Cnemidopyge* sp. A).
1997	*Cnemidopyge costata* (Boeck *in* Angelin 1854) – Krueger, pp. 147–156, table 1, pls 1–3.
2000a	*Cnemidopyge costata* (Angelin, 1854) – Månsson, pp. 324–326, fig. 8a–e.
2002	*Cnemidopyge costata* (Angelin, 1854) – Pålsson *et al.*, p. 42; figs 9 c, 10.

Norwegian material. – The Norwegian material mostly derives from the Sandbian Vollen Formation, but includes 14 cranidia (PMO 67589, 82361, 82845, 202.950b/4, 202.952/6, 202.959a/1, 202.965b/2, 203.288a/7, 203.288a/16, 203.316/2, 205.081b/2, 205.095 c/5, 205.173a/5, 207.410/1), 1 librigena (PMO 203.363), 1 thorax (PMO 202.959), 12 pygidia (PMO 33424, 33511, 72706 (two specimens), 72707, 82525, 105.964, 202.956/4, 202.959/2, 203.288a/17, 205.072, 206.065a/4) and one articulated specimen (PMO 36868) from the Elnes Formation.

Diagnosis. – Cephalon nearly semicircular in outline; glabella pyriform, width corresponding to or slightly more than one-third the cranidial width; occipital ring fairly wide and low; bacculae well developed; three pairs of distinct muscle scars moderately deeply impressed; glabellar front overhanging preglabellar border, continuing out into stout frontal spine approximately twice as long as cephalon, upwardly curved with oval cross-section and indistinct furrow on each proximal side; fixigena wide anteriorly. Librigena narrow with relatively broad border and continuing out into long cylindrical genal spine, its length corresponding to about three times the length of cephalon, thorax and pygidium.

Pygidium rounded triangular in outline; axis comparatively wide, occupying around 25 to 30% of the pygidial width; axis bearing about 15 axial rings;

pleural field with about 10 moderately deep to shallow pleural furrows; transition between pleural field and border not sharp, moderately broad border mainly defined by an increased inclination and closely spaced terrace-lines parallel to margin (modified from Størmer 1940; Krueger 1997 and Månsson 2000).

Remarks. – Nielsen (1995, pp. 350–352) described what he termed *C. costata* n. subsp. A from lower Darriwilian rocks from southern Scandinavia. It differs from the type form of *C. costata* in its short glabella, which does not overhang the anterior border; its smaller bacculae; its distinctly triangular frontal spine; its posterior cranidial furrow which has its widest point closer to the glabella; a relatively longer pygidium and by the weaker pleural ribs. The large differences indicate that it should be regarded as a separate species as suggested by Månsson (2000a, p. 326).

The material described from glacial deposits of northern Germany (Krueger 1997) differs slightly from the typical *C. costata* in the less wide (tr.) and longer (sag.) anterior cranidial border; the relatively wider pygidial axis corresponding to just over one quarter the pygidial width and by the more pronounced pleural ribs on the pygidial field. These differences are not considered sufficient enough to warrant splitting *C. costata* into subspecies and may simply be explained by ecological factors.

Occurrence. – In Norway *C. costata* is known from the lower Heggen and middle Sjøstrand members and up into the Vollen Formation. The following list is for the known occurrences within the Elnes Formation. Oslo-Asker: Djuptrekkodden and Elnestangen at Slemmestad (Sjøstrand to Håkavik Member, 2.19 to 51.92 m above datum in the Djuptrekkodden profile). Hadeland: Haugslandet and road at Grinaker, Gran (Elnes Fm.). Mjøsa: Hovinsholm on Helgøya (Elnes Fm.), Furnes Church (Helskjer Member), road-cut at Nydal (Heggen Member, 0.51 m above datum in sub-profile 2).

Outside Norway it is known from most of central to southwestern Baltoscandia. Sweden: Jämtland (lower part of Andersö Shale and Kogsta Siltstone) and Västergötland (Dalby Formation) (Jaanusson 1964; Månsson 2000a). Estonia and Germany (erratic builders) (Krueger 1997).

Cnemidopyge sp. A

Pl. 21, Fig 8

1995 *Cnemidopyge costatus* n. subsp. A Nielsen, pp. 350–352, fig. 253a–i.

Material. – One fragmentary cranidium (PMO 202.056C/6) from the Old Eternite Quarry at Slemmestad.

Remarks. – The cranidium, which is very poorly preserved, differs from *C. costata* in its short glabella, which does not overhang the preglabellar area; the more rounded glabellar front and by the possibly narrower distal part of the posterior fixigenal furrow. The general glabellar outline is closer to the slightly older *C. costata* subsp. A Nielsen, 1995 (see discussion above) but the present specimen differs on the distinctly less prominent muscle scars; the lack of a genal keel and the posterior cranidial furrow, which has its widest point situated more distally and furthermore appears narrower (exsag.).

Occurrence. – Collected 1.15 m above the base of the Helskjer Member at the Old Eternite Quarry just north of Slemmestad, Oslo-Asker district.

Genus *Lonchodomas* Angelin, 1854

Type species. – *Ampyx rostratus* Sars, 1835, by subsequent designation of Bassler (1915, p. 41).

Diagnosis. – Glabella carinate; frontal spine long, forwardly directed, subquadrate in cross-section; genal spines subquadrate in cross-section; thorax with five segments; pygidium with distinctly downward inclined posterior border, border width intermediate to narrow (modified from Moore 1959).

Remarks. – The genus was established in 1854 by Angelin, who assigned five species to the genus. All of them – except for *Ampyx rostratus* Sars, 1835 – were new. *L. domatus* Angelin, 1854, was selected as type species by Vodges in 1893. This was seemingly overlooked by Bassler (1915), who designed *L. rostratus* (Angelin, 1854) as the type species. Strictly, it is the first designator, which has precedence, but in this case the type material of *L. domatus* seems to be missing. There is also some doubt concerning the validity of the species as the type material may have consisted of more than one species (Balashova 1966; Nielsen 1995). In addition and since the time of Angelin no specimens belonging to this species have been found, even though extensive collecting from the type localities has been carried out (Nielsen 1995, p. 340). There are therefore weighty arguments for an acceptance of *L. rostratus* and not *L. domatus* as the type species. Whittington (1950) listed *L. rostratus* as type species, referring to Raymond (1925). The same was done by Moore (1959), although he referred to Bassler (1915). This has in practice resulted in *L. rostratus* being regarded as type species.

In his redescription of *L. rostratus* Whittington (1950) mentioned that among other things it differs from the genus *Ampyx* by a lack of the fourth glabellar muscle scar. However, all the muscle scars found on species belonging to *Ampyx* are actually also present on *L. rostratus*, and the composite constitution of F1 and F3 may actually appear slightly more obvious on the latter.

Lonchodomas rostratus (Sars, 1835)

Pl. 21, Figs 9–18; Text-Fig. 58; Table 35

1835	*Ampyx rostratus* nov. spec. Sars, pp. 334–335, pl. 8, fig. 3a–e.
1857	*Ampyx (Lonchodomas) rostratus* Sars – Kjerulf, p. 285.
1890	*Lonchodomas rostratus*, Sars – Brøgger, pp. 175–176.
? 1894	*Ampyx rostratus* Sars – Schmidt, pp. 85–86, pl. 6, figs 29–33.
1940	*Lonchodomas rostratus* (Sars) – Størmer, pp. 128–130, text-fig. 3, pl. 2, 1–4.
1950	*Lonchodomas rostratus* (Sars), 1835 – Whittington, pp. 556–557, pl. 74, figs 11–15.
1995	*Lonchodomas rostratus* (Sars, 1835) – Månsson, p. 103, fig. 5A–B.
2000a	*Lonchodomas striolatus* sp. nov. Månsson, pp. 325–326, fig. 9a–h.
2000a	*Lonchodomas* sp. Månsson, pp. 326–327, fig. 8f–i.
2000b	*Lonchodomas striolatus* Månsson, 2000 – Månsson, p. 14, fig. 9i–j.
2000b	*Lonchodomas* sp. – Månsson, p. 14.
2002	*Lonchodomas striolatus* Månsson, 2000b – Pålsson *et al.*, p. 42; fig. 9g–h, 10.
2002	*Lonchodomas* sp. – Pålsson *et al.*, p. 42; figs 9g–h, 10.

Type stratum and type locality. – The lectotype PMO 56407, selected from the type material by Størmer (1940), is from the Vollen Formation exposed at Huk on Bygdøy, Oslo-Asker district.

Table 35. Pygidial measurements on *Lonchodomas rostratus* (Sars, 1835)

PMO number	L	W	Wa	Wb1	Wb2
56401	5.3		3.4	1.4	1.3
33511	4.2		2.8		1
82476	1.9		1.4		
56407a	~4.5	11.3	3.5	1.1	1.1
56407b/1	~4.9	~12.5	4		
56407b/2	3.3	~8.9	3		
33589	3.8	9.3	2.8	0.9	0.8
33588	3.5	~8.8	2.3	0.7	0.7
82542	4	~9.4	2.6	1.2	0.9
82535	4	9.8	2.6	~1.2	~1.0
82852/1	2.1	5.8	1.5	~0.6	~0.6
82852/2	1.8	~5.1	1.3	~0.5	~0.5
3985	6.8	14.5	3.8	1.8	1.3
67276	5.1	~11.5	2.5	~1.1	1
33581	5.2	11.1	2.6	1.2	1.1
33530	3.2	7.6	2	0.9	~0.7
56406a	4.7	11	3.2	1.5	1.3
56406c/1	4.5	9.7	3	1.6	1.2
56406c/2	3.1	8.3	2	1	0.7
4072	4.5	~10.7	2.9	1.4	1.2
3983	4.8	10.4	3	1.5	1.3

Abbreviations: see Table 34.

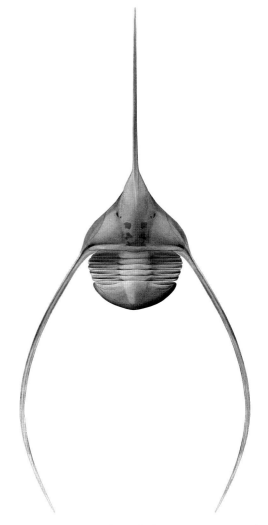

Fig. 58. Reconstruction of *Lonchodomas rostratus*. Length of specimen excluding spines approximately 2 cm.

Norwegian material. – The Norwegian material from the Elnes Formation and lowermost Vollen Formation consists of 1 cephalon (PMO 205.038), 75 cranidia (PMO H0540, H0541 (three specimens), S.1773, S.1876, 1999, 3984, 3993, 4072, 4073, 4084, 4755 (three specimens), 33334, 33512, 33523, 33527, 33532, 56267, 67275, 72075 (three specimens), 82052, 82054, 82472, 82636, 82640 (three specimens), 82644, 82866 (two specimens), 82877, 82878, 83008, 203.269b/5, 203.290b/3, 203.378a/41, 203.390/2–4, 203.394/2, 203.402a/1, 203.405/4, 203.406/3, 203.420, 203.426b/5, 203.426b/7, 203.433b/2, 203.462a/1, 203,480b/5, 205.025a/1, 205.035a/1, 205.043b/2–3, 205.046/1, 205.046/3, 205.081/15, 205.086/1, 205.087, 205.093a/1, 205.118a/5, 205.118/12, 205.177/3, 205.182b/5, 205.208/4, 206.065/5, 206.066c/9, 207.479a/6, 207.479/17, 207.481/4, 207.478/16), five librigenae (PMO 203.411/2, 203.462a/2, 203,480c/7, 205.140b/10, 206.068b/4), 41 pygidia (PMO H393, H490, 3983, 3985, 3988, 3993, 4072, 33511, 33581 (two specimens), 33587, 33588 (two specimens), 33589, 56267 (two specimens), 56406, 56406a (two specimens), 56406b, 56406c (three specimens), 67275, 67276, 82397, 82476, 82533, 82535, 82542, 82852, 83233, 203.123/9, 203.332/1, 203.390/1, 203.391/1, 203.488/2, 205.043/1, 205.045, 207.449/4, 207.479a/1) and six partly articulated specimens (PMO 59057, 82852, 205.044/1, 205.101/7, 205.205, 207.479/11).

Diagnosis. – Cranidium triangular in outline, the width generally greater than the exsagittal length (excluding frontal spine). Glabella diamond-shaped, distinctly carinate, extending for 20 to slightly more than 40% of its length in front of fixigena; maximal width corresponding to between 30 and 45% of cranidial width; posterior part of glabella low; anterior part in dorsal view with nearly straight flanks, continuing out into long frontal spine with strongly quadrate cross-section; spine nearly twice as long as rest of cranidium. Occipital ring flat, not separated by axial furrows from posterior fixigenal border, arched very gently rearwards. Anterior half of glabella bearing exsagittaly directed terrace-lines, which may continue back to posterior part. In a few cases the terrace-lines may turn out onto fixigena. Posterior border narrow, not widening abaxially. Facial sutures generally strongly sigmoid. Librigenal field moderately wide; librigena extended into long slender genal spine with quadrate cross-section; spine close to three times as long as cranidium (excluding frontal spine), reaching approximately 1.5 times the length of the animal (excluding the spines) behind the pygidium. Pygidium rounded subtriangular in outline; length corresponding to around 35 to 45% of the width; axis occupying between 25 and 35% of the pygidial width; pleural field with fairly deep and distinctly forward concave pleural furrow 2; posterior border strongly inclined, wide, sagittal width between 20 and 30% of pygidial length; terrace-lines on border moderately fine, semi-parallel to margin (modified from Månsson 2000a; Whittington 1950).

Remarks. – According to Månsson (2000a, p. 326) *Lonchodomas striolatus* Månsson, 2000 from the Andersö Shale formation of central Sweden differs from *L. rostratus* by its shorter glabella not reaching more than up to 40% of its length out in front of fixigena; in having concave glabellar flanks overhanging anterior part of fixed cheeks; by a more rounded and bulbous glabella and by having evenly convex outwards facial sutures. The present revision of *Lonchodomas* from the Elnes Formation shows *L. rostratus* to be an extremely variable species with regard to relative length of glabella; general outline of facial sutures and to some extent the degree of striation found on the cranidium. When comparing *L. striolatus* with *L. rostratus* only the slightly less sigmoid facial sutures and generally finer and more extensive cranidial striation differentiates the two, although most of the possible difference in striation may relate to the generally poor preservation of *L. rostratus* from the Elnes Formation. The identical stratigraphical occurrence in combination with few and small differences, especially when compared to the large variations seen within *L. rostratus*, clearly indicates that *L. striolatus* should be regarded as a junior synonym of *L. rostratus*. The same is the case for *Lonchodomas* sp. *sensu* Månsson (2000a).

The Estonian material assigned to *L. rostratus* by Schmidt (1894) may differ from the Norwegian material in having a slightly wider frontal glabellar spine; a generally much stronger glabellar keel and less inflated glabellar flanks; a slightly smaller pleural area on the pygidium and by a less steep pygidial border. Whether this is a local variation or perhaps a species related difference cannot be determined on the basis of the descriptions and photographs of the Russian-type material alone, and for the moment they are regarded as belonging to the same species.

Only a few specimens containing a complete frontal spine are found in the collections at the Natural History Museum of Oslo, but all of them show the same relative length of the spine compared to the rest of the cranidium, suggesting the spinal length was close to twice the cranidial length for all specimens belonging to this species. Specimens with librigenae still attached to each other are more common, espcially from the basal part of the Fossum Formation overlying the Elnes Formation in the southern Skien-Langesund district. Although most of the genal spines are broken, their uniform outline, and the width and length of the more complete specimens strongly supports the impression of a constant ratio between spinal length and cranidial size found for the frontal spines. This indicates that sexual dimorphism, if present, must be very small within *Lonchodomas rostratus*, suggesting that contrary to the proposed interpretation of Knell & Fortey (2005) the spines of at least a species like *Lonchodomas rostratus* did not function as a primary sexual feature, although their use as a secondarily sexual character is likely. Additionally the outline and horizontal, semi-axial orientation of the spines exclude a defensive purpose as the main explanation for their use. Perhaps the best explanation at the moment would therefore be the snow-shoe hypothesis proposed by Fortey & Owens (1978, p. 238), possibly as a stabilizing factor when the animal filtered the surface mud for food particles, but maybe also for raking up in the surface sediments like the walrus and narwhale today.

Occurrence. – Top Sjøstrand to Håkavik and upper Heggen Member of the Elnes Formation and at least the lower part of the Vollen Formation. Eiker-Sandsvær: Krekling and Vegutjern at Krekling. Oslo-Asker: Djuptrekkodden (22.70 to at least 52.48 m above base of profile), Elnestangen (Engervik Member), Engervik (Sjøstrand Member), Håkavik, Paradisbukta, Hukodden and Odden at Bekkebukta on Bygdøy (Engervik Member), Hjortnæs and central Oslo. Hadeland: Hofstangen and Haugslandet at Gran. Mjøsa: Håvesveen on Nes or between Hamar and Brumunddal,

Steinbergbakken in Ringsaker and along E6 between Nydal and Furnes.

Outside Norway it has been collected from the contemporaneous Andersö Shale formation and the Örån Shale of central Sweden (Månsson 2000a,b; Pålsson *et al.* 2002) and in the Estonian stages C_1 (upper Darriwilian to lower Sandbian) and C_2 (lower Sandbian) of Schmidt (1894).

Lonchodomas cuspicaudus n. sp.

Pl. 20, Figs 13–14, 16; Pl. 21, Figs 19–20; Pl. 28, Figs 1–7; Table 36

Derivation of name. – Referring to the pointed posterior part of the pygidium.

Type stratum and type locality. – Uppermost part of the Elnes Formation (or lower Vollen Formation) on Kojatangen between Vollen and Blakstad, Oslo-Asker.

Type material. – PMO 4044, a nearly complete, articulated specimen, is selected as holotype. Paratypes are PMO 81836A (cranidium), 40367 (cranidium), 206.271/8 (cranidium), 206.960 (cranidium), 3969 (pygidium), 40395 (pygidium), 97176 (pygidium) and 206.271/1 (pygidium).

Material. – 22 cranidia (PMO 4062, 4079, 4108, 40357, 40367 (two specimens), 40382, 40395 (two specimens), 44225, 44226, 44229, 44268, 60966 (two specimens), 81836 (two specimens), 81845, 82000, 82012 + 13/1, 206.271/8, 206.960), 27 pygidia (PMO 3709, 3710, 3950, 3951, 3969, 4061, 4068, 4107 (two specimens), 40375, 40376, 40377, 40384, 40395 (three specimens), 44227, 44228, 44267, 81999, 82000, 82002, 82004, 82053, 97176, 206.144/3, 206.271/1) and a single articulated specimen (PMO 4044).

Diagnosis. – Species of *Lonchodomas* defined by its rather compact, trapezoidal cranidium with its stout glabella only just overhanging anterior cranidial border; the exsagittal directed terrace-lines on the anterolateral flanks of the glabella and by the narrow and well-defined pygidial axis terminating just inside the distinctly pointed border.

Description. – Largest cranidium, PMO 82012, is more than 21 mm long (frontal spine excluded), while largest measured pygidium, PMO 44228, is 17.4 mm long.

Cranidium subtrapezoidal in outline and stoutly built for a *Lonchodomas*; glabella distinctly rhombic in outline, relatively short and wide, widening forwards markedly and only slightly overhanging anterior border, the length (frontal spine excluded)

Table 36. Pygidial measurements on *Lonchodomas cuspicaudus* n. sp.

PMO number	L	W	Wa	Wb1	Wb2
4044	~10.7	15.4		2.3	1.4
4068	13.3	23.1	7	3.3	~1.7
40395	16.2	23	7.5	~2.8	2
H487	5	~11.2	2.7	1.4	~1.0

Abbreviations: see Table 34.

corresponds to approximately 60% of the cranidial width, while the width corresponds to about one-third the cranidial width; the narrowest part of glabella, located at preoccipital lobes, corresponds to nearly 60% of the maximal glabellar width. Glabellar surface nearly smooth except for a slight swelling at the preoccipital lobes and an elongated impression just inside axial furrow and behind widest glabellar point; a distinct sagittal keel extends from preoccipital lobes all the way to the frontal spine, giving the glabella a rounded subtriangular frontal outline; posterior two principal cephalic axis muscle scars large, partly merged into one, the smaller posterior one, F1, V-shaped, situated on preoccipital lobes; F2 rounded, anterior margin located halfway between anterior and posterior cranidial border; F3 and F4 are small, situated in anterior part of elongated furrow just behind widest part of glabella. Frontal spine nearly horizontal, prismatic with distinct furrows, exact length unknown but surpassing 125% of cranidial length. Fixigena weakly convex with indistinct ridge directed in a gentle rearwardly directed curve from somewhere behind widest part of glabella and out to the posterior outward bend in the facial suture; anterior fixigenal border covering between 50 and 60% of anterior cranidial margin, abaxially terminated at fixigenal shoulder at which the facial suture continues posteriorly in a slightly adaxially convex curve down to a short distance in front of moderately broad (exsag.) posterior border furrow where it bends slightly outward. Test surface finely pitted and with exsagittal terrace-lines on anterior half of glabella.

Librigena and hypostome unknown.

Thorax consisting of five thoracic segments.

Pygidium subtriangular with distinctly pointed rear, giving it a bow-shaped outline, in cross-section moderately to weakly convex in both directions, length about 55% of the width. Axis narrow, occupying 25% of anterior pygidial width, fairly well defined by moderately deep axial furrows, terminating slightly in front of posterior border; articulating half ring short; anterior border furrow deep, gently curved, abaxial part V-shaped in cross-section; pleural field smooth without pleural ribs or furrows; border narrow but strongly developed, almost vertical, posteriorly narrowing, bearing around 7 to 11 terrace-lines largely parallel to outer margin. Test surface smooth.

Remarks. – *Lonchodomas cuspicaudus* n. sp. is readily distinguished from most of the other *Lonchodomas* species found on Baltoscandia by the distinctly pointed pygidium and the stout and relatively short glabella. The cranidium resembles somewhat that of the slightly older *Lonchodomas volborthi* (Schmidt,

1894) from Baltoscandia, but may be differentiated on the anterolateral glabellar furrow; stronger median keel; the presence of anterolateral terrace-lines on glabella and by the less well-developed pre-occipital lobes. It is distinguished from the contemporaneous *L. rostratus* (Sars, 1835) by the much shorter and stouter glabella, the widest part of which lies well behind anterior border furrow and by the relatively longer pygidium with its much narrower axis and pointed end. It shows some resemblance to *L. coagmentatum* Tripp, 1980, from approximately contemporaneous to slightly younger deposits in Scotland, but is readily distinguished on the well-defined and narrower pygidial axis and by the less forwardly reaching glabella. It differs from the Middle Ordovician *L. carinatus* Cooper, 1953 from eastern USA, which was refigured by Whittington (1959), by the complete anterior border and border furrow; the straight genal spines and by the more pointed pygidium with its narrower and clearly delimited axis. It is distinguished from both *L. clavulus* Whittington, 1965a, and *L. normalis* (Billings, 1865) refigured by Whittington (1965a), both from Western Newfoundland, by the more pointed posterior end of the pygidium; the sharper anterior corner formed by the anterior termination of the facial suture and by the shorter occipital ring. It may be differentiated from the Middle Ordovician *L. eximius* Andreyeva, 1986, from the Mongolian Altai Region by the wider glabella and pygidial axis.

Occurrence. – Upper part of the Heggen and Håkavik members of the Elnes Formation and up into the Vollen Formation. Eiker-Sandsvær: Råen at Fiskum (upper Heggen Member, 8.40 m below top of profile 1 of this study) and Krekling. Oslo-Asker: Håkavik (Vollen Fm.), exposure at Djuptrekkodden (Håkavik Member, 58.06 m above base of profile), Holmen (Vollen Fm.), Vollen (Vollen Fm.), Ny Drammensvei at Gyssestad (Vollen Fm.), Gråkammen in Vester Aker (Vollen Fm.), Kojatangen at Vollen (Uppermost Elnes Fm. or lower Vollen Fm.) and Huk on Bygdøy (Vollen Fm.). Mjøsa: Nydal-Furnes at Hamar.

Superfamily Remopleuridioidea Hawle & Corda, 1847

Family Remopleurididae Hawle & Corda, 1847

Subfamily Remopleuridinae Hawle & Corda, 1847

Genus *Sculptella* Nikolaisen, 1983

Type species. – *Sculptella scripta* Nikolaisen, 1983.

Diagnosis. – A remopleuridid with a cranidial outline like that of *Remopleurides*, but defined by a median area with densely spaced sculptural lines; three pairs of lateral glabellar furrows discernable as smooth stripes interrupting the sculptural lines; glabellar tongue very gently convex, less than one quarter as wide as median area; preglabellar field distinct and extending across anterior part of glabella; preglabellar furrow weakly defined or absent. Librigena bearing very long and strong genal spine; librigena very narrow anteriorly and anterolaterally, the lateral margin almost continous with the lateral margin of spine. Hypostome gently convex, median area with shallow V-shaped furrow separating the oval areas. Pygidium with distinct postaxial ridge, first pair of pleural spines clearly extending beyond tips of smaller second pair. Test relatively thin.

Sculptella scripta Nikolaisen, 1983

Pl. 22, Figs 1–8; Table 37

1887 *Remopleurides* sp. Brøgger, p. 17.
1983 *Sculptella scripta* n. gen., n. sp. Nikolaisen, pp. 268–270, pl. 9, figs 9–18, text-fig. 4–5.
1983 *Sculptella* aff. *scripta* n. gen., n. sp. Nikolaisen, pp. 270–271, pl. 13, figs 14–15.

Material. – At least 27 cranidia (PMO 3725, 3726, 4022, 4647, 59065, 74666, 74676, 74677, 74678, 74683, 74684, 74844 (two specimens), 74853, 74855, 74856, 82284, 95065, 95774, 203.134/1, 203.145, 206.215/6, 206.218/16, 208.487/3, 208.542/4, 208.557/10, 208.597/9), two hypostomes (PMO 74685, 95791), at three fragmentary thoraic segments (PMO 3724, 88125, 161.744), two fragmentary pygidia (PMO 3724, 88126).

Diagnosis. – Median area nearly 75% as long as wide, with sculptural lines transverse to gently convex forwards. Glabellar tongue 20% as wide as median area. Occipital furrow more than 40% as wide as median area. Pygidium with both pairs of pleural spines small (emended from Nikolaisen 1983).

Remarks. – The glabellar tongue is generally slightly broader than indicated by Nikolaisen (1983) averaging around 20% of that shown by the median area.

The two fragmentary specimens found in the Engervik Member do not seem in any way to differ from the type form belonging to the overlying Vollen Formation, though better preserved material is needed to confirm this

Specimen PMO 74683, which Nikolaisen (1983) denoted *Sculptella* aff. *scripta*, only differs from the typical form in its shorter glabellar tongue. The width of the glabellar tongue does not differ from the normal variation and the smooth surface is a result

Table 37. Cranidial measurements on *Sculptella scripta* Nikolaisen, 1983

PMO number	b1	b	e	j	j1	rem1	rem3	rem4	k00	o	g2	f1	f2	f3
59065	5.9	4.8	1.1	6.3	5.6	4	1.2	1.1	2.3	3.2				
74666	6.7	5.3	1.4	7.2	6.5	4.7	1.4	1.2	2.8	3.7				
74678	3.2	2.5	0.7	4	3.4	2.1	0.7	0.7	1.3	1.5				
4647			1.1	7.5	6.6	4.6			2.8		1.2	0.8	1.7	2.9
74677	3.3	2.6	0.7	4.7	4	2.2	0.8	0.9	1.4					
3725	7.3	5.7	1.6	7.9	7.1	5.3		1.3	3.3	4.1	1.1	1.1	2.1	3.2
4022	8.2	6.5	1.7	8.3	7.6	6.1	2.2	1.5	3.6		1.4	1.1	2.4	
206.215				6.4	5.6		1.3	1						
206.218	4.7	4.1	0.6	4.7	4.2	3.1	1.2	0.9	1.7					
203.145	5.8	4.7	1.1		5.4	3.8	1.1	1.1						
74683	4.3	3.3	1	5.2	4.6	3.2	1.1	1	1.8	2.5		0.8		2.1

Biometric abbreviations follow those of Nikolaisen (1983); b1, length of glabella; b, preoccipital length of glabella.
e, length (sag.) of occipital ring; j, cranidial width; j1, width of median area; rem1, length of median area.
rem3, length or height of glabellar tongue; rem4, width of glabellar tongue; k00, distance between posterior terminations of palpebral rim.
o, width (tr.) of occipital ring; g2, distance between f2 glabellar furrows; f1, distance from occipital furrow to posterior termination of f1 furrow.
f2, distance from occipital furrow to posterior termination of f2 furrow; f3, distance from occipital furrow to posterior termination of f3 furrow.

Table 38. Cranidial measurements on *Sculptella angustilingua* Nikolaisen, 1983

PMO number	b1	b	e	j	j1	rem1	rem3	rem4	k00	o	g2	f1	f2	f3
205.035.3				5.5	4.8	3.1		0.6						
205.035.2				5.8	5			0.6						
74679			1	5.8	5.3	3.4			1.8		1.1	0.7	1.3	2
4246	3.6	3	0.6	3.9	3.5	2.4	1.1	0.5	1.2	1.7		0.5		
74689	5.7	4.8	0.9	7.2	6.4	4.1	1.5	0.85	2.2	2.7		1	1.8	
205.101	4.2	3.5	0.7	5.6	4.8	3.2		0.6	1.7				1.3	
74691	4.6	3.7	0.9	5.5	5.1	3.5		0.5	1.8	2.3		0.7	1.4	
74682	5	4.2	0.8	6.6	6.1	4		0.8	2.3	2.5				
20314	5.6	4.6	1	6.9	6.1	4.3	1.2	0.55	2.4	2.9	1.4	0.8	1.8	2.8
74692	4	3.2	0.8	4.7	4.2	2.9			1.7	2.2		0.6	1.3	
203.269	3.9	3.25	0.65	4.7	4.1	2.8		0.6	1.7	1.9		0.6	1.3	1.9
4100		5.8		6.8	6	4.7	1.9	0.8	2.45					
74688				5.9	5.3		1.3	0.5			1.1			
4209				6.5	5.7			0.7						
95486	7.5	6.3	1.2		8.3	5.8	2.2	1.1			2.3	1.1	2.5	3.5
3948	4.3	3.5	0.8		4.8	3.2		0.7						

Abbreviations: see Table 37.

of weathering as also suggested by Nikolaisen (1983). The difference in the tongue length is not considered large enough for any taxonomic separation.

Sculptella scripta is easily distinguished from *Sculptella angustilingua* Nikolaisen, 1983 from Norway by its much broader glabellar tongue, transversely directed sculptural lines on median area and by the broader and less pronounced postaxial ridge on the pygidium (Nikolaisen 1983). In fragmentary or worn cranidia lacking the glabellar tongue, *S. scripta* may be differentiated on the less pronounced posterolateral bend on the palpebral furrow and on a generally wider occipital furrow.

Occurrence. – Elnes and Vollen formations. Skien-Langesund: Saltboden Light and Frierfjorden (Vollen Formation). Eiker-Sandsvær: profile 1 (Heggen Member, 12.07 to 22.22 m below datum) and profile 4 at Råen (Heggen Member, 4.23 m below datum). Modum: In the vicinity of Fure (probably Elnes Formation). Oslo-Asker: Bygdøy (Vollen Fm.), Semsvannet (Vollen Fm.), Vollen (Vollen Fm.), beach at Djuptrekkodden (Engervik Member, 28.78 to 29.28 m above datum) and Bøveien at Bø (Engervik Member, 13.40 to 13.43 m above datum). Hadeland: Hofstangen at Gran (Heggen Member).

Sculptella angustilingua Nikolaisen, 1983

Pl. 22, Figs 9–15; Table 38

1953   *Remopleurides* sp. – Størmer, p. 84.
1983   *Sculptella angustilingua* n. gen., n. sp. Nikolaisen, pp. 272–274, pl. 10, figs 9–17, pl. 11, figs 1–4, text-fig. 7, table 1.

Material. – 31 cranidia (PMO 3948 (two specimens), 4100, 4209, 4210, 4221, 4225 (two specimens), 4236, 4246, 20314, 74679, 74682, 74688, 74689, 74690, 74691, 74692, 74832, 74833, 82365, 95486, 95489, 203.269/7, 205.035/2, 205.035/3, 205.101/1, 206.088, 206.271/3 (two specimens), 208.553/1), three librigena (PMO 4234, 74680, 74681), one hypostome (PMO 4319), two thoracic segments (PMO 0551a, 95675) and one pygidium (PMO 0551b). Five of these cranidia belongs to the Elnes Formation. Most specimens are fragmentary and have been deformed by compaction.

Diagnosis. – Median area two-thirds as long as wide with sculptural lines V-shaped mesially. Glabellar tongue only about one-seventh to one-tenth as wide as median area, less than half as wide as long. Occipital furrow slightly more than one-third as wide as median area. Hypostome with large but low triangular median boss. Pygidium with narrow, sharply defined postaxial ridge (emended from Nikolaisen 1983).

Remarks. – The specimens belonging to the Elnes Formation differ somewhat from the typical form by a slightly narrower lateral border and are furthermore all characterized by a relatively broad glabellar tongue. Although there seem to be some stratigraphically related differences the present material is too limited and the specimens collected from the Elnes Formation too badly preserved for further speculation.

Occurrence. – Elnes and Vollen Formation. Eiker-Sandsvær: profile 1 at Råen (upper Heggen Member, 8.43 to 10.41 m below datum). Oslo-Asker: beach at Djuptrekkodden near Slemmestad (mid-Engervik Member to Håkavik Member, 36.88 to 53.71 m above datum); Håkavik (basal Vollen Fm.), Southern Kojatangen in Asker and on Bygdøy, Oslo (Vollen Formation, probably from the lower part (Nikolaisen 1983)). Ringerike: Gomnes, Gullerud, Kullerud and Norderhov. All from sandy and bioclastic beds at the base of the Vollen Formation (Nikolaisen 1983). Mjøsa: Hovinsholm on Helgøya (Elnes Fm.).

Sculptella sp. A

Pl. 22, Fig. 16; Table 39

1983 *Sculptaspis erratica* n. gen., n. sp. Nikolaisen, p. 279.

Material. – One fragmentary and somewhat compressed cranidium, PMO 74860.

Description. – Cranidium rounded spade-shaped in outline, 83% as long as wide. Median area 70% as long as wide with greatest width well behind transverse mid-line. The three pairs of glabellar furrows are gently impressed, although this may relate to taphonomy. Their exact length is impossible to discern. Glabellar tongue narrow, only 11% of cranidial width. Palpebral furrow deep and narrow in front, broadening posteriorly. Posteriorly the palpebral

Table 39. Cranidial measurements on *Sculptella* sp. A

PMO number	b1	b	e	j	j1	rem1	rem4	k00	f1
74860	4.4	3.7	0.7	5.3	4.4	3.1	0.6	2.1	0.7

Abbreviations: see Table 37.

furrow turns inwards and converges on the sagittal line with an angle of around 9° to the transverse mid-line. Palpebral rim convex anteriorly, flattening and broadening rearwards, becoming about twice as wide posterolaterally as at transverse mid-line. Occipital furrow forwardly convex, 48% as wide (tr.) as median area, though this could be lowered by as much as 5% with an intact test. Prominent occipital tubercle situated just behind occipital furrow.

The sculptural lines are unknown.

Remarks. – The specimen differs from *Sculptella angustilingua* Nikolaisen, 1983 on the slightly broader palpebral rim; the more posteriorly directed posterior palpebral furrow and on the much broader (tr.) occipital furrow corresponding to nearly half the width of the median area. Except for these differences they appear identical.

Occurrence. – The Elnes Formation. The cranidium was sampled from a railroad section at Flesberg Station in the Eiker-Sandsvær district.

Genus *Sculptaspis* Nikolaisen, 1983

Type species. – *Sculptaspis cordata* Nikolaisen, 1983.

Diagnosis. – A remopleuridid genus with a cranidial outline like that of *Remopleurides*, but defined by a gently convex (tr.) glabellar tongue less than one-third as wide as the median area; a median area with strong and closely spaced sculptural lines; three pairs of lateral glabellar furrows easily discernible as smooth fields interrupting the sculptural lines; preglabellar furrow distinct; preglabellar field distinct, extending across anterior end of glabellar tongue; anterior pits absent; palpebral furrows strongly incised. Librigena with short, stout and distinctly curved genal spine, very narrow laterally; a blunt but distinct angle is formed on lateral margin between spine and main librigena; posterior border furrow very broad (exsag.) but shallow. Test very thick (diagnosis from Nikolaisen 1983).

Sculptaspis insculpta Nikolaisen, 1983

Pl. 22, Figs 17–18; Pl. 23, Fig. 1; Table 40

1887 *Remopleurides* sp. Brøgger, p. 16.
1983 *Sculptaspis insculpta* n. gen., n. sp. Nikolaisen, pp. 281–282, pl. 11, figs 13–17, table 1.
1983 *Sculptaspis* sp. – Nikolaisen, pp. 286–287 (*partim*), (PMO 74694 only).

Type stratum and type locality. – Engervik Member of the Elnes Formation. Hukodden on Bygdøy, Oslo-Asker district.

Material. – 14 cranidia of which most are fragmentary and somewhat compressed (PMO S.3300, 3663, 3664, 3665, 3666, 60438, 72677, 72683, 74694, 74699, 74850, 74861, 203.141/1, 208.617/5).

Diagnosis. – Median area about two-thirds as long as wide, with sculptural lines concave forwards. Glabellar tongue around 20% as wide as median area. Preglabellar field at axial line one-fifth as long as glabellar tongue. Occipital furrow nearly half as wide as median area (diagnosis emended from Nikolaisen 1983).

Remarks. – Nikolaisen (1983, p. 282) stated the occipital furrow to be nearly 40% of the width of the median area. This is not supported by this study, which finds it to be nearly half as wide as the median area in five of the six specimens showing the width. The only specimen agreeing with the earlier description of a narrower occipital furrow is the holotype and which should be regarded as an extreme in the recorded normal variation. The glabellar tongue is also found to be slightly narrower than previous thought.

Occurrence. – Heggen and Engervik Member of the Elnes Formation. Eiker-Sandsvær: Muggerudkleiva (Heggen Member); profile 4 at Råen (upper Heggen Member, 6.74 m below datum). Oslo-Asker: Bygdøy (mid-Engervik Member); Elnestangen at Slemmestad (mid-Engervik Member) and at Djuptrekkodden (0.94 m above base of Engervik Member).

Sculptaspis erratica Nikolaisen, 1983

Pl. 23, Figs 2–8; Table 41

? 1909 *Remopleurides radians*, Barr., var. *angustata*, Törnq. – Holtedahl, pp. 7, 28, 40.
  1953 *Remopleurides* sp. – Størmer, p. 83.
  1983 *Sculptaspis erratica* n. gen., n. sp. Nikolaisen, pp. 279–281, pl. 12, figs 1–7, text-fig. 10, table 1.

Type stratum and type locality. – Elnes Formation. Road section between Nydal and Furnes Church, Nes-Hamar district.

Material. – 59 fragmentary cranidia (PMO 21929, 36899, 60393, 67194, 67230 (two specimens), 67245, 72690, 72694, 74701, 74702, 74703, 74704, 74756, 82372, 99089–90 (one specimen), 99095, 203.052/1 (two specimens), 203.062/1, 203.062/11, 203.077/1,

Table 40. Cranidial measurements on *Sculptaspis insculpta* Nikolaisen, 1983

PMO number	b1	b	e	j	j1	rem1	rem3	rem4	k00	o	g2	f1	f2	f3
74861				4.7	4.1			0.7						
72677				6.1	5.3	3.7								
3300	3.8	3.2	0.6	5.2	4.5	3		1						
203.141	5.2	4.3	0.9	6.6	6	3.8		1.3	3	3		0.8	1.6	2.5
74699	1.55	1.25	0.3	1.7	1.2	0.9	0.3	0.35	0.6	0.7				
3663	5	4.1	0.9	5.5	4.9	3.6	1.1	1.1	2.1	2.5	1.2	0.8	1.6	2.5
3666		4.8		7.35	6.6	4.4	1.6	1.25	3.2				2	2.9
72683	4	3.3	0.7	4.7	4.2	2.9	0.7	0.65						
3665	8.4	7	1.4	10.4	9.6	6.3		2.4	4.8	5.4				
74694				2	1.55	1.1		0.5	0.8		0.4	0.25	0.5	0.7

Abbreviations: see Table 37.

Table 41. Cranidial measurements on *Sculptaspis erratica* Nikolaisen, 1983

PMO number	b1	b	e	j	j1	rem1	rem3	rem4	k00	o	g2	f1	f2	f3
67194	5.3	4.4	0.9	5.8	5.2	3.6	1.1	1.7	2.6	3.1	1.6	0.8	1.7	2.5
74702	2.2	1.8	0.4	2.3	2	1.5	0.3	0.65	1	1.2	0.6	0.4	0.7	1.1
74701				4.8	4.2	3	0.9	1.3	2			0.8	1.4	
67245	4.9	4	0.9	5.2	4.7	3.5	1.2	1.4	2.2	2.4		0.9		
60393	5.1	4.3	0.8	6.6	5.9	3.6	1.3	1.7	3.2					
74703	5.6	4.7	0.9	6	5.3	3.9	1.2	1.5	2.7	3.2		0.9	1.8	2.7
74756	7.8	6.4	1.4	10	8.8	5.5	1.6	2	4.2		1.6	1	2	3.1
74704	5.6	4.6	1	6.4	5.7	3.7	1.7	1.4	2.6	2.8	1.6	0.9	1.7	2.7
36899	6.1	5.1	1	7.2	6.5	4.4	1.5	1.6	2.8					
21929	3.5	2.9	0.6	3.9	3.5	2.3	1	0.9	1.5					
99089	2	1.7	0.3	2.1	1.8	1.3		0.6						
203.226	3.7	3.1	0.6	4.5	3.9	2.5		1	1.9					

Abbreviations: see Table 37.

203.088/1, 203.092/1, 203.226, 203.346/4, 206.200/1, 207.432/2, 207.432/5, 207.435/1, 207.442/1, 207.445/5, 207.446/3, 207.447/1, 207.449/3, 207.450/2, 207.450/8, 207.454/3, 207.455/1, 207.456/1, 207.457/2, 207.458/1, 207.459/1, 207.460/1, 207.460/2, 207.460/3, 207.460/4, 207.460/5, 207.460/6, 207.460/9, 207.460/10, 207.462/2, 207.462/3, 207.463/3, 207.466/1, 208.624/2 (three specimens), 208.624/4); three librigenae (PMO 207.448/4, 207.450/3, 207.460/8) and one partly articulated specimen (PMO 203.341/1).

Diagnosis. – Median area slightly more than two-thirds as long as wide, with U-shaped to slightly V-shaped sculptural lines. Glabellar tongue nearly one-third as wide as median area. Preglabellar field of even width (sag.), one-eighth as long as glabellar tongue. Occipital furrow half as wide as median area (emended from Nikolaisen 1983).

Remarks. – Compressed specimens are characterized by the width of the glabellar tongue approximating 25% of the median area, and a small length/width ratio of the median area approximating 65%, contrasting with the 30 and more than 66% respectively in the undeformed cranidia. These values are surprisingly consistent between the deformed and undeformed specimens, resulting in two sharply defined morpho-groups.

Occurrence. – Specimens are found in the middle to upper part of the Elnes Formation. Eiker-Sandsvær: Muggerudkleiva in Heistad (Heggen Member); profile 4 at Råen (upper Heggen Member, 9.86 m below datum). Oslo-Asker: beach at Djuptrekkodden (upper Sjøstrand to Engervik Member, 20.56 to 31.57 m above datum); Bøveien (top Sjøstrand Member, 5.89 m above datum). Ringerike: Kullerud and western side of Røysetangen. Hadeland: Road section at Hvattum in Gran. Mjøsa: sub-profile 3 in road section between Nydal and Furnes Church (upper Heggen Member, 1.90 to 11.73 m above datum), Håvesveen, Stedrudstranda and at an unknown locality in Toten.

Sculptaspis sp. A

Pl. 23, Fig. 9; Table 42

1983 *Sculptaspis insculpta* n. gen., n. sp. – Nikolaisen, p. 281.

Material. – One fragmentary cranidium, PMO 74859.

Description. – Cranidium moderately convex both longitudinally and transversely. Median area oval to spade-shaped, moderately convex, 60% as long as wide with greatest width approximately 40% from posterior margin. The three pairs of lateral glabellar furrows are gently but distinctly impressed. S1 is rather broad exsagittally, describing a slight S-curve, turning more strongly rearwards distally. It is 28% as long as width of median area. S2 is very gently convex forwards, slightly narrower and less deep than S1, but of approximately the same length. S3 is very shallow, about 25% as long as S1. Glabellar tongue unknown.

Palpebral furrow moderately wide, deep laterally but shallowing rearwards. Palpebral rim convex, gradually flattening rearwards.

Occipital furrow is nearly half as wide as median area. Occipital ring convex transversely, almost flat longitudinally. Length of occipital ring less than 20% of the total cranidial length. Occipital tubercle small but prominent, located just behind the occipital furrow.

The sculptural lines on the median area are rather coarse, wavy and somewhat unorganized. They describe a gently concave forwards pattern on the posterior half, changing into a narrow V-shape anteriorly of the S2 furrows. The V-shaped pattern may continue onto the basal part of the glabellar tongue. The terrace-lines adjacent to the S2 furrows turn towards the furrows, where they terminate. The surface of the median area is densely and distinctly granulated along the margins, becoming sparser on the central part.

The sculptural lines on occipital ring are coarse and arranged in a concave forwards pattern with around 9 lines medially.

Other parts of the trilobite are unknown.

Remarks. – The specimen differs from *S. pannucea* Nikolaisen, 1983 from the Oslo Region in the sharply V-shaped anterior sculptural lines on the median area and the forwardly concave sculptural lines on the occipital ring. It differs from *S. insculpta* Nikolaisen, 1983 in the sculptural line-pattern and coarseness and perhaps in the denser granulation on the posterior part of the median area.

Occurrence. – The one cranidium is from the concretionary bed at 37.55 m above datum in the Djuptrekkodde profile or 9.5 m up in the Engervik Member.

Table 42. Cranidial measurements on *Sculptaspis* sp. A

PMO number	b1	b	e	j	j1	rem1	rem3	rem4	k00	o	g2	f1	f2	f3
74859	3.7	3.1	0.6	4.5	4.1	2.5			2		1	0.5	1.1	1.8

Abbreviations: see Table 37.

Table 43. Cranidial measurements on *Sculptaspis* sp. B

PMO number	j	j1	rem1	g2
74697	4.2	3.7	2.3	0.9

Abbreviations: see Table 37.

Sculptaspis sp. B

Pl. 23, Figs 10–11; Table 43

1983 *Sculptaspis* sp. Nikolaisen, pp. 286–287 (*partim*), pl. 13, figures 16–17, table 1 (PMO 74694 has been reassigned to *Sculptaspis insculpta*).

Material. – One cranidium (PMO 74697) and one librigena (PMO 74695).

Description. – Cranidium rounded rectangular in outline. Median area nearly two-thirds as long as wide, greatest width slightly more than 40% from posterior margin. Two posterior pairs of lateral glabellar furrows strongly impressed, the anterior pair more gently so. Furrows directed relatively strongly forwards, describing a slightly forwardly convex curve. S1 is rather broad exsagittally, about 28% as long as width of median area. S2 is slightly narrower and shallower but of similar length. S3 is shallow, less than half as long as S1. Glabellar tongue approximately 25% of the width of the cranidium. Palpebral furrows strongly incised. Posterior part of palpebral rim turned obliquely inwards and rearwards. Median area with fine concentric sculptural lines along rim. Central part lacks sculptural lines probably due to later dissolution.

Librigena with deep and wide eye socle furrow posteriorly. Base of spine situated in a relatively forward position, resulting in a deep genal notch. Genal spine is long, slender, slightly curved with a largely even transition to the lateral margin of the rest of the librigena.

Remarks. – The rectangular outline, moderately broad glabellar tongue and deep lateral glabellar furrows all point to these specimens belonging to *Sculptaspis* rather than *Sculptella*. Further support is found in the large resemblance in the sculptural line pattern with the much later *Sculptaspis cordata* Nikolaisen, 1983.

Occurrence. – Engervik Member. Oslo-Asker: The Old Eternite Quarry at Slemmestad.

Genus *Icelorobergia* n. gen.

Derivation of name. – The name refers to the several characters, which this genus has in common with the related genera *Robergia* and *Robergiella*.

Type species. – *Robergiella brevilingua* Fortey, 1980 from the Whiterockian Valhallfonna Formation of Spitsbergen, Norway (Fig. 59).

Diagnosis. – A remopleurid genus with very long and evenly curved eye lobes demarcating wide and moderately inflated median area; a broad tongue with only a slight forward expansion and bordered anteriorly by narrow horizontal rim; glabellar furrows on dorsal surface generally faint, convex forward,

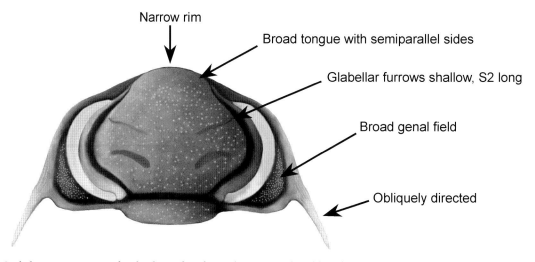

Fig. 59. Cephalic reconstruction of *Icelorobergia brevilingua* (Fortey, 1980) and based upon specimen illustrated *in* Fortey (1980, pl. 5). Some of the more important diagnostic features characterizing this genus are indicated. Length of cephalon is approximately 3.5 mm.

anterolaterally directed; S2 furrows narrow, extending out to palpebral furrow; relatively broad (tr.) area on the free cheek between the eye socle and borders; obliquely directed genal spines.

Included species. – *Robergiella brevilingua* Fortey, 1980, *Robergiella lundehukensis* Fortey, 1980 and probably *Remopleuridiella*? sp. of Burskij (1970) and *Robergiella*? sp. of Nikolaisen (1983) redescribed herein.

Remarks. – This new genus bears some resemblance to the Middle to Late Ordovician genus *Robergiella* to which the species have been assigned by various authors. *Robergiella* was established by Whittington (1959), based on the species *Robergiella sagittalis* Whittington, 1959 from the Porterfieldian Lower Edinburg Limestone of Virginia, USA. *R. sagittalis* is a remopleuridid trilobite with a distinctly expanding glabellar tongue, a relatively flat cranidium with marked glabellar furrows comparable to those found on *Robergia*, a broad librigenal field and posterior directed librigenal spines. Since 1972 several Middle to Late Ordovician remopleuridid trilobites have been assigned to *Robergiella* (Fortey 1980; Hoel 1999; Hunda *et al.* 2003; Ludvigsen 1975, 1980; Nikolaisen 1983; Owen 1981; Ross & Shaw 1972; Tripp 1980), though the justification at times has been quite weak. The first was *Robergiella* sp. from the Porterfieldian to Wildernessian (Middle Ordovician) Copenhagen Formation of Nevada, USA, described by Ross and Shaw (1972), who assigned it to *Robergiella* on strength of its large resemblance to the type species. It mainly differs in having granulate surface sculpture, but perhaps also in a slightly longer glabellar tongue.

Dean (1973) described some cranidia from the Darriwilian Keele Formation of Northwestern Yukon, Canada, which he named *Remopleurides* sp., though he found a large resemblance to the type species of *Robergiella*. They differ from the typical *Robergiella* on their broad (exsag.) glabellar furrows, but are in all other regards similar to this genus. This is one of the more important features found for *Robergiella*, and as the Yukon specimens also show the characteristic cranidial form with its long and expanding glabellar tongue, they should definitely be assigned to *Robergiella*.

Ludvigsen (1975) figured a *Robergiella* sp. from the Trentonian (Middle Ordovician) Road River Formation of Canada. This species is morphologically close to the type species but varies in the much more pronounced glabellar tongue and seemingly somewhat coarser granulation on the cranidium. The pygidium assigned to the species differs strongly from that of the type species of *Robergiella* with an axis consisting of three instead of two rings, laterally extending into three corresponding pairs of short

pleural spines without the deep dividing incisions observed on *R. sagittalis*.

Ludvigsen (1980) described specimens from upper Darriwilian to lowermost Sandbian limestones of Yukon, Canada, as *Robergiella* cf. *sagittalis*. He distinguished them from *R. sagittalis sensu stricto* by their longer (sag.) and narrower (tr.) glabellar tongue and relatively shorter (exsag.) palpebral lobes (Ludvigsen 1980, p. 105). Except for these characters they appear similar and they could just represent variations within one species.

Trip (1980) assigned *Remopleurides correcta* (Reed, 1903) from the Sandbian rocks of Girvan, Scotland to *Robergiella* together with the new species *R. neteorum* and *Robergiella* sp. from the same formation. They are all characterized by the typical long and broad glabellar tongue and broad area between eye and librigenal border and the assignment appears sound.

The two lower Darriwilian species *R. brevilingua* Fortey, 1980 (Fig. 59) and *R. lundehukensis* Fortey, 1980 from the Valhallfonna Formation of Spitsbergen, Norway, were assigned to *Robergiella* on the presence of a relatively broad (tr.) area on the free cheek between the eye socle and borders, which is absent on *Remopleurides*, and by the similar *Robergia*-like form of the glabellar furrows (Fortey 1980, p. 44). On the other hand, they differ significantly from *Robergia* in their more rounded cranidial outline, which is closer to that of *Remopleurides*; broad and comparable short glabellar tongue lacking the marked forward expansion; less distinct glabellar furrows; narrower though still broad area between eye socle and border and more laterally directed genal spines. As a consequence the assignment seems somewhat dubious, although the exclusion from *Remopleurides* seems valid enough. Unfortunately there are no pygidia assigned to the two species and it is therefore not possible to say if they have well-developed pleural furrows as seen on *Robergiella correcta*, the type species and to a lesser degree on *Robergiella insolitus* or the more shallow ones as is the general for *Remoplurides*. Neither are the thoracic segments known, which on *Robergiella* would be characterized by the lack of large axial articulating processes and sockets found on *Remopleurides*.

Nikolaisen (1983) followed the lead of Fortey (1980) and assigned a fragmentary cranidium from the Helskjer Member to *Robergiella* on strength of its large resemblance to the two species from the nearly contemporaneous beds on Spitsbergen. As the specimen lacks all other parts of the exoskeleton the assignment could only be tentative, but the narrow two anterior pairs of glabellar furrows are certainly suggesting a close relationship with at least the two species described by Fortey (1980).

The Lower Ordovician *Robergiella*? *tjernviki* described by Hoel (1999) from the Oslo Region is quite distinctive in having a narrow and nearly straight-sided and elongated glabellar tongue and broad librigena lacking a clear, narrow border as seen on the typical forms of *Robergiella*. On the other hand, it has the typical shape of the cranidium and glabellar furrows characterizing *Robergiella*. Until better material is known it therefore seems most prudent to retain Hoel's species in *Robergiella* with a question mark.

Hunda *et al*. (2003) described the new species *Robergiella insolitus* from the very top of the Ordovician in Canada. The species has a strong resemblance to the type species, but differs clearly in its smaller palpebral lobes and the lack of a median furrow on the second axial ring of the pygidium.

A general look through the species assigned to *Robergiella* reveals a clear grouping between cranidia of the typical form and those with a more rounded and *Remopleurides* like form characterized by a broad and short glabellar tongue lacking the marked forward expansion; less distinct glabellar furrows: a librigena with a narrower though still broad area between eye socle and border and relatively laterally directed genal spines. This second group comprise the species here assigned to *Icelorobergia* n. gen. and *Robergiella* is restricted to the following species: *R. sagittalis* Whittington, 1959, *Remoplurides* sp. (Dean 1973), *Robergiella correcta* (Reed, 1903), *R. insolitus* Hunda *et al*., 2003, *R. neteorum* Tripp, 1980, *R.* cf. *sagittalis* (Ludvigsen 1980), *R.* sp. (Ross & Shaw 1972), *R.* sp. (Ludvigsen 1975), *R.* sp. (Tripp 1980) and perhaps *R. tjernviki* Hoel, 1999.

The pygidium and thorax segments of *Icelorobergia* n. gen. are unknown, and hence its phylogenetic relationship to the other genera remains unclear, although the broad area on the librigena and the narrow (exsag.) anterior glabellar furrows suggest a closer relationship to *Robergiella* than to *Remopleurides*.

Icelorobergia? sp. A

Pl. 23, Figs 12–13; Table 44

1983 *Robergiella*? sp. Nikolaisen, pp. 287–288, pl. 13, figs 18–19.

Table 44. Cranidial measurements on *Icelorobergia*? sp. A

PMO number	b	j	j1	rem1	rem3	rem4	k00	f1
74686	2.2	3.05	2.6	1.65	1	1.2	1.2	0.35

Abbreviations: see Table 37.

Material. – One fragmentary cranidium lacking the occipital ring.

Description. – Cranidium gently convex, broadly elliptical in outline with a rather convex, strongly down-turned glabellar tongue. Glabellar tongue 40% as wide as cranidium or 46% as wide as the median area, slightly widening towards the front. Anterior margin slightly convex. The height, rem3, corresponds to 83% of the width. Median area around 65% as long as wide, anterolateral margin semielliptical, posterior margin more compressed. Glabellar furrows barely developed; posterior pair broad (exsag.), largely laterally directed. S2 narrow, nearly straight, laterally directed, reaching out to the palpebral furrow; S3 short. Preglabellar furrow strongly impressed. Palpebral furrow deep and narrow, turning strongly inwards posterolaterally, from where it continues in a gentle forwardly convex curve down to the occipital furrow. Palpebral rim strongly convex (tr.) with narrow median furrow posteriorly. Occipital furrow 46% as wide as median area; occipital ring missing. No sculptural lines are found but a moderately coarse granulation can be found on the glabellar tongue and along the rim of the median area. Test on the central part of the median area exfoliated.

Remarks. – With its overall form, the narrow S2 furrows extending out to the palpebral furrow and broad and long, down-turned glabellar tongue with a nearly straight frontal margin the cranidium bears a remarkable resemblance to those of the type species and other species of *Icelorobergia* n. gen.. On the other hand, all these characters may perhaps also be found in an extreme form of *Remopleurides* and a confident assignment to *Icelorobergia* n. gen. therefore has to wait until the librigena is known. The greatest resemblance with the Lower Ordovician *Remopleuridiella*? sp. of Burskij (1970) from Novaja Zemlya, suggesting that the species may have originated on the North Russian part of Baltoscandia and was not a Middle Ordovician migrant from Laurentia.

Occurrence. – From the Helskjer Member at the road section between Nydal and Furnes Church, Nes-Hamar.

Genus *Robergia* Wiman, 1905

Type species. – By monotypy, *Remopleurides microphtalmus* Linnarsson, 1875.

Diagnosis. – Length of palpebral lobes (exsag.) nearly half the cranidial length, extending from well behind S1 to well in front of S3; palpebral lobes with epipalpebral furrow. Librigenal spines originating from opposite mid-point of palpebral lobes or slightly behind. Pygidium with length equalling width; pleural fields flat with three pairs of pleural furrows and posterior spines, the third pair of spines may just be blunt rounded prolongations (modified from Moore 1959; Nikolaisen 1991).

Robergia sparsa Nikolaisen, 1983

Pl. 23, Figs 14–20; Table 45

1913   *Robergia microphthalma*? Linrs. sp. – Hadding, p. 78, pl. 8, fig. 18.
1950   *Robergia microphthalmus* (Linnarsson, 1875) – Whittington, pp. 543–544 (*partim*), pl. 71, fig. 7 (= figure of Hadding's pygidium).
1953   *Robergia microphthalma* – Størmer, p. 58.
1983   *Robergia sparsa* n. sp. Nikolaisen, pp. 292–294, pl. 15, figs 2–7.
1991   *Robergia sparsa* – Nikolaisen, 1983, pp. 45–45, 54, figs 7a–b, 12, 14k.
2000b *Robergia sparsa* Nikolaisen, 1983 – Månsson, p. 12.
2002   *Robergia sparsa* – Pålsson *et al.*, p. 42, fig. 10.

Type stratum and type locality. – Elnes Formation. Road section between Nydal and Furnes Church, Nes-Hamar district.

Norwegian material. – One cephalon (PMO 206.188/2), 20 cranidia (PMO S.1817, S.1818, 74848, 74982, 74986, 75055, 82214, 82220, 82221, 95542, 95551, 95792 (three specimens), 202.900/1, 202.900/2, 202.953/1, 203.108/2, 207.446/2, 207.450/4), one librigena (PMO 202.950/2), one hypostome (PMO 202.949/5), some thoracic segments and two pygidia (PMO 75056, 75058). Most specimens have been deformed by compaction.

Diagnosis. – Median area nearly three-fifths as long as wide, very finely striated. Glabellar tongue rounded in outline, 50 to 60% as wide as median area. Occipital furrow less than half as wide as median area. Pygidium subquadrate, four-fifths as long as wide and widening rearwards. First pair of pleural spines located somewhat behind rear end of pygidial axis (emended from Nikolaisen 1983).

Remarks. – The median cranidial furrow, which Nikolaisen (1983) used when separating *Robergia sparsa* from *R. microphthalma* (Linnarsson, 1875) is only present on some of the specimens and is also seen on several of the Norwegian specimens belonging to *R. microphthalma*. As a consequence it should not be used as a diagnostic character for differentiating these two species. The cranidium of *R. sparsa* differs from that of *R. microphthalma* in its narrower and more rectangular rounded glabellar tongue. In *R. sparsa* this is less than half as wide as the cranidial width, while it is greater in *R. microphthalma*. Another difference is the gentler curvature of the S1 furrows, which do not reach back to the posterolateral border distally.

Occurrence. – Elnes Formation. Oslo-Asker: Bøveien at Bødalen, Røyken (Sjøstrand or Engervik Member) and the Old Eternite Quarry middle Sjøstrand Member, 27.22 m above the Huk Formation) and beach profile at Djuptrekkodden, Slemmestad (middle to upper Sjøstrand Member, 2.17 to 22.65 m above datum). Mjøsa: Road section between Nydal and Furnes Church, where it has been taken from the Heggen Member, 6.40 to 6.41 m above datum in sub-profile 3.
Robergia sparsa has been found in the upper Darriwilian Lower Shale member of the Andersö Shale formation at Andersön in Jämtland, Sweden (Månsson 2000b, Nikolaisen 1991; Pålsson *et al.* 2002).

Table 45. Cranidial measurements on *Robergia sparsa* Nikolaisen, 1983 and *R. microphthalma* (Linnarsson, 1875)

Species	PMO number	b1	b	e	j	j1	rem1	rem3	rem4	k00	o	g2	f1	f2	f3
Robergia sparsa	S1817	12.4	11	1.4	13.2	11.8	6.9	2.1	6.1	4.9	4.4	1.4	1.5	2.9	4.7
	202.953	2.5	2.1	0.4	3.1	2.6	1.5		1.4	1.1	1.2	0.2	0.3	0.7	1.1
	95542	3.8	3.3	0.5	5.1	4.4	2.2		2.3	1.6	1.9	0.4	0.5	1	1.5
	95792.1	11.7	11.1	0.6	14.5	13	6.8		6.7			1.1	1.7	3.1	5
	S1818		4.6		6.4	5.5	3.1	0.9	3			0.7			
	203.108		2.1		3.1	2.5	1.4		1.4	1.1		0.3	0.3	0.6	1
	95792.2		1.4		2.5	2	1		1.2			0.2			
	75055		7.4		9.7	8.6	4.2	1.5	4.7	3.7		1	1	2.1	3
	95792.3		7.5		10.6	9.1	4.9						1.2	2.3	3.3
	74982	7.8	7	0.8	7.9	6.9	4.4					0.9	2.1	3.1
	74986				9.3	8.6	4.4					0.9	1	2	3.1
	95551	2.2	1.9	0.3	2.6	2.1	1.4			1	1.2	0.3	0.3	0.6	1
	202.900		4.3				2.4			2.2		0.5	0.5	1	1.7
R. micropht.	74978		6.1		7.7	6.2	3.4		4.4			0.8	0.9	1.7	2.5
	95647	8.7	7.7	1	8.3	7.4	4.5		5	4.2	4.5	0.7	1.1	2.1	3
	74876	8.5	7.4	1.1	9.4	8.1	4.2	1.1	5.4	4.4	4.7	0.7	0.9	1.9	3
	74869				6.66				3.63					
	74872	7.8	7	0.8	10.4	9.5	4.4		9.6	4.6	4.6		0.9	2	2.9

Abbreviations: see Table 37.

Remopleuridid genus et sp. indet.

Pl. 29, Fig. 12

Material. – Asingle fragmentary and partly exfoliated cranidium (PMO 67585).

Description. – Cranidium gently convex, broadly elliptical in outline. Glabellar tongue not preserved, but appears to have a very broad base, suggesting a total width of the tongue close to half the cranidial width. Median area just over half as long as wide, the posterior margin describing a nearly straight transverse line, while anterior margin approximates a more sub-elliptical outline. Glabellar furrows very weakly incised; posterior pair fairly short and narrow (exsag.), largely laterally directed. S2 and S3 moderately narrow, their distal terminations not seen. Surface of median area covered by rather dense and coarse pattern of sculptural lines forming into a V-shape on posterior half and a W anteriorly, resulting in a central quadratic pattern. Palpebral furrow distinctly impressed, turning strongly inwards posterolaterally, from where it continues in a gentle forward convex curve down to the occipital furrow. Palpebral rim broadly convex (tr.), covered by coarse, anterolaterally directed sculptural lines. Occipital furrow moderately deep but narrow; occipital ring very fragmentary, appearing moderately narrow (tr.) and covered by coarse sculptural lines forming a forwardly convex pattern. Test rather thick.

Remarks. – The dense pattern of sculptural lines; the effaced glabellar furrows; the relatively thick shell and the general outline of the cranidium suggest a close relationship with the genus *Sculptaspis* treated above, to which it probably should be allocated. The only deviating feature is found in the seemingly very broad glabellar tongue, which clearly sets it apart from other species belonging to this genus. The question remains as to whether the interpretation of the width of the glabellar tongue is correct or reflects later deformation and the general preservational state of the specimen.

Occurrence. – The cranidium has been collected from the Helskjer Member in the road-cut at Furnes Church, Mjøsa.

Order Corynexochida Kobayashi, 1935

Suborder Illaenina Jaanusson *in* Moore, 1959

Superfamily Illaenoidea Hawle & Corda, 1847

Family Illaenidae Hawle & Corda, 1847

Genus *Illaenus* Dalman, 1827

Type species. – *Entomostracites crassicauda* Wahlenberg, 1818.

Diagnosis. – Rather effaced exoskeleton with moderately sized eyes; a more or less hourglass-shaped glabella; a long rostral flange; broad and quadrangular anterior wings on hypostome; 10 thoracic segments; pygidial size nearly equivalent to the cephalon; pygidial axis short (modified from Moore 1959).

Illaenus aduncus Jaanusson, 1957

Pl. 17, Figs 14–19

1955 *Illaenus aduncus* Jaanusson – Bohlin, p. 115, fig. 3.
1957 *Illaenus aduncus* n. sp. Jaanusson, pp. 123–129 (*cum. syn.*), pl. 5, figs 1–8, pl. 6, fig. 6.
1984 *Illaenus aduncus* Jaanusson, 1957 – Wandås, p. 233, pl. 11M–Q.
1995 *Illaenus aduncus* Jaanusson, 1957 – Nielsen, p. 340 (*cum. syn.*), fig. 244.

Type stratum and type locality. – Holotype UM.Nr.Ar.4207 deposited at the Geological Museum in Uppsala, Sweden, was collected at Västenå in Östergötland, Sweden, from where it most probably came from the lower middle Darriwilian Raniceps Limestone (Jaanusson 1957).

Norwegian material. – The Norwegian material from the Elnes Formation is restricted to one cephalon (PMO 106.015), two cranidia (PMO 33524 (two specimens)), six pygidia (PMO 67187, 67597, 67598, 83107, 83243, 207.394/7) and two partly articulated specimens (PMO 33440, 207.521).

Diagnosis. – Sagitally strongly convex cephalon with nearly vertical front; strongly forwardly convex terracelines centrally on glabella; cuspate inner doublural margin on the pygidium; delicate and very dense, radial terrace-line pattern on the external pygidial surface (based on Jaanusson 1957).

Description. – Largest cranidium from the Elnes Formation, PMO 33440, approximately 29 mm long (forelengthened specimen), while largest pygidium PMO 83107 measures 15 mm in length.

The species was thoroughly described by Jaanusson (1957), and nothing new is added by the Norwegian material.

Remarks. – The contemporaneous to slightly older *Illaenus incisus* Jaanusson, 1957 is nearly identical to *I. aduncus*, but can be distinguished by its coarser and sparser terrace-lines on the cephalon, by the more rounded and downward projected librigenal corners and by the lateral direction of the terrace-lines on pygidial pleural fields, only turning forward into a forward pointing 'V' behind the axis. *Illaenus aduncus* is also morphologically close to the lower Darriwilian *I. sarsi* Jaanusson, 1954 from Baltoscandia, but differs in the characters mentioned in the diagnosis above. It differs from the Swedish *I. schuberti* Nielsen, 1995 of the lower middle Darriwilian *Asaphus raniceps* Zone on the distinctly U- to V-shaped central glabellar terrace-lines; the relatively wider and more sharply rounded (sag.) cranidium and by the dense radial terrace-line pattern on the pygidium.

Occurrence. – In Norway *I. aduncus* has been found from the Svartodden Member of the Huk Formation and up into the basal part of the Elnes Formation (Nielsen 1995; Wandås 1984). Hadeland: Lynne (basal Elnes Fm.). Mjøsa: Helskjer on Helgøya (Helskjer Member, 5.10 m above the Huk Fm.), road-cut along highway E6 at Nydal just north of Hamar (middle Helskjer Member, somewhere between 5 and 6.5 m above the Huk Fm. and from level 7.3 m).

Outside Norway it has been collected from the lower middle Darriwilian *Asaphus raniceps* Zone and possibly also from the lower Darriwilian *A. expansus* Zone in southern to central Sweden (Jaanusson 1957; Nielsen 1995).

Family Styginidae Vodges, 1890

Subfamily Stygininae Vodges, 1890

Genus *Raymondaspis* Přibyl, 1948

Subgenus *Raymondaspis* (*Cyrtocybe*) Holloway, 2007

Type species. – *Raymondaspis turgidus* Whittington, 1965a.

Diagnosis. – Cephalon highly convex (sag.); anterior part of glabella overhanging cephalic rim in dorsal view; preglabellar furrow and cranidial portion of lateral border furrow effaced and cephalon with an anterior branch of facial suture diverging weakly forwards over most of its length. Pygidium around twice as wide as long, lacking postaxial ridge (emended from Holloway 2007).

Remarks. – Holloway (2007) erected *Cyrtocybe* to include parts of the former genus *Turgicephalus* Fortey, 1980, which he convincingly rejected as a subjective junior synonym of *Raymondaspis*. *Cyrtocybe* was found to differ from *Raymondaspis* on characters such as the preglabellar furrow and anteromedial

part of lateral border furrow being effaced; front of glabella overhanging marginal cephalic rim in dorsal view; an axial furrow, which becomes shallow from S1 to anterior pit of fossula; a relatively inflated L1 lobe; an anterior branch of facial suture diverging weakly forwards; absent genal spine, and a pygidium with width more than twice the length and carrying no postaxial ridge. Some of the characters, though, do not hold up when comparing the various Scandinavian species belonging to *Raymondaspis* sensu Holloway (2007) with *Cyrtocybe*. This applies to the postaxial ridge on the pygidium, which is absent on pygidia of *Raymondaspis insignus* Nielsen, 1995 and *R.* sp. B *sensu* Nielsen (1995, fig. 225F) and the weak forward divergence of the anterior branch of the facial suture, which also may be found in *R.* sp. B *sensu* Nielsen (1995, fig. 225D). The diagnostic difference between the two genera is further diluted by the former *Turgicephalus* cf. *turgidus* (Whittington, 1965a) of Wandås (1984) described below and here assigned to *Cyrtocybe* on strength of characters like the cephalic convexity and the highly effaced to completely absent preglabellar and lateral border furrows. The species features a number of *Raymondaspis*-like characters such as a genal spine or at least a pointed genal angle; a deep axial furrow in front of indistinct S1 furrow; a strongly effaced L1 and a shallow occipital furrow. The rest of the characters used in separating the two genera may possibly also be problematic and are at least not as clear cut as originally thought. *Cyrtocybe* is here assigned subgenus status of *Raymondaspis*.

Raymondaspis (*Cyrtocybe*) seems to have originated from Laurentia as it is here found down into lower Middle Ordovician deposits, while the first Baltoscandian representatives are from the lower Darriwilian of Norway (Holloway 2007; this study). The geographical distributions may indicate that the genus was able to cross the Iapetus Ocean already at an early stage when the ocean was still at its widest, possibly with the help of an easterly directed oceanic current. It could also be indicative of the problematic status of the subgenus.

Raymondaspis (*Cyrtocybe*) sp.

Pl. 17, Figs 20–23

1984 *Turgicephalus* cf. *turgidus* (Whittington, 1965) – Wandås, pp. 232–233, pl. 11G–I.

Material. – The material consists of four exfoliated cranidia (PMO 67578, 83182, 83239, 207.398/1) and a single fragmentary librigena (PMO 67578).

Description. – Largest cranidium, PMO 83239 about 7 mm long. Cranidium highly convex sagitally, resulting from a downward deflection of the cranidium just in front of the palpebral lobes; transverse convexity moderate posteriorly, flattening anteriorly. The length corresponds to just over two-thirds the width. Axial furrows posteriorly deep and moderately wide, becoming shallow in front of eye ridge before terminating at anterior pits slightly behind cranidial front. Occipital ring low and fairly wide (tr.), occupying between 33 and 40% of the cranidial width, bounded anteriorly by shallow and nearly straight occipital furrow. Glabella somewhat pestle-shaped, forward expansion increasing in front of the palpebral lobes; largest width located just in front of anterior pits corresponding to 140% of the smallest width found between the palpebral lobes. Muscle scars rather faint, located on lateral flanks of glabella at occipital furrow, between palpebral lobes and slightly behind anterior pits. Lateral glabellar lobes and furrows rather effaced. Anterior part of glabella bearing faint sagittal ridge. Anterior glabellar margin sharply delimited, fluctuating anterior cranidial margin and only slightly overhanging marginal cephalic rim in dorsal view. Preocular area is relatively narrow, only widening forward slightly, giving the cranidial front a distinctly rectangular appearance. Palpebral lobes moderately sized, continuing into strong, forward curving eye ridges. Librigena poorly preserved, but showing the presence of a genal spine or point.

Remarks. – The specimens most closely resemble the North American *R.* (*C.*) *turgidus* Whittington, 1965a from the Middle Ordovician Table Head Formation of Western Newfoundland. Unfortunately, Whittington (1965a) only figured specimens with preserved test, making it difficult to compare them with the internal moulds from the Elnes Formation. Even so it is clear that the Norwegian specimens are separated by the extremely shallow occipital furrow and the presence of a genal spine or point and possibly also by a more pronounced eye ridge; a slightly more sagitally convex cranidium and by a straighter transverse course of the occipital furrow. These differences are supported by a cranidial mould, which has been described from the contemporaneous Otta Conglomerate of south central Norway (Bruton & Harper 1981). That specimen, which was tentatively assigned to *Turgicephalus turgidus*, shows the same deep occipital furrow observed for the external test surface of the North American specimens. It further differs from the Oslo Region specimens by the less pronounced eye ridges and the relatively wider glabella, but is similar in its less forwardly curved

occipital furrow and strongly effaced posterior lateral glabellar lobes.

Occurrence. – The species is very rare and is only known from the uppermost Helskjer Member in the northern part of the Oslo Region. Mjøsa: Road-cut at Furnes Church (7.75 m above the Huk Formation); northern road-cut along E6 just south of Nydal (10.06 m above the Huk Formation).

Genus *Bronteopsis* Nicholson & Etheridge, 1879

Type species. – *Bronteopsis scotica* Nicholson & Etheridge, 1879 (by monotypy; junior subjective synonym of *Ogygia? concentrica* Linnarsson, 1869 (see Skjeseth 1955)).

Diagnosis. – Preglabellar field absent; four pairs of glabellar furrows; genal spines present and are longer and broader based than seen in *Stygina*. Thorax has eight segments. Pygidium with seven or eigth axial rings; unfurrowed axial tip; broad and gently concave border on pleural field; seven pairs of shallow pleural furrows separated by low and broad, posteriorly turning pleural ribs. External surface covered with terrace-lines (Moore 1959).

Bronteopsis holtedahli Skjeseth, 1955

Pl. 18, Figs 1–4

1955 *Bronteopsis holtedahli* sp. n. Skjeseth, pp. 17–18, pl. 5, figs 4, 7.
1963 *Bronteopsis holtedahli* – Skjeseth, p. 63.
1984 *Bronteopsis holtedahli* Skjeseth, 1955 – Wandås, p. 233, pl. 11E–F.

Type stratum and type locality. – The holotype PMO 67011 is a fragmentary cranidium from the lower Elnes Formation at a road-cut at Furnes Church, Mjøsa.

Material. – The material consists of two cranidia (PMO 67011b, 67011 c) and four pygidia (PMO 9039, 33108 (+ 33106), 64458 (67011a)).

Diagnosis. – Species of *Bronteopsis* characterized by a narrow anterolateral border, corresponding to between 15 and 20% of the cranidial length; a moderately wide and not strongly forward widening glabella with well-defined S1 furrows and fairly shallow S2 and S3 impressions. Pygidium with fairly distinct pleural ribs and an axis taking up around two-thirds the pygidial length (emended from Skjeseth 1955).

Description. – See Skjeseth (1955).

Remarks. – The species appears very close to the possibly contemporaneous *B. panderi* Schmidt, 1904 from Estonia, but is differentiated on the slightly wider glabella; the less deeply impressed lateral glabellar furrows and by the more pronounced pleural ribs on the pygidium. It is differentiated from *B. concentrica* Linnarsson, 1869 from Baltoscandia on the strength of the distinctly narrower anterolateral cranidial border; the clearly less forwardly widening glabella and possibly by slightly deeper lateral glabellar furrows or impressions, though this may relate to different preservation. The pygidial axis is somewhat longer than that in *B. concentrica*, which following Whittington (1950, p. 546) only occupies around half the pygidial length and not the two-thirds seen on *B. holtedahli*. Whether some of this could relate to ontogeny is open for question.

Occurrence. – Hadeland: Lundebakken (Elnes Fm.), Hofstangen (Elnes Fm.). Mjøsa: road-cut at Furnes Church (probably from the Elnes Formation).

Order Proetida Fortey & Owens, 1975

Superfamily Proetoidea Salter, 1864

Family Proetidae Salter, 1864

Subfamily Proetinae Salter, 1864

Genus *Proetus* Steininger, 1831

Subgenus *Proetus* (*Proetus*) Steininger, 1831

Type species. – *Calymene concinna* Dalman, 1827.

Diagnosis. – Glabella elongate, commonly weakly constricted laterally; field of free cheek with fine pitting, generally inconspicuous; prominent genal spine and distinct lateral occipital lobes always present; no preglabellar field. Pygidium without border and with occasional incurved marginal terrace ridges; surface smooth (from Owens *in* Helbert *et al.* 1982).

***Proetus* (*Proetus*) sp.**

Pl. 28, Figs 10–11

Material. – Two librigenae (PMO 36617, 67604).

Description. – The largest librigena (PMO 36617) is 2.7 mm long (spine excluded). It is characterized by a large librigenal field with a small, posteriorly located eye-socle; length of eye approximating two-thirds the distance from eye to anterior librigenal margin; border furrow moderately deep and narrow, delineating narrow border; lateral border continuing evenly into short and slender posterolateral genal spine of unknown length.

Remarks. – The material at hand is too insufficient for a closer identification or comparison.

Occurrence. – Specimen PMO 67604 was found in the Helskjer Member in the upper quarry at Helskjer on Helgøya, Mjøsa district, while the other (PMO 36617) has been collected from a locality named Tømte from the central or northern part of the Oslo Region (upper Elnes Formation).

Superfamily Bathyuroidea Fortey *in* Kaesler, 1997

Family Telephinidae Marek, 1952

Genus *Telephina* Marek, 1952

Type species. – *Telephus fractus* Barrande, 1852.

Remarks. – The genus was subdivided by Nikolaisen (1963) on strength of the presence or absence of frontal glabellar spines, but a rejection of the subgeneric division was presented by Tripp (1976) and Ahlberg (1995a,b) and is followed here.

Diagnosis. – Cephalon wider than long with forward tapering glabella; glabella broadly rounded to truncate anteriorly; posterior glabellar furrows may be represented by short longitudinal depressions; occipital ring typically with rearward directed median spine; convex eye forming large part of librigena, enclosing crescentic palpebral lobe with prominent ring; narrow cephalic border outside eye lobes; short pair of spines on anterior cephalic border and generally long and slim genal spines. Thorax with wide axis. Pygidium small, semicircular with two axial rings; pleural field narrow, unfurrowed. Surface tuberculate with pattern of anastomosing raised lines; the glabella having three or four pairs of smooth muscle attachment areas (modified from Moore 1959; Ahlberg 1995a).

Telephina bicuspis (Angelin 1854)

Pl. 24, Figs 1–2, 15–16; Pl. 25, Fig. 1; Text-Figs 60–61; Tables 46–47

1854   *Telephus bicuspis.* n. sp. Angelin, p. 91, pl. 41, fig. 22.

1963   *Telephina (Telephina) bicuspis* (Angelin, 1854) – Nikolaisen, pp. 364–367 (*cum. syn.*), fig. 2, pl. 1, figs 1–10.

1963   *Telephina (Telephina) vesca* n. sp. Nikolaisen, pp. 379–381, pl. 3, figs 5–10.

1963   *Telephina (Telephina)* sp. no. 1 Nikolaisen, p. 384, pl. 3, fig. 12.

1995a   *Telephina bicuspis* (Angelin 1854) – Ahlberg, pp. 264–268 (*cum. syn.*)

1995b   *Telephina bicuspis* (Angelin 1854) – Ahlberg, p. 51.

2000b   *Telephina bicuspis* (Angelin 1854) – Månsson, p. 12, fig. 7g.

2006   *Telephina bicuspis* (Angelin 1854) – Bruton & Høyberget, pp. 359–364.

Type stratum and type locality. – The neotype selected by Thorslund *in* Thorslund & Asklund (1935) originates from the Elnes Formation in Oslo, Oslo-Asker.

Material. – The Natural History Museum of Oslo has a relatively large collection of mainly fragmentary cranidia from the Elnes Formation, which have been studied here. The following examination and revision of the species was mainly based on 26 cranidia (PMO I.0245, 3772a, 20291, 21920, 36457, 36738, 36854, 67198a, 67200, 67230b, 67238, 72668, 72692a, 72693a, 72693c, 72693d, 72694a, 72694b, 72694c, 72695, 72696, 72724a, 72728a, 72731, 82545, 105.966).

Diagnosis. – Cranidium of fully grown specimens about three-fifths as long as wide; preoccipital part of glabella parabolic in outline, wider than long, tapering

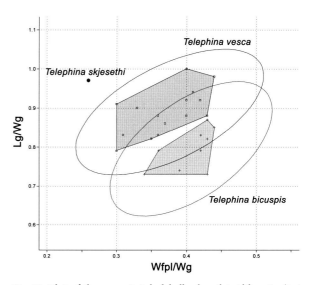

Fig. 60. Plot of the preoccipital glabellar length/width ratio (Lg/Wg) against the ratio between the width of the fixigenal field and the preoccipital glabellar width (Wfpl/Wg) selected of species of *Telephina* from the Elnes Formation of the Oslo Region. The 95% confidence intervals of *Telephina bicuspis* (Angelin 1854) and *T. vesca* Nikolaisen, 1963 overlap significantly, whereas *T. skjesethi* Nikolaisen, 1963 plots well outside of the 95% confidence intervals of the two others.

moderately forward; frontal glabellar lobe broadly rounded to truncate anteriorly; lateral glabellar furrows indistinct; palpebral area with narrow raised rim along posterolateral margin, becoming slightly wider anteriorly; fixigena rather large and wide; glabellar surface, except on large muscle attachment areas, covered by moderately dense tuberculation. Pygidium wide with broadly rounded posterior margin (modified from Ahlberg 1995a; Nikolaisen 1963).

Discussion. – *Telephina vesca* Nikolaisen, 1963 from the Elnes Formation of the Oslo Region, was

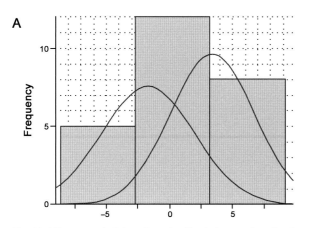
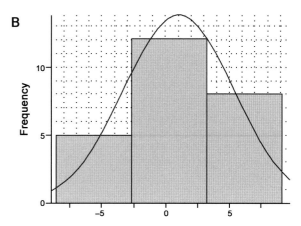

Fig. 61. Mixture analysis based on the discriminant values for the two taxa *T. bicuspis* and *T. vesca*. Graph A investigates the likelihood of two taxa illustrated by the two curves on top of the actual 'size'-grouping in the form of a three-partitioned histogram. Graph B examines the possibility that the data set represents a single taxon only. The two possible scenarios are very close, but with a slightly higher probability for the one-group scenario. The investigation is carried out with the statistical program PAST (Hammer *et al.* 2001).

Table 46. Cranidial measurements on *Telephina bicuspis* (Angelin 1854), including specimens formerly grouped under the junior synonym *T. vesca* Nikolaisen, 1963

Species	PMO number	Lg	Wg	Wfpl	Lf	Wf
Telephina vesca	72694b	2.2	2.4	1		
	36457	3	3.2	1.3		
	20291	5	5.1	2.3		
	82545	4.5	~5.5	1.9		
	67230b	3.6	4.1	1.7		
	I.0245	2.8	3	1.2		
	67238	3	3.6	1.1		
	72694a	4.4	4.9	1.6		
	72731	4.9	5.5	2.4	1.8	5.3
	72695	2.9	3.3	1.2		
	72694c	3.7	4.3	1.6		
	67200	6.2	~7.5	2.7		
	72693a	2.1	2.3	0.7		
	72728a	1.8	2	0.9		
	3772a	1.3	1.7	0.5		
	72724a	1.8	~1.8	0.7		
Telephina bicuspis	72668	3.7	4.5	1.9		
	67198a	4.5	~6.1	2.4		
	105966	3.5	4.7	2	1.1	4.5
	72692a	1	1.2	0.5		
	72693c	1.2	1.4	0.6		
	36854	1.6	2.2	0.8		
	72693d	2.2	2.7	1.1		
	21920	1.5	1.9	0.8		
	72696	1.6	2	0.7		
	36738	2.5	2.9		0.9	2.85

Lg, length of preoccipital part of glabella; Wg, width of glabella.
Wfpl, width of palpebral area taken as the line running perpendicularly from the axial furrow out to the distal corner of the palpebral furrow.
Lf, distance from occipital furrow to lateral glabellar furrow or pit; Wf, glabellar width at glabellar furrows.

Table 47. List of preoccipital glabellar length/width ratio (Lg/Wg); ratio between width of fixigenal pleural field (Wfpl) and preoccipital glabellar width (Wg) for a number of telephinid specimens. The last column shows the resulting discriminant values when examining *Telephina vesca* against *T. bicuspis*

Species	PMO number	Lg/Wg	Wfpl/Wg	Discriminant
Telephina vesca Nikolaisen	72694b	0.92	0.42	2.8632
	36457	0.94	0.41	4.6355
	20291	0.98	0.44	5.1928
	82545	0.82	0.35	1.1712
	67230b	0.88	0.4	1.7085
	I.0245	0.92	0.4	4.0581
	67238	0.83	0.31	4.1482
	72694a	0.9	0.33	7.0652
	72731	0.89	0.43	0.5036
	72695	0.88	0.36	4.0981
	72694c	0.86	0.37	2.3259
	67200	0.83	0.36	1.1611
	72693a	0.91	0.3	9.4449
	72728a	0.88	0.43	−0.083798
	3772a	0.79	0.3	2.3961
	72724a	1	0.4	8.7573
T. bicuspis (Angelin)	72668	0.82	0.43	−3.6082
	67198a	0.74	0.39	−5.9177
	105966	0.73	0.43	−8.8948
	72692a	0.87	0.43	−0.6712
	72693c	0.85	0.44	−2.4434
	36854	0.73	0.34	−3.518
	72693d	0.83	0.42	−2.4234
	21920	0.79	0.42	−4.773
	72696	0.79	0.36	−1.1885

T. bicuspis: N: 9
  Chi^2: 2.1111
P(normal): 0.14623
T. vesca: N: 16
  Chi^2: 1
P(normal): 0.31731

considered to differ from *T. bicuspis* by the narrower glabella and palpebral lobes; a more well-rounded fixigena and by densely tuberculation throughout the glabellar surface (Nikolaisen 1963, pp. 379–380). A comparison of the two based on the material in the Natural History Museum, University of Oslo, does not support any differentiation on the strength of the glabellar tuberculation or the roundness of the fixigenae or width of palpebral lobes. The length/width ratio (Lg/Wg) of the glabella and the relative width of the pleural field on fixigena relative to the glabellar width (Wfpl/Wg) do separate the two taxa (Fig. 60). On the other hand the separation between the two groups appears rather small and a large overlap is seen on the 95% confidence intervals. The question is therefore whether the observed separation is merely an artifact of subjective subdivision. In order to investigate this, the discriminant values were found for the two taxa (Table 47) and were examined for a normal distribution by using the chi-square method.

The hypothesis was accepted for both groups. The discriminant values were then examined in a Mixture analysis (Fig. 61), which estimates the likelyhood of of the presence of one or more groups with normal distribution in the material (for further explanation see Hammer & Harper 2006). The Akaike Information Criterion (AIC) value for the hypothesis of one and two groups respectively were:

Two groups: Log l.hood: −71.24
  AIC: 154.5
One group: Log l.hood: −72.64
  AIC: 151.3

Although the difference is small, the AIC value is slightly smaller for the one group hypothesis indicating that *T. vesca* and *T. bicuspis* should be regarded as a single species and thus *T. vesca* is considered a junior subjective synonym of *T. bicuspis*.

T. sp. no. 1 *sensu* Nikolaisen (1963) (Pl. 25, fig. 1) from the Elnes Formation in the northern Oslo Region was thought to differ from the former *T. vesca* by the coarser glabellar tuberculation and narrower and more convex fixigena. The width and convexity of fixigena largely relate to preservation, and the suggested difference in tuberculation is here regarded as lying inside the variance expected for a natural population.

Occurrence. – The species is known from the Heggen, Sjøstrand and Engervik members of the Oslo Region and the corresponding Lower Andersö Shale of Jämtland, Sweden and the Folkeslunda Limestone on Öland. In the East-Baltic area it has been found in beds from the lowermost Uhaku Stage (~Engervik Member) at Engure in Latvia (Ahlberg 1995b; Månsson 2000b; Nikolaisen 1963; this study).

Telephina intermedia (Thorslund, 1935)

Pl. 24, Figs 3, 5–7; Table 48

1935   *Telephus intermedius* n. sp. Thorslund, pp. 22, 61, pl. 2, figs 7–8.
1963   *Telephina* (*Telephina*) *furnensis* n. sp. Nikolaisen, pp. 367–368, pl. 1, fig. 11.
1963   *Telephina* (*Telephina*) *intermedia* (Thorslund 1935) – Nikolaisen, pp. 369–370, pl. 2, figs 1–5.
1995a *Telephina furnensis* Nikolaisen, 1963 – Ahlberg, p. 266.
1995a *Telephina intermedia* (Thorslund 1935) – Ahlberg, p. 269, fig. 3.

Type stratum and type locality. – Lower Andersö Shale member (Upper Darriwilian) at Raftan in Jämtland, Sweden (Thorslund & Asklund 1935).

Material. – A material of 22 cranidia was selected from the numerous, mostly fragmentary specimens found in the collection of the Natural History Museum of Oslo. The specimens are listed in Table 48.

Diagnosis. – Glabella rather forward reaching, rounded sub-trapezoid, narrowing distinctly forward with rounded to slightly indented frontal margin; preoccipital part of glabella distinctly wider than long with marked pair of lateral glabellar furrows; surface granulation fairly dense, absent on rather small muscle attachment areas anterolaterally of occipital furrow and in front of lateral glabellar furrows (modified from Nikolaisen 1963).

Remarks. – *T. furnensis* Nikolaisen, 1963 from the Elnes Formation of the northern Oslo Region, represented by only two cranidia, and is here regarded as being a junior synonym of *T. intermedia*. The shorter glabella with its subquadratic outline and the more abaxial location of the lateral glabellar furrows of the

Table 48. Cranidial measurements on *Telephina intermedia* (Thorslund 1935)

PMO number	Lg	Wg	Wfpl	Lf	Wf
72732c		3.4	1.6	1.5	~3.2
82404	4.4	4.9		1.8	4.5
67199	2.7	~3.5		1.1	~3.3
67230	2.4	2.6		1	2.4
72732b	4	4.9	1.8	1.5	4.6
3809	4.1	4.5	1.6	1.5	4.1
72732a	2.9	3.8	1.6	1.1	3.6
72727b	1.4	1.7	0.6	0.55	1.6
72782	~2	2.5	1.1	0.8	2.4
33541	2.6	3.4	1.3	1.1	3.2
33105	2.5	3.2	1.3	1	3
4247	1.7	1.95	0.7	0.7	1.9
72725	3.8	~4.5	1.6	1.6	~4.3
36649	3	3.8	1.5	1.3	3.5
72727a	1.6	2	0.65	0.7	1.9
36759	4.4	5.1	1.7	1.9	5
72690_1	2.4	2.85	1.1	0.9	2.7
72690_2	2.15	2.7	0.95	0.8	2.55
72690_3	1.2	1.37	0.63	0.45	1.31
72690_4	2.5	3.65	1.1	1	3.45
105966	3.6	3.9	1.7	1.6	3.6
73013a	1.45	1.9	0.75	0.65	1.85

Abbreviations: see Table 46.

Table 49. Cranidial measurements on *Telephina mobergi* (Hadding, 1913)

PMO number	Lg	Wg	Wfpl	Lf	Wf
33103	3.5	5.2	2	1.3	4.9

Abbreviations: see Table 46.

specimens previously placed in *T. furnensis* do distinguish them from the typical *T. intermedia* but their differences are small in comparison with the overall resemblance to Thorslund's species (compare Pl. 24, fig. 3 and Pl. 24, fig. 5). The composite outline of the occipital furrow and the moderately distinct lateral glabellar furrows provide evidence against Ahlberg's (1995a) assignment of *T. furnensis* to *T. bicuspis* (Angelin 1854).

Occurrence. – Heggen Member in the Mjøsa District and in the Lower Andersö Shale member at Raftan in Jämtland (Nikolaisen 1963; Thorslund & Asklund 1935).

Telephina mobergi (Hadding, 1913)

Pl. 24, Figs 8–9; Table 49

1963   *Telephina* (*Telephina*) *mobergi* (Hadding, 1913) – Nikolaisen, pp. 371–373 (*cum. syn.*), pl. 2, figs 6–12.
1995a *Telephina mobergi* (Hadding, 1913) – Ahlberg, pp. 269–272 (*cum. syn.*), pl. 3, figs 3–14.

Type stratum and type locality. – Lectotype selected by Nikolaisen (1963, p. 371) from the Lower Andersö Shale member (middle to upper Darriwilian) at Andersön in Jämtland, Sweden (Nikolaisen 1963).

Material. – The collections at the Natural History Museum of Oslo includes a large number of Norwegian specimens belonging to this species. The 22 best preserved cranidia (PMO 33103, 33251a–b, 36493, 36505, 36506, 36526, 57199, 67202, 72071a–d, 72101, 72685a–b, 72686, 72717, 72728c, 72730b–d, 72768) were examined more closely.

Diagnosis. – Glabella comparatively short and distinctly triangular in outline with rounded front and pair of deep lateral glabellar furrows; preoccipital part of glabella clearly wider than long and characterized by inflated posterolateral corners; anterior part of palpebral lobe wide, directed nearly perpendicular to sagittal line and slightly in front of the glabella; glabella smooth (modified from Ahlberg 1995a).

Occurrence. – Heggen Member and middle Sjøstrand to Engervik Member in central to northern Oslo Region and contemporaneous Lower Andersö Shale member of the Andersö Shale formation in Jämtland, Sweden (Nikolaisen 1963; Ahlberg 1995a; Pålsson *et al.* 2002; This study).

Telephina norvegica Nikolaisen, 1963

Pl. 24, Figs 4, 10; Table 50

1963   *Telephina* (*Telephina*) aff. *furnensis* n. sp. Nikolaisen, p. 369, text-fig. 4.
1963   *Telephina* (*Telephina*) *norvegica* n. sp. Nikolaisen, pp. 373–375, pl. 3, figs 1–2.
1963   *Telephina* (*Telephina*) sp. no. 2 Nikolaisen, p. 384–385, pl. 3, figs 15–16.
1995a *Telephina norvegica* Nikolaisen, 1963 – Ahlberg, p. 266.

Type stratum and type locality. – Heggen Member in the road section at Furnes Church, Mjøsa.

Material. – Four cranidia are known (PMO 33550, 72721, 72730a, 205.197a/1).

Diagnosis. – Preoccipital part of glabella clearly wider than long, subtrapezoidal, distinctly forward narrowing with gently rounded to distinctly truncate front and indistinct pair of lateral glabellar furrows

located midway to anteriorly on preoccipital part of glabella; glabella fairly short, not reaching far in front of palpebral furrows; Anterolateral fixigenal corners sharp; anterior cranidial margin nearly straight; glabella bearing moderately dense tuberculation (emended from Nikolaisen 1963).

Remarks. – The fragmentary cranidium PMO 72725 assigned by Nikolaisen (1963) to this species should rather be placed in *T. intermedia* treated above.

T. sp. no. 2 *sensu* Nikolaisen (1963) from the Engervik Member, Elnes Formation, in the Oslo-Asker district was described as having a distinctly longer glabella relative to the width and a narrower fixigena than *T. norvegica*. The latter feature is merely because a large part of fixigena is covered by matrix, while the difference in the relative length of the glabella is quite small and most probably relates either to size or later deformation.

T. norvegica was proposed a junior synonym of *T. bicuspis* (Angelin 1854) by Ahlberg (1995a), but may differ on features such as the more triangular fixigena with a possibly slightly wider anterior palpebral rim and by the more pronounced lateral glabellar furrows.

Occurrence. – The species is known from the Heggen and Engervik members of the central to northern Oslo Region (Nikolaisen 1963; This study). Oslo-Asker: Djuptrekodden at Slemmestad (uppermost Engervik Member, 42.82 m above datum in beach profile) and Huk on Bygdøy (Engervik Member). Hadeland: South of Skiaker in Gran (upper Elnes Fm.). Mjøsa: Road-cut at Nydal-Furnes Church (Elnes Fm.).

Telephina skjesethi Nikolaisen, 1963

Pl. 24, Figs 11–12; Table 51

1963   *Telephina* (*Telephina*) *skjesethi* n. sp. Nikolaisen, pp. 375–376, pl. 3, figs 3–4.

Type stratum and type locality. – Heggen Member at Dyste in Toten, Mjøsa.

Material. – Holotype cranidium PMO 72722 and associated two thoracic segments.

Diagnosis. – Preoccipital part of glabella approximately as long as wide, parabolic in outline with

Table 50. Cranidial measurements on *Telephina norvegica* Nikolaisen, 1963

PMO number	Lg	Wg	Wfpl	Lf	Wf
33550	1.2	1.85	0.7	0.55	1.7
72730a	2.3	2.8	1.2	1.2	2.6

Abbreviations: see Table 46.

Table 51. Cranidial measurements on *Telephina skjesethi* Nikolaisen, 1963

PMO number	Lg	Wg	Wfpl
72722	1.9	2	0.5

Abbreviations: see Table 46.

rounded to slightly concave front and indistinct or completely effaced lateral glabellar furrows; occipital ring with short spine; glabella relatively long, reaching well out in front of palpebral furrows; palpebral lobe ridges narrow; glabellar surface densely tuberculate except on medium sized muscle attachment areas (modified from Nikolaisen 1963).

Remarks. – The species has a strong affinity to *T. bicuspis*, but plots at present far outside the 95% confidence interval of that species on a plot on the glabellar length/width ratio against the width of the fixigenal field relative to the glabellar width (Fig. 60). For this reason it is upheld for the time being.

Occurrence. – Heggen Member in the northern Mjøsa district (Nikolaisen 1963).

Telephina sulcata Nikolaisen, 1963

Pl. 24, Figs 13–14; Table 52

1953   *Telephus mobergi* – Størmer, p. 102.
1963   *Telephina* (*Telephina*) *sulcata* n. sp. Nikolaisen, pp. 377–379, pl. 3, figs 13–14.

Type stratum and type locality. – Erratic boulder from the uppermost part of the Helskjer Member in the road-cut at Furnes Church, Mjøsa.

Material. – Nine cranidia PMO 67011d–e, 72706, 72707, 72756b, 72757, 72758a–b, 73014.

Diagnosis. – Preoccipital part of glabella broadly rectangular to subtrapezoidal in outline with broad flattened front and deeply impressed pair of lateral glabellar furrows; glabella not extending in front of broad palpebral furrow; palpebral lobe moderately wide, delimiting rather narrow fixigena. Occipital ring covered by moderately dense tuberculation (emended from Nikolaisen 1963).

Remarks. – The cranidium appears to have a smooth surface, but this could easily relate to taphonomy.

Occurrence. – Helskjer Member, 7.7 m above the Huk Formation in the road-cut at Furnes Church (Nikolaisen 1963).

Table 52. Cranidial measurements on *Telephina sulcata* Nikolaisen, 1963

PMO number	Lg	Wg	Wfpl	Lf	Wf
73014	2.4	3.4	1	1.2	3.1

Abbreviations: see Table 46.

Telephina viriosa Nikolaisen, 1963

Pl. 24, Fig. 17

1963   *Telephina* (*Telephina*) *viriosa* n. sp. Nikolaisen, pp. 381–383, pl. 3, fig. 11.

Type stratum and type locality. – Top Helskjer Member, 8.2 m above the Huk Formation in the road-cut at Furnes Church, Mjøsa (Nikolaisen 1963).

Material. – The holotype is the only known specimen.

Diagnosis. – Preoccipital part of glabella broadly subrectangular with broad and slightly truncated front and shallow lateral glabellar furrows; glabella not extending in front of palpebral furrows; anterior palpebral lobe wide. Occipital ring covered by dense tuberculation (emended from Nikolaisen 1963).

Remarks. – The taxon, which comes from practically the same horizon and locality as *T. sulcata*, appears morphologically very close to it. The main difference is the lack of deep lateral glabellar furrows or pits, but *T. viriosa* may merely represent a subjective synonym of *T. sulcata* (Nikolaisen 1963).

The cranidium appears to have a smooth surface, but this could easily relate to taphonomy.

Occurrence. – See for type specimen.

Telephina aff. *granulata* (Angelin, 1854)

Pl. 25, Figs 3–4

1963   *Telephina* (*Telephops*) *granulata* (Angelin, 1854) – Nikolaisen, pp. 381–383, pl. 3, fig. 11.
1995a   *Telephina* sp. – Ahlberg, p. 274.

Diagnosis. – Glabella subparabolic in outline with well-rounded front and fairly closely spaced frontal glabellar spines; length of preoccipital part of glabella approximating the width; occipital spine much longer than cranidium; lateral glabellar furrows moderately distinct (modified from Nikolaisen 1963).

Material. – Two cranidia (PMO 33331, 72726b).

Remarks. – This material was originally assigned to *T. granulata* by Nikolaisen (1963), but a later comparison with the type material of that species convinced Ahlberg (1995a) that the Norwegian specimens most probably should be assigned their own species. The most pronounced differences seem to be the presence of lateral glabellar furrows and possibly by a slightly coarser granulation.

Occurrence. – Heggen Member and perhaps the 'Cephalopod Shale'. South of Skiaker in Gran, Hadeland and road-cut at Furnes Church, Mjøsa (Nikolaisen 1963).

Telephina invisitata Nikolaisen, 1963

Pl. 25, Figs 5–6

1963 *Telephina* (*Telephops*) *invisitata* n. subgen., n. sp. Nikolaisen, pp. 392–394 (*cum. syn*), fig. 6, pl. 4, figs 11–13.

Type stratum and type locality. – Heggen Member. Håve in Håvesveen at Ringsaker, Mjøsa.

Material. – Six fragmentary cranidia (PMO 62112, 67198, 72691b, 72723, 72728b, 72729) and possibly a single librigena (PMO 72729a).

Diagnosis. – Glabella subparabolic in outline, preoccipital part about as long as wide with well-developed, compound lateral glabellar furrows and heavily tuberculation; frontal glabellar spines rather closely spaced; occipital spine long; fixigena narrow, crescentic (modified from Nikolaisen 1963).

Occurrence. – Heggen and Engervik members in the central to northern Oslo Region.

Order Ptychopariida Swinnerton, 1915

Suborder Ptychopariina Richter, 1933

Family Elviniidae Kobayashi, 1935

Genus Indet.

Remarks. – The taxon is closest to *Carolinites* Kobayashi, but is distinguished by the presence of strongly developed cranidial tuberculation; a rather wide occipital furrow and the rather convex and extremely wide fixigenae with small but distinctly developed genal spines. The rare occurrence in Baltoscandia (one specimen) may suggest an origin outside Baltica, probably reflecting an oceanic/cosmopolitan lifestyle like that of *Carolinites*.

Genus et sp. indet.

Pl. 25, Figs 7–9

1962 Komaspidid – Nikolaisen, p. 14.
1984 aff. *Carolinites* – Wandås, p. 211, pl. 2I.

Description. – Cranidium 10.4 mm long and strongly convex in transverse view. Glabella expanding forward, sub-ovoid, slightly bulbous; occipital ring wider than posterior part of preoccipital part of glabella, length approximating 20% the width, not sharply delimited from posterior fixigenal border; occipital furrow deep and wide (exsag.), the width nearly equaling the length of the occipital ring; preglabellar part of glabella widening slightly forwards, sub-ovoid, somewhat inflated, the length only slightly larger than the width; lateral glabellar furrows present but extremely faint. Fixigena highly convex, wide, the width equalling the glabellar width; posterior border narrow, continuing out into short and slender, posteriorly directed genal spine; posterior border furrow wide; palpebral lobes reaching from 25% of the cranidial length in front of the posterior margin and forward to somewhere around the anterolateral glabellar corners; bacculae large and strongly developed, their length corresponding to slightly over 20% of the cranidial length. Glabella and fixigena covered by large tubercles centrally, the tubercles becoming sparser or disappearing on the flanks.

Occurrence. – The specimen was found in the lowermost Helskjer Member, 2 m above the Huk Formation at Hovindsholm on Helgøya, Mjøsa.

Suborder Olenina Fortey, 1990

Family Olenidae Burmeister, 1843

Genus *Porterfieldia* Cooper, 1953

Type species. – *Triarthrus caecigenus* Raymond, 1920.

Diagnosis. – Preglabellar furrow and anterior border converge frontally; small anteriorly located eyes; 11 thoracic segments; a pygidium bearing seven axial rings (modified from Moore 1959; Månsson 1998).

Porterfieldia humilis (Hadding, 1913)

Pl. 25, Fig. 10

1998 *Porterfieldia humilis* (Hadding, 1913) – Månsson, p. 61 (*cum. syn.*), figs 2, 14.

Diagnosis. – Cranidium lacking preglabellar field; glabella subquadrate with three slightly impressed lateral glabellar furrows; S1 and S2 gently curved; S3 faint; occipital furrow almost straight medially, curving forward laterally; surface of internal mould covered by very faint granules; fixigena narrow. Pygidium with five axial rings (after Månsson 1998).

Occurrence. – Heggen Member in the road-cut at Furnes Church, Mjøsa. Outside Norway it is known from the contemporaneous Lower Andersö Shale member at Andersön and Raftan in Jämtland, Sweden (Månsson 1998; Nikolaisen 1965).

Suborder Harpina Whittington *in* Moore, 1959

Family Harpididae Hawle & Corda, 1847

Genus *Scotoharpes* Lamont, 1948

Type species. – *Scotoharpes domina* Lamont, 1948.

Diagnosis. – Cephalic outline suboval to subcircular, prolongations almost straight to adaxially curving; glabella longer than wide; occipital, preglabellar and axial furrows strong; eye tubercles opposite front quarter of glabella; gena and fringe with pits separated by smooth genal caeca that appear to branch from region of axial furrow just behind eye ridge; genal caeca dominantly radial on outer part of gena; single rows of slightly coarser pits are ususally developed against girder and upper and lower rims. Pygidium with strongly curved pleural furrows (modified from Norford 1973).

Scotoharpes rotundus (Bohlin, 1955)

Pl. 18, Figs 5–9

1955 *Aristoharpes? rotundus* n. sp. Bohlin, pp. 129–131, pl. 3, figs 5–7.
1984 *Scotoharpes rotundus* (Bohlin, 1955) – Wandås, pp. 233–234, pl. 11J–L, pl. 12A–C.

Type stratum and type locality. – The holotype cephalon U.M. No. Ar. 4225 was collected from the upper Asaphus raniceps Limestone at Gunnarslund on the parish of Persnäs, Öland, southern Sweden.

Norwegian material. – The Norwegian material consists of nine cephala (PMO S.1682, S.1683, 83216, 83242, 83243, 90575, 106.152, 107.385, 211.307) and two fringes (PMO 67147, 90575).

Diagnosis. – A species of *Scotoharpes* characterized by subcircular cephalon with short posterior prolongations reaching less than the central cephalic length posteriorly; a brim width which does not change anteriorly to laterally; relatively small alae and with poorly developed or absent caeca on girder (modified after Wandås 1984).

Description. – The species is fully described by Bohlin (1955) and Wandås (1984).

Remarks. – The cephala from the Huk Formation may vary slightly from the ones from the Helskjer Member in the more oval outline; slightly more posterior position of the eyes in relation to the glabella and possibly also by slightly larger alae. These differences are considered to be too small for taxonomic separation from *S. rotundus*.

Occurrence. – In Norway *S. rotundus* has been found from the middle Lysaker Member of the Huk Formation and up into the uppermost Helskjer Member of the Elnes Formation. Skien-Langesund: Skinnvikstangen at Rognstrand (Helskjer Member). Eiker-Sandsvær: Muggerudkleiva (lower Elnes Formation). Oslo-Asker: Grundvik (house) at Hiken just south of Slemmestad (Lysaker Member, Huk Fm.), Hukodden on Bygdøy (Svartodden Mb., Huk Fm.). Mjøsa: Road-cut at Furnes Church (top Helskjer Member, 7.75 m above the Huk Formation).
  Outside Norway it has been described from Öland in southern Sweden, where it occurs from the upper Asaphus raniceps Limestone and the succeeding Megistaspis gigas Limestone (Bohlin 1955).

Order Phacopida Salter, 1864

Suborder Phacopina Struve *et al. in* Moore, 1959

Family Pterygometopidae Reed, 1905

Subfamily Pterygometopinae Reed, 1905

Genus *Pterygometopus* Schmidt, 1881

Type species. – *Calymene sclerops* Dalman, 1827.

Diagnosis. – Preglabellar furrow distinct, joining dorsal furrow laterally. Anterior branch of facial suture running just in front of preglabellar furrow; posterior branch situated in deep furrow. Eyes of moderate size, anteriorly reaching the axial furrow. Genal angles rounded. Vincular furrow distinct. Pygidial pleurae commonly faintly concave peripherally, with five to six pleural furrows (after Jaanusson & Ramsköld 1993).

Pterygometopus sp.

Pl. 26, Fig. 3; Pl. 29, Fig. 11

Material. – A single external mould of a compressed cephalon (PMO 207.522) preserved in marly mudstone and a small pygidium (PMO 83192).

Description. – Cephalon sub-triangular, the length corresponding to 50% of the width on flattened specimen; glabella widening strongly forward, the maximal width at anterior margin corresponding to 220% of the smallest width found outside S1 or 46% of the total cranidial width; occipital ring bearing distinct mesial tubercle; lateral glabellar furrows short but deep, directed slightly rearwards abaxially; S1 bifurcating with one the side-branch reaching slightly rearwards; S3 located slightly behind anterior margin of eye, very gently posteromedially curved; lateral glabellar lobes rectangular in outline, the length increasing forward; frontal lobe distinctly rhomboidal. Axial furrow deep, trench-like. Eyes large, situated close to each other, the distance between inner margins corresponding to just over the width of the occipital ring; length of eye approximating 45% of total cephalic length, the posterior margin nearly reaching posterior fixigenal furrow. Posterior branch of facial suture located in deep furrow situated just in front of the posterior border furrow on adaxial part of fixigena, continuing in a large, forward reaching curve abaxially, the anteriormost part at level with the anterior half of L2. Genal corners rounded.

Pygidium 3.5 mm long, the length corresponding to half the width; axis strongly delimited, the length corresponding to 83% of the pygidial length, while the width takes up approximately 30% of the pygidial width; axis bearing five strong anterior and two effaced posterior axial rings; pleural field with five strongly developed pleural ribs bearing moderately distinct interpleural furrows; the ribs extend to a point slightly over 10% of the pygidial width from the posterolateral margin, where they fade out, although faint traces of the interpleural and pleural furrows may be followed out to just inside the margin.

Remarks. – *Pterygometopus* sp. resembles the type species *P. sclerops* Dalman from slightly older rocks of Baltoscandia, but is readily differentiated on the larger eyes corresponding to nearly half the cephalic length; the much stronger abaxial lengthening (exsag.) of the posterolateral fixigenal projection; the lack of an anterior prolongation of the palpebral lobe and possibly by a relatively longer frontal glabellar lobe. Many of the same differences also separate *P.* sp. from *P. sclerops* var. *angulata* Schmidt, 1881 from the East-Baltic area.

Occurrence. – Both specimens have been collected from the Helskjer Member in the road-cut at Furnes Church, Mjøsa; one in an erratic boulder and the other from approximately 3.7 m above the base of the Helskjer Member.

Suborder Calymenina Swinnerton, 1915

Family Calymenidae Burmeister, 1843

Subfamily Calymeninae Burmeister, 1843

Genus *Gravicalymene* Shirley, 1936

Type species. – *Gravicalymene convolva* Shirley, 1936.

Diagnosis. – Glabella with three pairs of lateral glabellar lobes, none papillate; preglabellar furrow deep, its anterior edge merging into thick (sag.) rolled anterior border (Moore 1959).

Gravicalymene capitovata Siveter, 1977

Pl. 26, Figs 5–8

1953 *Calymene* – Størmer, p. 79.
1977 *Gravicalymene capitovata* sp. nov. Siveter, pp. 377–385, figs 11A–H, 12A, E–I, K–L.

Type stratum and type locality. – Heggen Member in the road-cut at Muggerudkleiva, Eiker-Sandsvær.

Material. – cranidia (PMO 60414, 60418 (three specimens), 60421, 66598 (two specimens), 80844, 82305, 82681, 91032–3, 91042, 91043, 91045, 91047–8, 91060, 91061–2, 91064, 121.644, 208.537a/3), three rostral pl.s (PMO 91036 (two specimens), 91056), five librigenae (PMO 91034, 91036, 91037, 91055, 91063), two hypostomes (PMO 91038, 91051–2), eight pygidia (PMO 60404, 60410, 82705, 91039, 91049–50, 91058–9, 91061, 121.644) and one partly articulated specimen (PMO 60606).

Diagnosis. – Glabella strongly bell-shaped in outline with large ovate L1 lobe corresponding to three times the L2 lateral glabellar lobe; frontal glabellar lobe situated behind anterior part of fixigena; anterior border slopes steeply forwards and upwards from the nearly vertical anterior side of short preglabellar furrow; eye ridge strongly delineated abaxially of axial furrow opposite S2 lateral glabellar furrow; posterior border furrow narrow. The pygidium has six complete and two incomplete axial rings and six pleural furrows (Siveter 1977).

Description. – The species was fully described by Siveter (1977) and no new data may be added here.

Occurrence. – The species occur from the middle to upper part of the Heggen Member and up into the basal Vollen Formation in the central Oslo Region. Eiker-Sandsvær: Muggerudkleiva (Heggen Member); waterfall in the Ravalsjøelv and in profile 1 at Råen (18.94 m below datum). Oslo-Asker: Oslo (lower Vollen Formation together with *Reedolithus carinatus* (Angelin)).

Gravicalymene sp.

Pl. 26, Fig. 4

1984 *Pterygometopus* sp. Wandås, p. 237, pl. 13M–N.

Material. – A single pygidium (PMO 83016).

Description. – Pygidium with long, strongly vaulted and sharply delimited axis reaching nearly all the way back to the posterior margin; the width correponds to around 35% of the pygidial width; axis bearing six complete and strongly developed axial rings and one nearly effaced ring in front of short terminal piece. Pleural field with six pleural ribs, each with a distinct interpleural furrow reaching all the way from the axial furrow to the lateral margin, slightly deepening and expanding posterolaterally.

Remarks. – Although having only seven axial rings, the pygidium shows a strong resemblance to that of *G. capitovata* treated above, but it is currently excluded from that both on the basis of the missing axial ring and its early stratigraphic occurrence.

Occurrence. – The pygidium derives from the Helskjer Member in the road-cut at Furnes Church.

Genus *Sthenarocalymene* Siveter, 1977

Type species. – *Sthenarocalymene lirella* Siveter, 1977.

Diagnosis. – Glabella rather inflated with an overall bell-shaped outline, becoming sub-quadrate in front of lateral glabellar lobe L1; frontal lobe anteriorly steep; lateral glabellar furrows short; anterior border narrow, running smoothly into preglabellar furrow and lying well below dorsal glabellar surface; axial furrow deep; posterior border furrow narrow (exsag.) and surface sculpture consisting of small, closely spaced tubercles and granules (Siveter 1977).

Sthenarocalymene lirella Siveter, 1977

Pl. 26, Figs 9–12

1977 *Sthenarocalymene lirella* gen. et sp. nov. Siveter, pp. 388–392, figs 12B–C, J, 13A–J.

Type stratum and type locality. – Vollen Formation on western side of Bygdøy, Oslo-Asker.

Material. – The material consists of 13 cranidia (PMO 81919, 82465–6, 82467 (two specimens), 82528, 82581, 91065, 91067–8, 91069–71 (three cranidia), 91072–3, 206.170a/3) and two partly articulated specimens (PMO 57471, 82521).

Diagnosis Preglabellar furrow very short and shallow; anterior border short with gently inclined dorsal surface; anterior margin of frontal glabellar lobe straight or very slightly convex forwards; mid-length of palpebral lobe opposite centre of posterior part of lateral glabellar lobe L2; eye ridge absent from palpebral lobe to axial furrow. Thorax with 13 segments (Siveter 1977).

Remarks. – *S. lirella* is readily differentiated from *Gravicalymene capitovata* treated above by the less forwardly narrowing glabella; the smaller L1 lateral glabellar lobes and the much shorter anterior border, which is shorter (sag.) than the occipital ring. At present no differences have been observed on the librigena and like *G. capitovata*, *S. lirella* has six pleural ribs and probably seven axial rings on the pygidium, not five of each as thought by Siveter (1977, p. 391).

Occurrence. – Basal Sjøstrand and upper Heggen members, Elnes Formation, and up into the Fossum and Vollen formations. Skien-Langesund: 100 m NW of the lighthouse at Saltboden in Frierfjorden (Fossum Fm.). Eiker-Sandsvær: Muggerudkleiva (top Heggen Member). Oslo-Asker: Bøveien south of Slemmestad (top Sjøstrand Member, 3.14 m above datum), Elnestangen (basal Engervik Member) and western side of Bygdøy (Vollen Fm.).

Suborder Cheirurina Harrington & Leanza, 1957

Family Cheiruridae Salter, 1864

Subfamily Deiphoninae Raymond, 1913b

Genus *Sphaerocoryphe* Angelin, 1854

Type species. – *Sphaerocoryphe dentata* Angelin, 1854.

Diagnosis. – Cheirurid genus with bulbous frontal glabellar lobe; glabellar S2 and S3 apodemes in ventral view reduced to a low knob in the axial furrow with no continuation adaxially onto the glabella on the dorsal side and well-developed fixigenae bearing a profixigenal spine in front of strong genal spines. Thorax with nine segments (Moore 1959; Pärnaste 2004a,b).

Sphaerocoryphe sp.

Pl. 27, Figs 1–4

Material. – One exfoliated glabella (PMO 143.563) and one fragmentary cephalon (PMO 143.562).

Description. – The largest cephalon is approximately 3.5 mm long with a broad subtriangular outline. Sagittal length equals around one-third the width (genal spine included).

Glabella semicircular in frontal view, widest part corresponding to one-third the cephalic width located opposite adaxial part of anterior librigenal border. Occipital ring and furrow not preserved. Posterior lateral glabellar lobes low, appearently only slightly isolated from the rest of the glabella. Frontal glabellar lobe bulbous, strongly delineated with an even semicircular outline sagitally and overhanging anterior border. Lateral flanks steep, but not overhanging. Lateral glabellar furrows, except S1, absent on internal mould at least. S1 deep, curving around L1, adaxial termination located halfway back to the occipital furrow.

The axial furrows moderately deep, forwardly diverging.

Palpebral lobes lying close to axial furrow, their posterior edge situated just in front of S1. The posterior fixigenal field is short but wide with transverse curvature strong adaxially, decreasing distally on genae; exsagittal curvature small. Posterior border and furrow exsagittally narrow, only widening slightly abaxially. Lateral border continuing out into a short profixigenal spine situated opposite posterior margin of fixigenal field. Genal spines short and stout, curving slightly posteriorly, the length corresponding to half the frontal glabellar lobe length. Anterior branches of facial sutures diverge slightly forwards; posterior branches weakly sigmoid, resulting in a very low increase in posterior fixigenal field width distally, turning strongly rearwards on lateral border, where they reach the margin just in front of the profixigenal spine. The librigena is strongly triangular in outline. The specimens are both strongly exfoliated, but the fixigena bears strong pitting.

Remarks. – The species may possibly be the oldest known member of *Sphaerocoryphe* with its appearence at the top of the *Didymograptus artus* graptolite Zone or just above. This makes the specimens very significant as all the other Middle Ordovician species are from Laurentia, occurring in Scotland and North America. The earliest Baltoscandian species previously recorded is from the Upper Ordovician Sandbian Stage, suggesting a non-

Baltoscandian origin of the genus. The new findings together with the recently described and closely related *Hemisphaerocoryphe platinflata* Hansen, 2005, from the lowermost Darriwilian of Baltoscandia indicate that the subfamily Deiphoninae probably originated on Baltica.

The specimens differ from the nearly contemporaneous *S. saba* Tripp, 1962 from the Confinis Flags in the Girvan district, Scotland, by the less bulbous frontal glabellar lobe; the short glabellar 'neck' between occipital furrow and frontal glabellar lobe and by the shorter profixigenal spine. The slightly younger *S.* sp. A of Tripp (1962) from the same unit as *S. saba*, is differentiated on the larger posterior L1 lobes; the long glabellar 'neck'; the more bulbous frontal glabellar lobe and by the distinctly pustulate glabella. *Sphaerocoryphe* sp. C of Tripp (1962) from Scotland may be separated on its more posteriorly directed and slender genal spines.

The upper Darriwilian *Sphaerocoryphe ludvigseni* Chatterton, 1980 from the Esbataottine Formation in Canada is easily differentiated on the nearly effaced L1 lobes situated behind a well-developed glabellar 'neck'; the presence of two profixigenal spines and by the clearly more curved outline of the genal spines. The upper Darriwilian *S. akimbo* Tripp, 1967 from the upper Stinchar Limestone of the Girvan district may be separated on the much longer and more curved genal and profixigenal spines.

Occurrence. – The two specimens are both collected from the limestone bed 7.75 m above the Huk Formation in the road-cut at Furnes Church, Mjøsa.

Subfamily Cheirurinae Hawle & Corda, 1847

Genus *Cyrtometopus* Angelin, 1854

Type species. – *Calymene? clavifrons* Dalman, 1827.

Diagnosis. – Anterior cranidial border angulate in outline anterolaterally, connected to palpebral lobe by sutural ridge. Glabella moderately inflated, elongate, with isolated, ovate L1 lobes; S2 and S3 short; eyes anteriorly, situated opposite L3; rostral pl. with a pair of short, anteriorly directed spines. Hypostome with narrow borders and anterior wings located far forward. Pygidium with three pairs of pygidial spines, anterior pair very long and posterior two short; terminal piece of axis forming slight median projection posteriorly (Lane 2002).

Cyrtometopus clavifrons (Dalman, 1827)

Pl. 27, Figs 5–9

2002   *Cyrtometopus clavifrons* (Dalman, 1827) – Lane, pp. 155–164 (*cum. syn.*), fig. 1, pls 1–3.
2003   *Cyrtometopus clavifrons* – Hansen & Nielsen, pp. 108, 110, fig. 3.

Type stratum and type locality. – Lectotype cranidium RM Ar.17907 selected by Lane (2002) comes from the Expansus or lower Raniceps Limestone of the Swedish Komstad Limestone at Skarpåsen, Östergötland.

Norwegian material. – The examined material consists of one cephalon (PMO 102.648), 32 cranidia (PMO 1994, 90587, 101.699, 101.770, 102.457, 102.474, 102.549, 102.550, 102.557, 102.568, 102.569, 102.575, 102.591, 102.593, 102.598, 102.603, 102.652, 102.677, 105.984, 106.055, 106.065, 106.073, 106.074, 106.077, 106.376, 106.444, 106.454, 106.526, 106.533–34 (one specimen), 106.539, 142.203, 206.947/1) and two partly articulated specimens (PMO 1664, 83587).

Diagnosis. – Largest glabellar width and height located centrally on cranidium; genal spines diverging posterolaterally with approximately 30° to exsagittal line (modified from Angelin 1854; Schmidt 1881).

Description. – The species was thoroughly redescribed by Lane (2002).

Remarks. – The cranidium seems in general to have been very morphologically conservative through time but a change in sagittal curvature of the cranidium, resulting in a stronger convexity in the stratigraphically youngest specimens is possible (own observation).

Occurrence. – In Norway *C. clavifrons* occurs from the Lysaker Member or perhaps even the underlying Hukodden Member of the Huk Formation and up into the lower Elnes Formation (Brøgger 1882; Nielsen 1995; Wandås 1984). Skien-Langesund: Skinnvikstangen at Rognstrand (Helskjer Member). Eiker-Sandsvær: Stavlum at Fiskum (Helskjer Member, 0.32 to 1.16 m above the Huk Formation). Modum: road-cut at Vikersund skijump (Helskjer and lower Heggen Member, 0.40 to 15.35 m above the Huk Fm.). Oslo-Asker: Djuptrekkodden north of Slemmestad (Hukodden to middle Lysaker Member, Huk Fm. (Nielsen 1995)), Huk on Bygdøy (Huk Fm.), Tøyen in Oslo (Lysaker Member). Mjøsa: road-cut at Furnes Church (Helskjer Member), Helskjer on Helgøya (Helskjer Member, 4.5 m above the Huk Formation).
  Cyrtometopus clavifrons is relatively common in the Baltoscandian deposits from around the Volkhov-Kunda Regional Stage. In the East-Baltic it is known from the top of zone $B_{II}\beta$ and up into $B_{III}\alpha$ of Lamansky (1905), corresponding to the Glauconite Limestone and succeeding Vaginatum Limestone of Schmidt (1881) (Schmidt 1881, 1907; Brøgger 1882; Lamansky 1905; Hansen & Nielsen 2003). It is found from the *Megistaspis simon* trilobite Zone and up into the *Asaphus raniceps* trilobite Zone of the Swedish Komstad Limestone; the Komstad Limestone on the Danish island of Bornholm and in the Polish Miedzygórz bed (Linnarsson 1869; Brøgger 1882; Törnquist 1884; Poulsen 1936; Bohlin 1949; Tomczykowa & Tomczyk 1970; Nielsen 1995).

Family Pliomeridae Raymond, 1913a

Subfamily Pliomerinae Raymond, 1913a

Genus *Pliomera* Angelin, 1854

Type species. – *Asaphus fischeri* Eichwald, 1825 (by subsequent designation, Vodges 1925)

Diagnosis. – Genus defined by gonatoparian facial sutures; anterior lateral glabellar furrows reaching anterior border furrow, situated in front of anterolateral angles of glabella; frontal glabellar lobe with mesial indentation; anterior cephalic border bearing seven to nine denticules; fixigena L-shaped, pitted; eyes small, situated posteriorly outside L2 lobes; palpebral lobes raised; eye ridges absent. Thorax with 12 to 18 segments. Pygidium bearing four to five axial rings; terminal piece minute; pleurae ending in very short spines or blunt points, last pair completely embracing piece (after Moore 1959).

Pliomera fischeri (Eichwald, 1825)

Pl. 27, Figs 10–13

1825   *Asaphus Fischeri*, m. – Eichwald, pp. 52–53, pl. 3, figs 2a–b.
1857   Pliomera Fischeri Eichw. – Kjerulf, p. 285.
1878   *Pliomera fischeri*. Eichw. – Angelin, pp. 30–31, pl. 20, fig. 2.
1881   *Amphion Fischeri* Eichw. sp. – Schmidt, pp. 191–195, pl. 13, figs 1–8.
1882   *Amphion Fischeri*, Eichw. – Brøgger, pp. 135–136, pl. 6, figs 3–3a.
1955   *Pliomera fischeri* – Bohlin, pp. 114–115, fig. 3, table 1.
1955   *Pliomera fischeri* (Eichw.) – Jaanusson & Mutvei, p. 30.
1957   *Pliomera fischeri* (Eᴵᴄʜᴡᴀʟᴅ) – Harrington & Leanza, pp. 36, 215.
1981   *Pliomera fischeri* (Eichwald, 1825) – Bruton & Harper, pp. 171–172, pl. 5, figs 8–13.
1984   *Pliomera fischeri* (Eichwald, 1825) – Wandås, p. 236 (*cum. syn.*), pl. 13G–L.
2003   *Pliomera fischeri* (Eichwald, 1825) – Hansen & Nielsen, p. 108, fig. 3.

Type stratum and type locality. – Stergötland in Sweden.

Norwegian material. – The examined material consists of 41 cranidia (PMO 83171, 83237, 90512, 101.783 (two specimens),

101.799, 101.805, 102.398, 102.438, 102.441, 102479, 102.483, 102.485, 102.513, 102.522, 102.523 (five specimens), 102.529, 102.530, 102.536 (two specimens), 102.552, 102.570, 102.582 (two specimens), 102.628, 102.795, 105.661, 105.927, 106.076, 106.395, 106.400, 106.423, 106.530, 106.543 (two specimens), 201.996/2, 201.997/2), one hypostome (PMO 102.814), 24 pygidia (PMO 101.783 (two specimens), 102.410, 102.429 (two specimens), 102.467, 102.479, 102.480, 102.486, 102.512, 102.523 (five specimens), 102.548, 102.611 (three specimens), 102.814, 106.073, 106.406, 106.409, 202.275/1) and 12 partly articulated or complete specimens (PMO H2627, 2259, 2276, 2279, 2280, 102.416, 102.489, 102.497, 102.537, 106.428, 106.431, 202.117/3).

Diagnosis. – Species characterized by 18 thoracic segments and five axial rings on pygidium (based on Harrington & Leanza 1957).

Description. – The species is thoroughly described by Schmidt (1881, pp. 192–193).

Remarks. – The differences inferred by Bruton and Harper (1981, p. 171) between the pygidia of early and late forms can not be verified by the present study, which was based on the large collection of Norwegian specimens stored at NHM. The axial width was thus found to approximate 65 to 75% of the anterior pleural rib width (tr.) on both older and younger specimens, while the posterior distance from axis to pygidial margin corresponds to 60% of the axial length. The proposed differences are solely related to different grades of deformation and weathering.

Occurrence. – In the Oslo Region *P. fischeri* is known from the Lysaker Member of the Huk Fm. and up into the basal Sjøstrand and Heggen Members. Modum: road-cut at Vikersund skijump (Helskjer to lower Heggen Member, 1.0 to 13.9 m above the Huk Fm.). Oslo-Asker: Slemmestad (Lysaker Member), Gertungsholmen (Lysaker Member), the Old Eternite Quarry (Helskjer Member, 0.42 to 5.63 m above the Huk Fm.), Tøyen in Oslo (upper Lysaker Member, Huk Fm.). Hadeland: Hovodden at Randsfjorden (8.3 to 13.6 m above the Huk Fm.). Mjøsa: road-cut at Furnes Church (upper Helskjer Member, 7.75 m above the Huk Fm.).
In south central Norway it has been described from the middle Darriwilian Otta Conglomerate of probable island arc origin (Bruton & Harper 1981).
Outside Norway it is known from the Asaphus limestones of Vestergötland, Östergötland, Öland, Småland and in a few cases also from Scania (Angelin 1878; Bohlin 1955). It is quite common in the East Baltic area. In the NW Russian St Petersburg province it occurs from the basal Volkhov Formation of the Volkhov Regional Stage and up into the succeeding Vaginatum Limestone from Kunda Regional Stage (Hansen & Nielsen 2003; Schmidt 1881). In Estonia it is known from the Vaginatum Limestone (Schmidt 1881).

Family Encrinuridae Angelin, 1854

Subfamily Cybelinae Holliday, 1942

Genus *Cybelurus* Levitskiy, 1962

Type species. – *Cybelurus planus* Levitskiy, 1962.

Diagnosis. – Glabella expanding evenly forward; three pairs of transversely directed lateral glabellar furrows long, deepest adaxially; S3 lateral furrow forked; frontal lobe deeply indented anteriorly by median furrow; palpebral lobe situated more than half way out across fixigenal field opposite L2 lobe; prominent eye ridge connecting palpebral lobe with axial furrow at S3. Thorax consisting of 12 or more segments; sixth segment from rear prolonged by long, posteriorly directed spines. Pygidium with approximately 14 axial rings; pleural region composed of four segments, posterior pleural bands continuing out into short spines (modified from Whittington 1965a).

Cybelurus cf. *mirus* (Billings, 1865)

Pl. 27, Figs 14–17

1961   *Pliomerops* sp. Nikolaisen, p. 295, pl. 2, fig. 4.
1965a   *Miracybele* sp. – Whittington, p. 424.
1968   *Miracybele* sp. Nikolaisen, pp. 2–3 (comments on Norwegian material).
1973   *Cybelurus* sp. – Dean, p. 13.
? 1980 *Cybelurus* cf. *mirus* (Billings 1865) – Fortey, p. 99, pl. 23, figs 7–9.
1984   *Cybelurus* cf. *mirus* (Billings, 1865) – Wandås, pp. 236–237, pl. 13 K–L.

Norwegian material. – Two cranidia (PMO 36600, 82991).

Description. – Cranidium just over 2.5 times as wide as long. Glabella gently convex (sag., tr.), only slightly elevated above cheeks, moderately expanded forward, maximum width at anterolateral corners around 85% of sagittal length or just over 1.6 times the width of the occipital ring. Anterior margin gently rounded. Occipital ring longest (sag.) medially, the length taking up around 12% of the glabellar length. Occipital furrow curving forwards and shallowing medially. Three pairs of lateral glabellar furrows extend one-third across the glabella, S1 and S2 deepening and widening adaxially, the first directed slightly rearwards adaxially, while the latter crosses glabella transversely. S3 bifurcates just outside its midpoint, the posterior branch gently concave forward, strongly deepening adaxially; main branch directed obliquely forward, dying out on frontal glabellar lobe. Glabellar lobes progressively larger anteriorly, slightly inflated. Axial furrows narrow, fairly deep, describing a gently outwards concave curve, deepening slightly in front of eye ridge. Preglabellar furrow shallow laterally, medially deepening into deep, elongate (tr.) pit from which a rather deep mesial furrow runs rearwards, bisecting the frontal glabellar lobe; mesial furrow distinctly Y-shaped with wide anterior opening. Fixigena triangular, posterolaterally very short (exsag.), highest point located slightly adaxially and posteriorly to palpebral lobe. Palpebral lobe far back on fixigena, opposite L2 or one-fourth the cranidial length from posterior margin and nearly two-thirds the glabellar width from axial furrow. Thin eye ridge reaches inwards and forwards from palpebral lobe, meeting axial furrow opposite anterior edge of L3 lobe. Genal spine short and slender, directed posterolaterally. Posterior border furrow deep but narrow. Posterior border convex, narrow (exsag.), widening slightly laterally. Anterior border rather narrow, only slightly expanded medially, sagittal length corresponding to less than 10% of the cranidial length.

Fixigena characterized by strongly developed fine reticulation. Glabella and borders are covered by fine granulation, most strongly developed on frontal part of glabella and on anterior border.

Remarks – As earlier noted by Wandås (1984, pp. 236–237) this taxon has a strong affinity with the approximately contemporaneous *Cybelurus mirus* from the Table Head Formation of Newfoundland and is clearly more closely related to this species than any other described species. The Norwegian specimens differ in their distinctly Y-shaped anterior glabellar furrow and by the narrower (exsag.) posterior fixigenal border continuing into rather small and slender genal spines. The differences though are rather small and could easily reflect local variation. Fortey (1980, p. 99) described a single specimen from mid-Darriwilian rocks of Spitsbergen, which he tentatively assigned to *C. mirus*. This seems even closer to the Norwegian specimens than the type form and differs merely on a stronger forward expansion of the glabella and a slightly wider and more flattened distal part of the posterior fixigenal border. In contrast to the type form of *C. mirus* it resembles the Oslo material in the wider anterior opening of the mesial furrow frontally on the glabella and in the more slender genal spines.

Occurrence – The two Norwegian specimens are coming from the top of the lower half of the Elnes Formation of the northern Oslo Region. Mjøsa: road-cut at Furnes Church (top Helskjer Member, 7.75 m above the Huk Formation), Redalen south of Biri (Heggen Member). A single specimen, which may be identical with the Norwegian material, has been found in the mid-Darriwilian Profilbekken Member on Svalbard (Fortey 1980).

Genus *Cybellela* Reed, 1928

Type species – *Zethus rex* Nieszkowski, 1857.

Diagnosis – Glabella with subparallel to slightly forwardly diverging sides, widest point located outside

frontal glabellar lobe; three pairs of deep, slightly pit-like lateral glabellar furrows; anterior cranidial border with five, forward directed short spines or thorns; eyes stalked, situated opposite to anteriorly of L2; fixigena generally lacking a genal spine. Thorax with strongly effaced tuberculation. Pygidium with four pairs of pleural ribs; each rib bearing one or two tubercles (based on Moore 1959; Krueger 2003).

Remarks – Henningsmoen *in* Moore (1959) regarded *Cybellela* Reed, 1928 as a junior synonym of *Atractopyge*, but the genus was resurrected by Whittington (1965b, p. 44), who referred to the diagnostic characters presented by Reed (1928): a parallel-sided glabella with three to four pairs of regularly disposed tubercles; distinctly forward eyes; produced genal corners; the fine pitting on the cheeks with few scattered tubercles. Except perhaps for the degree of forward glabellar expansion, where the glabella generally may be described as having either markedly forward diverging sides or parallel sides, all the characters appear gradual and may occur in various combinations, making the differentiation very subjective. On the other hand Krueger (2003, 2004) uses additional features such as the general eye length and pygidial outline, where *Cybellela* is characterized by having long, stalked eyes and a posteriorly sharply pointed hypostome. The hypostome is still only known for a few species, making this a slightly problematic diagnostic character, but the length of the eye stalk may actually be a sound character, especially in connection with the forward expansion of the cranidium and generally much sparser tuberculation found on the cranidia of *Cybellela*. For these reasons *Cybellela* and *Atractopyge* are here regarded as two separate, but closely related genera.

Cybellela sp.

Pl. 27, Figs 19–20; Pl. 28, Fig. 8

Material. – The material consists of one fragmentary cranidium (PMO 83173), two librigenae (PMO 83178, 207.394/8) and two pygidia (PMO 207.395/3, 207.395/7).

Description. – Cranidium only represented by one very fragmentary specimen (PMO 83173, Pl. 28, fig. 8). Glabella widening forwardly. Lateral glabellar furrows relatively long and deep, *Cybele* like, only slightly posteriorly directed adaxially; length of furrows forwardly increasing from posterior to anterior pair. Lateral glabellar lobes slightly inflated and widen abaxially. Axial furrows deep but fairly wide. Fixigena with well developed, posterolaterally directed eye ridge; base of eye stalk located outside S2 to posterior part of L3 at a distance corresponding to just below half the posterior preoccipital width of glabella from the axial furrow. Strong, posterolaterally directed genal spine.

Glabella smooth except for at least three pairs of large and strongly developed median tubercles and an additional, but slightly smaller tubercle situated just in front of S3 furrow. Fixigenal field covered by pits and sparse, but evenly distributed tubercles.

Librigena strongly arched with well rounded margin.

Pygidium fairly wide with shoulder-like posterolateral corners and slightly convex lateral margin; axis occupying 80% of the pygidial length (anterior half ring excluded) and 40% of the pygidial width; 13 to 16 axial rings, the anterior three or four distinct throughout, while the posterior ones become effaced mesially; pleural field with four pairs of ribs ending in short, club-like spines at pygidial margin. Pleural ribs with no or only a few indistinct tubercles.

Remarks. – The species show marked similarities to the ancestral genus *Cybele* Lovén, 1846 including the long, lateral glabellar furrows and strong genal spine, but is here assigned to *Cybellela* because of the lack of a fifth pleural rib on the pygidium. It is differentiated from the slightly younger *Cybellela wollinensis* Krueger, 2003 from Germany by the longer and less anterolaterally directed lateral glabellar furrows; the longer (exsag.) posterolateral fixigenal projection and by the finer and denser fixigenal tuberculation. The species is clearly distinguished from the approximately contemporaneous *Cybele woehrmanni* Schmidt, 1907 and *Cybele ruegensis* Krueger, 2003 by the number of pleural ribs; the posterior position of the eye stalk and by the sparse, but evenly distributed fixigenal tuberculation.

Occurrence. – The species is rare in the uppermost Helskjer Member from the northern Oslo Region. Mjøsa: road-cut at Furnes Church (top Helskjer Member, 7.75 m above the Huk Fm.) and road-cut along E6 at Nydal just north of Hamar (top Helskjer Member, 7.30 to 7.32 m above the Huk Fm.).

Genus *Atractopyge* Hawle & Corda, 1847

Type species. – *Calymene verrucosa* Dalman, 1827.

Diagnosis. – Glabella widening forward with three pairs of lateral glabellar furrows; row of tubercles or spines in front of glabella; glabella moderately to densely tuberculated; eyes sessile or shortly stalked. Hypostome with rounded to moderately pointed posterior margin. Pygidium with four pairs of ribs (modified from Krueger 2003, 2004; Moore 1959).

Atractopyge dentata (Esmark, 1833)

Pl. 27, Figs 18, 21–24; Pl. 28, Figs 12

1833	*Trilobites dentatus* Esmark, p. 269–270, pl. 7, fig. 10.
1854	*Cybele dentata.* Esm. – Angelin, p. 89, pl. 41, fig. 12.
1878	*Cybele dentata.* Esm. – Angelin, p. 89, pl. 41, fig. 12.
1940	*Cybele dentata* (Esmark) – Størmer, p. 125, text-fig. 2: 10, pl. 1, fig. 5.
1961	*Atractopyge dentata* (Esmark, 1833) – Nikolaisen, pp. 298–302 (*partim*), pl. 4, fig. 4 (specimens from stage 4b belong to *Cybellela grewingki*).
non 1965b	*Cybellela dentata* (Esmark, 1833) – Whittington, pp. 44–46, text-fig. 3D, G, pl. 13: 3–12, 14.
non 1971	*Atractopyge dentata* (Esmark, 1833) – Neben & Krueger, pl. 37, figs 4–6, pl. 46, figs 18–19 (belongs to *Cybellela grewingki*).
non 2003	*Cybellela dentata* (Esmark, 1833) – Krueger, pp. 41–44, pl. 2, figs 10–11, 8, pl. 9, fig. 12 (belongs to *Cybellela grewingki*).

Type locality and type stratum. – Lectotype pygidium PMO 56161a selected by Størmer (1940) originates from the upper Elnes Formation or less likely the Vollen Formation at the Royal Palace in Oslo.

Material. – The material consists of two cranidia (PMO 121.644, 211.428–29), one fixigena (PMO 56161a) and four pygidia (PMO 56161a, 33473, 60607, 62111).

Diagnosis. – Glabella widening strongly forward with two distinct rows of large tubercles on each side of the glabella; occipital ring short, bordered anteriorly by narrow and straight occipital furrow; stalked eyes situated centrally on the coarsely granulated fixigena opposite S2 and posterior part of L3; anterior border with short spines mesially and genal corners continuing out into well-developed genal spines.

Description. – Cranidium 40% as long as wide (width excluding genal spines), gently convex in both directions. Glabella strongly forward widening, the anterior width corresponding to nearly 1.4 times the occipital ring width on examined specimen, while the anterior width approximates 90% of the glabellar length. Occipital ring taking up 20% of the glabellar length, widening moderately mesially, anteriorly bordered by distinct but rather narrow and straight occipital furrow ending in apodeme laterally. Lateral glabellar furrows short and deeply pit-like, directed obliquely anterolaterally, the anteriormost located around 60% of the glabellar length from the posterior margin; S1 reaching just over halfway towards the sagittal line. Frontal glabellar lobe expands abruptly, broadly rounded anteriorly and lacking independent convexity. Axial furrows deep with deep fossula just behind largest glabellar width. Preglabellar furrow shallow laterally, becoming slightly deeper mesially. Anterior border only preserved on one specimen (PMO 121.644) on which it continues out into a broad, six-fingered and somewhat hand-like extension reaching more than half the glabellar length out in front of glabella; the anterior half has a width corresponding to half the glabellar width. Only the base of the extension is preserved on cranidium PMO 211.429, but the outline of the proximal part of the extension appears identical. Fixigena moderately narrow, the width corresponding to 40% of the cranidial width, laterally continuing out into stout, posteriorly directed genal spines. Eye stalk located opposite S2 and posterior half of L3 or nearly 25% of the cranidial length in front of the posterior margin and around 12% of the cranidial width from the axial furrow. Eyes situated on short stalks, the length corresponding to nearly 25% of the maximum glabellar width. Eye ridge strongly developed, reaching axial furrow in front of S3 lateral glabellar furrow. Posterior border expands strongly outwards and bordered anteriorly by moderately broad and deep posterior furrow.

Glabella bears three pairs of large tubercles mesially. Another row consisting of five tubercles is found laterally on each side of the glabella, the tubercles distributed with one on each lateral glabellar lobe and last two describing an anterolaterally directed line on the frontal glabellar lobe. Smaller granules cover the area between the tubercles. Fixigena is coarsely granulated with scattered tubercles throughout; a pair of especially large tubercles is located in continuation of the lateral glabellar row of tubercles anterolaterally on the preglabellar border.

Librigena, hypostome and thorax unknown.

Pygidium with shoulder-like posterolateral corners and slightly convex lateral margin; axis occupying up to 80% of the pygidial length (anterior half ring excluded) and one-third the pygidial width; about 13 to 16 axial rings, the anterior three or four distinct throughout, while the posterior ones become effaced mesially; pleural field with four pairs of ribs ending in short, club-like spines at pygidial margin. Axis and pleural ribs with few indistinct tubercles.

Remarks. – Because of the sparse and rather poor material from the type stratum this species was thought conspecific with Upper Ordovician specimens from the Oslo Region, which has produced some confusion regarding its diagnostic features and

its stratigraphical and geographical occurrence, and species such as the somewhat younger *Cybellela grewingki* (Schmidt, 1881) were regarded as its junior synonyms (see Nikolaisen 1961; Whittington 1965b). The new cranidia described above make it clear that all the specimens formerly assigned *A. dentata* from the Upper Ordovician of the Oslo Region and contemporaneous part of the Bala Series of southwestern United Kingdom should be assigned to the East-Baltic *Cybellela grewingki* rather than to *A. dentata*. *A. dentata* differs from *C. grewingki* in the strong forward widening of the glabella; the less pit-like lateral glabellar furrows; the hand-like forward extension of the preglabellar border, the shorter eye-stalks and by the better developed axial furrows on the cranidium. *Atractopyge* aff. *dentate* of Owen & Parkes (2000) from the lower Sandbian of Ireland has the beginning of the robust, hand-like mesial spines on the anterior cranidial border, but *A. dentata* from the Oslo Region differs in a number of characters like the sparser tuberculation on the cranidium; the presence of a strong pair of preglabellar tubercles; shorter lateral glabellar furrows and in the more posterior location of the palpebral stalk.

Occurrence. – The species is extremely rare in the Elnes Formation, where it has been found in the Heggen and possibly Engervik members throughout the Oslo Region. A single specimen, PMO 62111, may be from the overlying Vollen Formation. Eiker-Sandsvær: Muggerudkleiva at Hedenstad (Heggen Member). Oslo-Asker: Tørtberg at Majorstuen (Vollen Fm.? or upper Elnes Fm.). Hadeland: Haugslandet at Gran (Elnes Fm.?).

Order Lichida Moore *in* Moore, 1959

Superfamily Lichoidea Fortey *in* Kaesler, 1997

Family Lichidae Hawle & Corda, 1847

Subfamily Lichinae Hawle & Corda, 1847 [*sensu* Pollitt, Fortey & Wills 2005]

Genus *Metopolichas* Gürich, 1901

Type species. – *Metopias hübneri* Eichwald, 1842.

Diagnosis. – Longitudinal furrow usually terminating at base of bullar lobe but may extend weakly to occipital furrow; bullar lobe sometimes circumscribed but usually confluent with L1b abaxially. Middle body of hypostome circumscribed by relatively small posterior lateral lobes; lateral borders broad.

Pygidial axis approximately one-third sagittal length of pygidium, widening at back and bearing two axial rings; pleural area with three pairs of furrowed pleurae, the posterior margin of the third pair rounded or indented (emended from Moore 1959; Thomas & Holloway 1988).

Metopolichas celorrhin (Angelin, 1854)

Pl. 25, Figs 15–17; Pl. 28, Figs 13; Pl. 29, Figs 8, 10

1854 (1878) *Lichas celorrhin*. n. sp. Angelin, p. 69, pl. 35, figs 1–1c.

1882 *Lichas celorrhin*, Ang. – Brøgger, pp. 128–130 (*partim*), pl. 5, figs 11–13 (*non* Pygidium H.2553).

1885 *Lichas celorhin* Ang. – Schmidt, pp. 56–59, pl. 1, figs 4–9.

1885 *Lichas pachyrhina* Dalm. – Schmidt, pp. 59–65, pl. 1, figs 10–12.

1907 *Lichas celorhin* Ang. [*sic*] – Schmidt, pp. 29–32 (*partim*), text-fig. 4 (*non* pl. 2, figs 5–5c).

1937 *Lichas celorrhin* ANG. – Warburg, p. 212.

1939 *Lichas celorhin* Angelin 1854 [*sic*] – Warburg, pp. 24–34 (*cum. syn.*), pl. 1–2.

1971 *Lichas celorhin* Angelin 1854 [*sic*] – Neben & Krueger, pl. 9, figs 1–2, 5.

1984 *Metopolichas celorrhin*? (Angelin, 1854) – Wandås, pp. 237–238, pl. 13O–Q.

1988 *Metopolichas celorrhin* (Angelin, 1854) – Thomas & Holloway, p. 214, pl. 11, figs 222–228.

Type stratum and type locality. – The lectotype cranidium (RM.-Ar.2237) selected by Warburg (1939) comes from the Asaphus Limestone at Humlenäs, Sweden.

Norwegian material. – The material is very fragmentary, but 23 cranidia (PMO 2376, H.2587, H.2588, H.2589, H.2590, H.2594, H.2601, H.2602, 33205, 33217, 82706, 101.725, 102.622, 102.632, 102.644, 106.051, 106.067, 106.389 + 394, 106.391, 106.392–3, 106.447–7a, 112.134, 212.657), two hypostomes (PMO 106.418, 106.421) and a single pygidium (PMO 106.052) has been identified.

Diagnosis. – Glabella wider than long with sub-conical and strongly overhanging anteromedian lobe; bicomposite glabellar lobes relatively narrow, uninflated; most cranidial furrows rather strong and wide, the preglabellar furrow being very wide; pygidium with narrow posterior indention and an axial width corresponding to just over 25% of the pygidial width; surface densely covered by intermingled coarse and fine tuberculation.

Remarks. – Although the Norwegian material from the basal Elnes Formation is rather poor, the approx-imately identical stratigraphical level of occurrence with the lectotype for *M. celorrhin* (Helskjer Member ≈ Gigas Limestone, which is suggested as the type stratum by Warburg (1939)) together with a strong resemblance in the prominent, potato-like and broad frontal glabellar lobe covered by a mixture of fine and rather coarse granulation all point toward them being conspecific. The possible differences lined up by Wandås (1984) are not supported here and are merely related to the strong compression and defor-mation of the Norwegian material examined by him.

Occurrence. – In Norway it is recorded from the upper Lysaker Member, Huk Fm. (Brøgger 1882), but no material older than the Svartodden Member, Huk Fm. has been identified in connec-tion with the present study. The species continues up into the basal Heggen Member of the Elnes Formation and occurs throughout most of the Oslo Region. Skien-Langesund: Rogn-stranda (Svartodden (?) and Helskjer Member). Eiker-Sandsvær: Krekling, Stubberud and Muggerudkleiva (Svartodden Member of the Huk Fm. and up into the Heggen Member, Elnes Fm.). Modum: road-cut at Vikersund skijump (basal Helskjer to lower Heggen Member, 0.2 to 4.0 m above the Huk Formation). Oslo-Asker: Kampen in Oslo (Svartodden Member of the Huk Fm. or possibly the Helskjer Member, Elnes Fm.). Hadeland: Grinaker near Gran (Helskjer Member).

Outside Norway it is reported from the upper Komstad Limestone and possibly Gigas Limestone of Sweden and from the whole Kunda to perhaps basal Aseri Regional Stage in the East Baltic area (Schmidt 1907; Warburg 1939).

Subfamily Trochurinae Phleger, 1936

Genus et sp. indet.

Pl. 25, Fig. 18; Pl. 26, Figs 1–2; Text-Fig. 62

Material. – two small cranidial external moulds (PMO 211.733, 211.734) preserved in turbiditic siltstone.

Description. – Largest specimen, PMO 211.733, measuring 4.6 mm in length and 6.5 mm in width. Cranidium broadly triangular in outline with some-what hourglass-shaped glabella having its widest point at occipital ring; Occipital ring extremely well-developed, composite and strongly delineated anteriorly, less so laterally, the width (tr.) corresponding to 80% of the total cranidial width, while the length takes up slightly more than 25% of the cranidial length; occipital furrow very deep and wide, making an incision into occipital ring behind L1 lobes, turning posteriorly laterally; preoccipital part of glabella narrow just in front of occipital furrow, the width corre-sponding to 40% of the occipital ring width, strongly forward widening, the widest point centrally between occipital furrow and the front of the glabella corre-sponding to approximately two-thirds the occipital

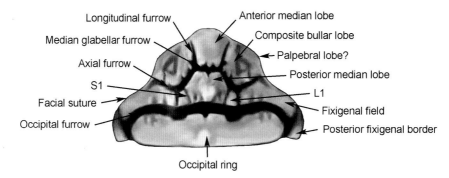

Fig. 62. Tentative reconstruction of the cranidium of the indeterminate trochurine from the Håkavik Member showing the terminology used in the description.

width; median lobe narrow, divided into two nearly equal halves by strong transverse furrow, posterior half broadly sub-hexagonal, the length equalling the occipital ring length; anterior median lobe slightly longer but narrower, moderately forward widening, anteriorly melting together with bullar lobes; L1 lobes moderately sized, triangular, situated as lateral extensions of posterior median lobe from which they are separated by shallow S1 furrow; bullar lobes sub-pentagonal in outline with central depressions and a width nearly equalling that of the posterior median lobe; longitudinal furrows deep, reaching from posterior corner of bullar lobes and forward to anterior part of glabella, where they fade out; posterior half of longitudinal furrows converges forwards, bending abruptly outwards at transverse furrow from where they diverge moderately forward; axial furrows broad along lateral occipital ring margin, becoming deep and well defined along abaxial margin of L1 and posterior part of bullar lobe before fading out anterolaterally; fixigenal field short and wide, the width corresponding to nearly 40% of the cranidial width, laterally terminating in lateral border furrow; posterior fixigenal border rudimentary, separated from occipital ring by wide axial furrow.

Cranidial surface gnarled or coarsely granulated, the granulation becoming finer and more even on anterior and anterior part of posterior median lobe.

Librigena, hypostome, thorax and pygidium unknown.

Remarks. – The taxon appears quite different from the known species within this subfamily with its strongly developed transverse furrow dividing the median lobe into two approximately equally sized halves; the well-developed and large occipital ring and furrow and by the L1 lobes partly merged with the posterior median lobe and reaching forward to the bullar lobe. These and the general combination of other characters like the rather weakly convex cranidium and glabella, the lack of large spines on median lobes and the hourglass-shaped outline of the median lobes when combined clearly indicate that it represents a new genus.

The taxon is probably closest related to genera such as *Hemiarges* Gürich and *Uripes* Thomas & Holloway, both of which have been treated in detail by Thomas & Holloway (1988). The Norwegian taxon differs, though, on features like the presence of only one L1 lobe; the much wider occipital ring; the more deeply incised occipital furrow and by the fading out of the anterior branch of the longitudinal furrows, which continues strongly on *Hemiarges* and *Uripes*.

Occurrence. – The two specimens belongs to the same siltbed from the Håkavik Member on the south-eastern side of the small island Killingen west of Bygdøy, Oslo-Asker.

Superfamily Odontopleuroidea (Whittington *in* Moore, 1959)

Family Odontopleuridae Burmeister, 1843

Subfamily Odontopleurinae Burmeister, 1843

Genus *Primaspis* R. & E. Richter, 1917

Type species. – *Odontopleura primordialis* Barrande, 1846.

Diagnosis. – Glabella with small L3 lateral glabellar lobes; occipital ring not greatly lengthened or inflated; median tubercle or small paired occipital spines present; occipital lobes small; eye lobes located opposite basal glabellar lobes and about halfway across genal region; lateral cephalic margin bearing short spines. Thorax with 10 segments; posterior pleural bands inflated at fulcra; posterior pleural spines stout; anterior pleural spines small (modified from Moore 1959).

Primaspis multispinosa Bruton, 1965

Pl. 25, Figs 11–14

1965 *Primaspis multispinosa* n. sp. Bruton, pp. 349–352, pl. 2, figs 1–6.

Type stratum and type locality. – Heggen Member. Road-cut at Furnes Church, Mjøsa.

Material. – The material consists of five cranidia, 12 librigenae, two hypostomes, four thorax segments and five pygidia (PMO 72772a–f, 72773, 72773c, 72774a–b, 72776, 72776a–d) coming from one single concretion and one additional cranidium (PMO 36494).

Diagnosis. – Primaspid with wide librigena; a square hypostome with short curved middle furrows running from deep V-shaped anterolateral depressions and an asymmetrically multispinose pygidium bearing seven secondary spines between and five outside of major pleural spines.

Remarks. – The species is found in an allochthonous storm deposit (shell bed) together with the cephalopod *Trilacinoceras* sp. and the gastropod *Sinuites* sp. The associated fauna would suggest that the deposit belongs to the latest Darriwilian, corresponding with the uppermost Sjøstrand Member to top Elnes Formation in the Oslo-Asker district.

Occurrence. – Heggen Member near Kjølja at Vesttorpen (Nordre Land) and in the road-cut at Furnes Church, Mjøsa.

Order Agnostida Salter, 1864

Suborder Agnostina Salter, 1864

Superfamily Agnostoidea M'Coy, 1849

Family Metagnostidae Jaekel, 1909

Subfamily Metagnostinae Jaekel, 1909

Genus *Geragnostus* Howell, 1935

Type species. – *Agnostus sidenbladhi* Linnarsson, 1869.

Diagnosis. – Glabella with semi-ovate anterior lobe; F3 effaced or with straight median portion, lateral

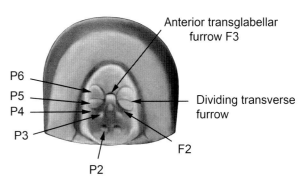

Fig. 63 Reconstruction of the cephalon of *Geragnostus hadros* Wandås, 1984 showing the terminology used in the description.

portions curving forward and outward; posterior lobe with weak F2 furrows; glabellar node immediately behind F3 furrow. Pygidial axis of moderate length, commonly constricted across M2; posterior lobe nearly equidimensional, semi-ovate to subrectangular, commonly with weak terminal node (from Kaesler 1997).

Geragnostus hadros Wandås, 1984

Pl. 26, Figs 13–14; Text-Fig. 63

1984 *Geragnostus hadros* n. sp. Wandås, pp. 216–217, pl. 1A–G.

Type stratum and type locality. – Basal Helskjer Member, 0.2 m above the Huk Formation in the road-cut below Vikersund skijump, Modum.

Material. – The examined material contains 15 cephala (PMO 36207, 67186, 67187, 82660, 82847, 90482, 90496, 95313, 102.656a, 102.567, 102.793, 103.711, 105.633–34, 106.082, 106.086) and seven pygidia (PMO 82175, 101.701, 102.576, 102.606, 106.076, 106.083, 106.084) and one partly articulated specimen consisting of pygidium and a thoracic segment (PMO 82660).

Diagnosis. – Species of *Geragnostus* defined by the nearly equal size of 5p and 6p muscle impressions separated by transverse furrow; the location of transverse furrow F2 well behind anterior transverse furrow F3 and by the indistinct axial node. Pygidium with slightly rearward narrowing anterior part of axis and by the fairly deeply impressed axial and border furrows; the somewhat rectangular and posteriorly rounded terminal piece bearing a small terminal node (modified from Wandås 1984).

Occurrence. – Basal Helskjer Member to lower Heggen Member. Skien-Langesund: Rognstranda (Heggen Member). Eiker-Sandsvær: Krekling (Helskjer Member, 0.76 m above Huk Fm.), Stavlum (Helskjer Member, 0.76 m above the Huk Fm.) and Muggerudkleiva (?) (Heggen Member). Modum: road-cut below Vikersund skijump (Helskjer to basal Heggen Member, 0.2 to 5.45 m above the Huk Formation). Oslo-Asker: Slemmestad (Sjøstrand Member). Mjøsa: road-cut at Furnes Church (top Helskjer Member 0.0–0.55 m below the Heggen Member).

Geragnostus sp.

Pl. 26, Figs 15–16

Material. – The material is very fragmentary and rather poorly preserved, but consists of four cephala (PMO 60373, 60449, 82080, 203.287) and three pygidia (PMO 36617, 82672, 206.060/6).

Description. – The largest cephalon (PMO 203.287) measures nearly 5.0 mm. Glabella fairly short and narrow, the length corresponding to between 50 and 60% of the cranidial length, while the length approximates 60% of the length; glabellar surface smooth with small mesial tubercle situated 40% of the glabellar length from the posterior margin; axial furrow fairly deep and narrow; cheeks smooth, curving downwards and outwards to the border; border moderately wide anteriorly, narrowing posteriorly and adaxially delineated by strong border furrow; no marginal spines present.

Pygidium very fragmentary, but characterized by smooth surface; a short and distinctly delineated axis and a wide, flattened border with a single pair of marginal spines.

Remarks. – The material is too small and poorly preserved for a confident classification, but shows a strong resemblance to *Geragnostus hadros* treated above. Most likely it is closely related to this species.

Occurrence. – Heggen, Engevik and Håkavik members of the Elnes Formation and up into the Vollen Formation. Eiker-Sandsvær: Muggerudkleiva (Heggen Member to possibly basal Fossum Fm.). Oslo-Asker: Beach-profile at Djuptrekkodden near Slemmestad (Engervik and Håkavik Member, 36.86–51.78 m above datum in profile). Ringerike: Gomnes (Vollen Fm.).

Acknowledgements. – During the preparation of this monograph I have been helped by many persons and institutions. First of all I wish to thank the Natural History Museum, University of Oslo, for housing and financing my PhD project and the publication of this monograph. I want to express my sincere gratitude to my first supervisor Professor David L. Bruton for his continual help and advice in my work with the doctoral thesis and to my second supervisor Dr. Arne T. Nielsen, Geological Museum, University of Copenhagen, for much helpful advice and the help in outlining the project. Professor Per Ahlberg, University of Lund, and especially Dr. Alan W. Owen, University of Glasgow, are thanked for refereeing the paper. Professor Felix Gradstein and Mr. Erik Anthonissen are thanked for their help with stratigraphical questions. Thanks are also due to Dr. Øyvind Hammer for his ready help with statistical problems; to Mr. Per Aas for skilfully photographing the larger specimens; to Mrs. Inger Eide, librarian at NHM for help locating books and obtaining literature; to Dr. Hans Arne Nakrem and Franz-Joseph Lindemann for their help with the databases and collections; to Dr. Kristina Månsson, Professor Kent Larsson and Professor Per Ahlberg at the GeoBiosphere Science Centre at Lund for making the Andersö Shale collections available for closer examinations; to Dr. Sven Egenhoff, Department of Geosciences, Colorado State University, and Dr. Jörg Maletz, Department of Geology, State University of New York at Buffalo, for their advice and help regarding the sedimentological interpretation and graptolite identification; to Dr. Gunnar Juve for his translational work; to Mr. Leif Koch for his help in identifying old locality names; to the people at Heimtun Internat at Fiskum for renting me a room during their holiday time; to Statens Vegvesen, Region øst for their ready help, and last but not least to Jesper Hansen, Tromsø Universitets Museum, University of Tromsø, for interesting discussions on aspects of palaeoecology and faunal distribution.

References

Ahlberg, P. 1995a: Telephinid Trilobites from the Ordovician of Sweden. *Palaeontology 38(2)*, 259–285.

Ahlberg, P. 1995b: Telephinid trilobites from the Ordovician of the East Baltic. *Geologiska Föreningens i Stockholm Förhandlingar 117*, 49–52.

Andreyeva, O.N. 1986: Middle Ordovician Brachiopods and Trilobites of Tuva and Altai Region. *Paleontological Journal 19* (for 1985), *2*, 34–45.

Angelin, N.P. 1851: Palæontologia Svecica. I. Iconographia Crustaceorum Formationis Transitionis. Fasciculi I. Berlingianus, Holmiæ [Stockholm], 1–24, pls 1–24.

Angelin, N.P. 1854: *Palæontologia Scandinavica. I. Crustacea formationis transitionis.* Fasciculi II. Lipsiæ, 21–92, pls 25–41.

Angelin, N.P. 1878: *Palæontologia Scandinavica. Pars I. Crustacea formationis transitionis.* Fasciculi I and II. I–IX, 1–92. Appendix (ed. G. Lindström), 93–96, pls I–III, XLII, Stockholm.

Balashova, E.A. 1966: Trilobity ranneordovikskikh otlozhenij Russkoj platformy. [Trilobites of the Lower Ordovician rocks of the Russian platform.] *Voprosi Paleontologii 5*, 3–22, pls 1–2. [In Russian.]

Balashova, E.A. 1971: K ustanovleniju novovo podsemeistva Pseudobasilicinae [To the erection of the new subfamily Pseudobasilicinae]. *Voprosi Paleontologyi 6*, α. Igdatelstvo Leningradskovo universiteta, 52–60, pls 1–2. [In Russian.]

Balashova, E.A. 1976: *Sistematika trilobitov Asaphina i ikh predstaviteli v SSSR.* [Systematics of Asaphine Trilobites and their Representatives in the USSR.] 215 pp. Nedra, Leningrad. [In Russian.]

Bassler, R.S. 1915: Bibliographic Index of American Ordovician and Silurian Fossils. *Bulletin of the United States National Museum 92(1)*, 718 pp.

Bergström, J. 1982: Scania. *In* Bruton, D.L. & Williams, S.H. (eds): Field Excursion Guide. IV International Symposium on the Ordovician System. *Paleontological Contributions from the University of Oslo 279*, 184–197.

Bergström, S.M., Larsson, K., Pålsson, C. & Ahlberg, P. 2002: The Almelund Shale, a replacement name for the Upper *Didymograptus* Shale and the Lower *Dicellograptus* Shale in the lithostratigraphical classification of the Ordovician succession in Scania, Southern Sweden. *Bulletin of the Geological Society of Denmark 49*, 41–47.

Berry, W.B.N. 1964: The Middle Ordovician of the Oslo Region, Norway. 16. Graptolites of the *Ogygiocaris* series. *Norsk Geologisk Tidsskrift 44*, 61–170.

Billings, E. 1861–1865: *Palæozoic Fossils 1.* 426 pp. Geological Survey of Canada, Ottawa, ON, Canada.

Bjørlykke, K. 1965: The Middle Ordovician of the Oslo Region, Norway. 20. The geochemistry and mineralogy of some shales from the Oslo Region. *Norsk Geologisk Tidsskrift 45*, 435–456.

Bjørlykke, K. 1974: Depositional history and geochemical composition of Lower Palaeozoic epicontinental sediments from the Oslo Region. *Norges Geologiske Undersøkelse 305*, 1–81.

Bockelie, J.F. 1978: The Oslo region during the early Palaeozoic. *In* Ramberg, I.B. & Neumann, E.R. (eds): Tectonics and Geophysics of Continental Rifts; Volume two of the proceedings of the NATO Advanced Study Institute Paleorift Systems with Emphasis on the Permian Oslo Rift. *NATO ASI Series. Series C: Mathematical and Physical Sciences 37*, 195–202.

Bockelie, J.F. & Nystuen, J.P. 1985: The southeastern part of the Scandinavian Caledonides. *In* Gee, D.G. & Sturt, B.A. (eds): *The Caledonide Orogen – Scandinavia and Related Areas*, 69–88. John Wiley, Chichester, UK.

Boeck, C. 1838: Übersicht der bisher in Norwegen gefundenen Formen der Trilobiten-Familie. *Gæa Norvegica 1*, 138–145.

Bohlin, B. 1949: The Asaphus Limestone in Northernmost Öland. *Bulletin of the Geological Institution of the University of Upsala 33*, 529–569, pls 1–2.

Bohlin, B. 1955: The Lower Ordovician Limestones between the *Ceratopyge* Shale and the *Platyurus* Limestone of Böda Hamn. *Bulletin of the Geological Institution of the University of Upsala 35*, 111–151, pls 1–6.

Bohlin, B. 1960: Trilobites of the genus *Megistaspis* from the Vaginatum Limestone of Scandinavia. *Bulletin of the Geological Institutions of the University of Uppsala 38(2)*, 155–205, pls 1–12.

Bruton, D.L. 1965: The Middle Ordovician of the Oslo Region, Norway. 19. The trilobite family Odontopleuridae. *Norsk Geologisk Tidsskrift 45(3)*, 339–356, Chart 1, pls 1–3.

Bruton, D.L. & Harper, D.A.T. 1981: Brachiopods and trilobites of the early Ordovician serpentine Otta Conglomerate, south central Norway. *Norsk Geologisk Tidsskrift 61(2)*, 153–181.

Bruton, D.L. & Høyberget, M. 2006: A reconstruction of *Telephina bicuspis*, a pelagic trilobite from the Middle Ordovician of the Oslo Region, Norway. *Lethaia 39*, 359–364.

Bruton, D.L., Lindström, M. & Owen, A.W. 1985: The Ordovician of Scandinavia. 273–282. *In* Gee, D.G. & Sturt, B.A. (eds): *The Caledonide Orogen – Scandinavia and Related Areas*. John Wiley & Sons Ltd. Chichester, UK.

Brünnich, M.T. 1781: Beskrivelse over Trilobiten, en Dyreslægt og dens Arter, med en nye Arts Aftegning. *Konglige Danske Videnskabernes Selskab Skrifter, Nye Saml., Del 1*, 384–395, 1 pl. Kiøbenhavn [Copenhagen].

Brøgger, W.C. 1882: *Die silurischen Etagen 2 und 3 im Kristianiagebiet und auf Eker, ihre Gliederung, Fossilien, Schichtenstörungen und Contactmetamorphosen 1882*. 376 pp. Universitätsprogramm für 2. Sem, Kristiania [Oslo].

Brøgger, W.C. 1884: Spaltenverwefungen in der Gegend Langesund-Skien. *Nyt Magazin for Naturvidenskaberne 28*, 253–419.

Brøgger, W.C. 1886: Ueber die Ausbildung des Hypostomes bei einigen Skandinavischen Asaphiden. *Sveriges Geologiska Undersökning C. Afhandlingar och uppsatser 82*, 78 pp., pls 1–3.

Brøgger, W.C. 1887: Geologisk kart over øerne ved Kristiania. *Nyt Magazin for Naturvidenskaberne 31(2)*, 1–36. 1 map. Special offprint for same paper published in 1890.

Brøgger, W.C. 1890: Geologisk kart over øerne ved Kristiania. *Nyt Magazin for Naturvidenskaberne 31(2)*, 162–195, 1 map [published as an offprint in 1887].

Brongniart, A. & Desmarest, A.-G. 1822: *Histoire Naturelle Crustacés Fossiles, sous les Rapports zoologiques et géologiques*. 154 pp., pl. 1–11. Paris.

Burmeister, H. 1843: *Organisation und Uebersicht der Trilobiten*. I–XII, 1–148, 4 pls. Georg Reimer, Berlin.

Burskij, A.Z. 1970: Ranneordoviskie trilobity severa Paj-Khoja [Early Ordovician Trilobites from northern Pai-Khoya]. 96–138. *In* Bondarev, V.I. (ed.): *Opornyj razrez ordovika Paj-Choja, Vajgaca i juga Novoj Zemli (sbornik stat'ej)* [Ordovician key section in Pai-Khoya, Vangach Islands and S. Novaja Zemlya (collections of papers)]. Institut Geologii Arktiki, Leningrad [in Russian].

Chatterton, B.D.E. 1980: Ontogenetic studies of Middle Ordovician trilobites from the Esbataottine Formation, Mackenzie Mountains, Canada. *Palaeontographica. Abteilung A, Paläozoologie-Stratigraphie 171*, 1–74, pls 1–19.

Cocks, L.R.M. & Fortey, R.A. 1982: Faunal evidence for oceanic separations in the Palaeozoic of Britain. *Journal of the Geological Society of London 139*, 465–478.

Cocks, L.R.M. & Fortey, R.A. 1990: Biogeography of Ordovician and Silurian faunas. *Geological Society of London, Memoir 12*, 97–104.

Cocks, L.R.M. 2000: The Early Palaeozoic geography of Europe. *Journal of the Geological Society, London 157*, 1–10.

Dahll, T. 1857: Profile durch die Gegend von Skien, Porsgrund und Langesund. 306–333. *In* Kjerulf, T. (ed.): *Ueber die Geologie des südlichen Norwegens 9*, 3. Nyt Magazin for Naturvidenskaberne.

Dalman, J.W. 1827: Om Palaeaderna eller de så kallade trilobiterna. *Kongliga Svenska Vetenskaps-Akademiens Handlingar för år 1826*, 2, 113–162, 226–294, pls 1–6. Stockholm.

Dalman, J.W. 1828: *Über die Palæaden oder die sogenannten Trilobiten*. Translation by F. Engelhart. Johan Leonhard Schrag. Nürnberg. 84 pp., 6 pls.

Dean, W.T. 1973: Ordovician Trilobites from the Keele Range, Northwestern Yukon Territory. *Geological Survey of Canada, Bulletin 223*, 1–43.

Dons, J.A. & Larsen, B.T. 1978: The Oslo Palaeorift. A review and guide to excursions. *Norsk Geologisk Undersøkelse Bulletin 337*, 199 pp.

Eichwald, E. von 1825: *Geognostico – zoologica per Ingriam marisque Baltici provincias, nec non de Trilobitis observationes*. 58 pp. Casani.

Eichwald, E. von 1855–1860: *Letheae Rossica, ou paléontologie de la Russie. 2. volume*. 1657 pp. Stuttgart.

Esmark, H.M.T. 1833: Om nogle nye Arter af Trilobitter. *Magasin for Naturvidenskaberne, 2. række 1*, 268–270, pl. 7.

Fortey, R.A. 1975: The Ordovician trilobites of Spitsbergen. II. Asaphidae, Nileidae, Raphiophoridae and Telephinidae of the Valhallfonna Formation. *Norsk Polarinstitutt Skrifter 162*, 207 pp.

Fortey, R.A. 1980: The Ordovician trilobites of Spitsbergen. III. Remaining trilobites of the Valhallfonna Formation. *Norsk Polarinstitutt Skrifter 171*, 163 pp., fig. 1.

Fortey, R.A. & Cocks, L.R.M. 2003: Palaeontological evidence bearing on the global Ordovician–Silurian continental reconstructions. *Earth-Science Reviews 61*, 245–307.

Fortey, R.A. & Owens, R.M. 1975: Proetida: a new order of trilobites. *Fossils and Strata 4*, 227–239.

Fortey, R.A. & Owens, R.M. 1978: Early Ordovician (Arenig) stratigraphy and faunas of the Carmarthen district, south-west Wales. *Bulletin of the British Museum of Natural History, Geology 30(3)*, 225–294.

Gradstein, F.M., Ogg, J.G. & Smith, A.G. (eds). 2004: A Geologic Time Scale 2004. Cambridge University Press, 1–610 pp.

Grahn, Y. & Nõlvak, J. 2007: Remarks on older Ordovician Chitinozoa and biostratigraphy of the Oslo Region, southern Norway. *Geologiska Föreningens i Stockholm Förhandlingar (GFF) 129*, 101–106.

Grorud, H.-F. 1940: Et profil gjennem Ogygiaskifer og Ampyxkalk på Tørtberg, Frogner ved Oslo. *Norsk Geologisk Tidsskrift 20*, 157–160.

Hadding, A. 1913: Undre dicellograptusskiffern i Skåne jämte några därmed ekvivalenta bildningar. *Acta Universita (N. F. Avd. 2) 9(15)*, 1–91, 8 pls. Lund & Leipzig.

Hamar, G. 1966: The Middle Ordovician of the Oslo Region, Norway. 22. Preliminary report on conodonts from the Oslo-Asker and Ringerike districts. *Norsk Geologisk Tidsskrift 46(1)*, 27–83.

Hammer, Ø. & Harper, D.A.T. 2006: *Paleontological Data Analysis*. Blackwell Publishing, Malden, MA. 351 pp.

Hammer, Ø., Harper, D.A.T. & Ryan, P.D. 2001: PAST: Paleontological Statistics Software Package for Education and Data Analysis. *Palaeontologia Electronica 4(1)*, 9 pp.

Hansen, T. 2005: A new trilobite species of *Hemisphaerocoryphe* from the Arenig of the St. Petersburg area, Russia. *Norwegian Journal of Geology 85(3)*, 203–208.

Hansen, T. & Nielsen, A.T. 2003: Upper Arenig trilobite biostratigraphy and sea-level changes at Lynna River near Volkhov, Russia. *Bulletin of the Geological Society of Denmark 50*, 105–114.

Hansen, T., Bruton, D.L. & Jacobsen, S.L. 2005: Starfish from the Ordovician of the Oslo Region, Norway. *Norwegian Journal of Geology 85*, 3, 209–216.

Harrington, H.J. & Leanza, A.F. 1957: Ordovician trilobites of Argentina. *Department of Geology, University of Kansas, Special Publications 1*, 1–276.

Hawle, I. & Corda, A.J.C. 1847: *Prodrom einer Monographie der böhmischen Trilobiten*, 176 pp., 7 pls. J. G. Calve, Prague.

Helbert, G.J., Lane, P.D., Owens, R.M., Siveter, D.J. & Thomas, A.T. 1982: Lower Silurian Trilobites from the Oslo Region. *In*

Worsley, D. (ed.): *IUGS Subcommission on Silurian Stratigraphy. Field Meeting, Oslo Region 1982 278*, 129–148. Paleontological Contributions from the University of Oslo.

Henningsmoen, G. 1960: The Middle Ordovician of the Oslo Region, Norway 13, Trilobites of the family Asaphidae. *Norsk Geologisk Tidsskrift 40(3–4)*, 203–257.

Henningsmoen, G. 1960a: Cambro-Silurian deposits of the Oslo Region. *Norges Geologiske Undersøkelse 208*, 130–169.

Heuwinkel, J. & Lindström, M. 2007: Sedimentary and tectonic environment of the Ordovician Föllinge Greywacke, Storsjön area, Swedish Caledonides. *Geologiska Föreningens i Stockholm Förhandlingar (GFF) 129(1)*, 31–42.

Hoel, O.A. 1999 Trilobites of the Hagastrand Member (Tøyen Formation, lowermost Arenig) from the Oslo Region, Norway. Part II: Remaining non-asaphid groups. *Norsk Geologisk Tidsskrift 79*, 259–280.

Holliday, S. 1942: Ordovician trilobites from Nevada. *Journal of Paleontology 16*, 471–478.

Holloway, D.J. 2007: The trilobite *Protostygina* and the composition of the Styginidae, with two new genera. *Paläontologische Zeitschrift 81*, 1, 1–16.

Holtedahl, O. 1909: Studien über die Etage 4 des Norwegischen Silursystems beim Mjösen. *Videnskabs-Selskabets Skrifter. I. Mathematisk-Naturvidenskabelig Klasse 1909(7)*, 76 pp.

Hughes, C.P. 1979: The Ordovician trilobite faunas of the Builth-Llandrindod Inlier, central Wales. Part III. *Bulletin of the British Museum (Natural History). Geology series 32(3)*, 109–181.

Hughes, C.P., Ingham, J.K. & Addison, R. 1975: The morphology, classification and evolution of the Trinucleidae (Trilobita). *Philosophical Transactions of the Royal Society of London. B, Biological Sciences 272*, 537–604.

Hughes, C.P., Rickards, R.B. & Williams, A. 1980: The Ordovician fauna from the Contaya Formation of eastern Peru. *Geological Magazine 117(1)*, 1–21.

Hunda, B.R., Chatterton, B.D.E. & Ludvigsen, R. 2003: Silicified Late Ordovician trilobites from the Mackenzie Mountains, Northwest Territories, Canada. *Palaeontographica Canadiana 21*, 1–87.

Ivantsov, A.Y. 2003: Ordovician trilobites of the subfamily Asaphinae of the Ladoga Glint. *Paleontological Journal 37(3)*, 229–337.

Ivantsov, A.Y. 2004: *Classification of the Ordovician trilobites of the subfamily Asaphinae from the neighbourhood of St. Petersburg*, 60 pp. Palaeontological Institute, Russian Academy of Sciences, Moscow. [In Russian].

Jaanusson, V. 1953a: Untersuchungen über baltoskandische Asaphiden I, Revision der Mittelordovizischen Asaphiden des Siljan-Gebietes in Dalarna. *Arkiv för Mineralogi och Geologi. Kungliga Svenska Vetenskapsakademien 1(14)*, 377–464, pls 1–10.

Jaanusson, V. 1953b: Untersuchungen über baltoskandische Asaphiden II, Revision der Asaphus (Neoasaphus)-arten aus dem Geschiebe des südbottnischen Gebietes. *Arkiv för Mineralogi och Geologi. Kungliga Svenska Vetenskapsakademien 1(15)*, 465–499, pls 1–6.

Jaanusson, V. 1954: Zur Morphologie und Taxonomie der Illaeniden. *Arkiv för Mineralogi och Geologi 1(20)*, 545–583, pls 1–3.

Jaanusson, V. 1956: Untersuchungen über baltoskandische Asaphiden. III. Über die Gattungen *Megistaspis* n. nom. und *Homalopyge* n. gen. *Bulletin of the Geological Institutions of the University of Uppsala 36*, 59–77, pl. 1.

Jaanusson, V. 1957: Unterordovizische Illaeniden aus Skandinavien. *Bulletin of the Geological Institutions of the University of Uppsala 37*, 79–165, pls 1–10.

Jaanusson, V. 1960: The Viruan (Middle Ordovician) of Öland. *Bulletin of the Geological Institutions of the University of Uppsala 38*, 207–288, 5 pls.

Jaanusson, V. 1963: Lower and Middle Viruan (Middle Ordovician) of the Siljan District. *Bulletin of the Geological Institutions of the University of Uppsala 42*, 1–40, pl. 1.

Jaanusson, V. 1964: The Viruan (Middle Ordovician) of Kinnekulle and Northern Billingen, Västergötland. *Bulletin of the Geological Institutions of the University of Uppsala 43*, 1–73.

Jaanusson, V. 1973: Aspects of carbonate sedimentation in the Ordovician of Baltoscandia. *Lethaia 6*, 11–34.

Jaanusson, V. 1982: Ordovician in Västergötland. *In* Bruton, D.L. & Williams, S.H. (eds): Field Excursion Guide. IV International Symposium on the Ordovician System. *Palaeontological Contributions from the University of Oslo 279*, 164–183.

Jaanusson, V. & Mutvei, H. 1951: Ein Profil durch den Vaginatum-Kalkstein im Siljan-Gebiet, Dalarna. *Geologiska Föreningens i Stockholm Förhandlingar 73*, 630–636.

Jaanusson, V. & Mutvei, H. 1955: Stratigraphie und Lithologie der unterordovizischen *Platyurus*-Stufe im Siljan-Gebiet, Dalarna. *Bulletin of the Geological Institution of the University of Upsala 35*, 7–34, pls 1–5.

Jaanusson, V. & Ramsköld, L. 1993: Pterygometopine trilobites from the Ordovician of Baltoscandia. *Palaeontology 36(4)*, 743–769.

Kaesler, R.L. (ed.) 1997: *Treatise on Invertebrate Paleontology. Part O. Arthropoda 1*. Trilobita, Revised. Volume 1: Introduction, Order Agnostida, Order Redlichiida. The Geological Society of America, Boulder, Colorado, and The University of Kansas, Lawrence, Kansas. 530 pp.

Karis, L. 1982: The sequence in the lower allochton of Jämtland. *In* Bruton, D.L. & Williams, S.H. (eds): *Field Excursion Guide. IV International Symposium on the Ordovician System 279*, 55–63. Palaeontological Contributions from the University of Oslo.

Karis, L. & Strömberg, A.G.B. 1998: Beskrivning till berggrundskartan över Jämtlands län. Del 2: Fjälldelen. *Sveriges Geologiska Undersökning Ca 53(2)*, 363 pp.

Kjerulf, T. 1857: Ueber die Geologie des südlichen Norwegens. *Nyt Magazin for Naturvidenskaberne 9(3)*, 193–333 [with text by T. Dahll on the geology of the Skien-Langesund area, 306–333].

Knell, R.J. & Fortey, R.A. 2005: Trilobite spines and beetle horns: sexual selection in the Palaeozoic? *Biology letters 1*, 196–199.

Krueger, H.-H. 1997: Die Trilobitengattung *Cnemidopyge* aus Geschieben des Ludibundus-Kalkes. *Archiv für Geschiebekunde 2(3)*, 147–156, pls 1–3.

Krueger, H.-H. 2003: Die Trilobitengattungen *Cybele* und *Cybellela* aus baltoskandischen Geschieben und ihrem Anstehenden sowie ein Nachtrag zu *Atractopyge. Archiv für Geschiebekunde 4(1)*, 15–48.

Krueger, H.-H. 2004: Die Trilobitengattung *Atractopyge* (Ordovizium) aus baltoskandischen Geschieben. *Archiv für Geschiebekunde 3(8)*, 12 [Schallreuter-Festschrift], 747–766.

Lamansky, W. 1905: Die ältesten silurischen Schichten Russlands (Etage 8). *Mémoires Comité Géologuque. Nouvelle Série 20*, 1–223.

Lane, P.D. 2002: The taxonomic position and coaptative structures of the Lower Ordovician trilobite *Cyrtometopus. Special Papers in Palaeontology 67*, 153–169.

Levitskiy, E.S. 1962: On a new trilobite genus *Cybelurus* gen. nov. Izvestiya Vysshikh Uchebnykh Zavedeniy. *Geologiya i Razvedka 1962(7)*, 129–131.

Linnarsson, J.G.O. 1869: Om Vestergötlands cambriska och siluriska aflagringar. *Kongliga Svenska Vetenskaps-Akademiens Handlingar 8(2)*, 1–89, pls 1–2.

Linnarsson, J.G.O. 1875: En egendomlig Trilobitfauna från Jemtland. *Geologiska Föreningens i Stockholm Förhandlingar 2*, 491–497, *pl.* 22.

Ludvigsen, R. 1975: Ordovician Formations and Faunas, Southern Mackenzie Mountains. *Canadian Journal of Earth Sciences 12*, 663–697.

Ludvigsen, R. 1980: An unusual trilobite faunule from Llandeilo or lowest Caradoc strata (Middle Ordovician) of northern Yukon Territory. *In* Current Research B (ed.): *Geological Survey of Canada, Paper 80–1B*, 97–106.

Maletz, J. 1995: The Middle Ordovician (Llanvirn) graptolite succession of the Albjära core (Scania, Sweden) and its implication for a revised biozonation. *Zeitschrift für geologische Wissenschaften 23*, 249–259.

Maletz, J. 1997: Graptolites from the *Nicholsonograptus fasciculatus* and *Pterograptus elegans* Zones (Abereiddian, Ordovician) of the Oslo Region, Norway. Greifswalder *Geowissenschaftliche Beiträge 4*, 5–100.

Maletz, J. & Egenhoff, S.O. 2005: Dendroid graptolites in the Elnes Formation (Middle Ordovician), Oslo Region, Norway. *Norwegian Journal of Geology 85*, 217–221.

Maletz, J., Egenhoff, S., Böhme, M., Asch, R., Borowski, K., Höntzsch, S. & Kirsch, M. 2007: The Elnes Formation of southern Norway: A key to late Middle Ordovician biostratigraphy and biogeography. *Acta Palaeontologica Sinica 46* (Suppl.), 298–304.

Marek, L. 1952: Contribution to the stratigraphy and fauna of the uppermost part of the Králův Dvůr Shales (Ashgillian). *Sborník Ústředního Ústavu Geologického, Svazek 19*, 429–455, pls 1–2.

McCormick, T. & Fortey, R.A. 1999: The most widely distributed trilobite species; Ordovician *Carolinites genaceinaca. Journal of Paleontology 73*, 2, 202–218.

Moore, R.C. 1959 (ed.): *Treatise on Invertebrate Paleontology O. Arthropoda 1*, 560 pp. Geological Society of America, Boulder, Colorado, and University of Kansas Press, Lawrence, Kansas.

Männil, R.M. 1966: *Istoriya pazvitiya Baltijskogo bassejna v ordovike*, [*History of the evolution of the Baltic basin in the Ordovician.*] 1–201. Eesti NSV Teaduste Akadeemia Toimetised, Tallinn. [In Russian with English summary.]

Månsson, K. 1995: Trilobites and stratigraphy of the Middle Ordovician Killeröd Formation, Scania, Sweden. *Geologiska Föreningens i Stockholm Förhandlingar 117*, 97–106.

Månsson, K. 1998: Middle Ordovician olenid trilobites (*Triarthrus* Geen and *Porterfieldia* Cooper) from Jämtland, central Sweden. *Transactions of the Royal Society of Edinburgh: Earth Sciences 89*, 47–62.

Månsson, K. 2000a: Dionidid and raphiophorid trilobites from the middle Ordovician (Viruan Series) of Jämtland, central Sweden. *Transactions of the Royal Society of Edinburgh: Earth Sciences 90*, 317–329.

Månsson, K. 2000b: Trilobites from the Middle and Upper Ordovician Andersön Shale Formation in Jämtland and the equivalent Killeröd Formation in Skåne, Sweden. Doctoral Thesis. *Lund Publications in Geology 152*, 21 pp. Lund.

Neben, W. & Krueger, H.H. 1971: Fossilien ordovicischer Geschiebe. *Staringia 1, Nederlandse Geologische Vereniging 1971*, 1–7, pls 1–50.

Nielsen, A.T. 1995: Trilobite systematics, biostratigraphy and palaeoecology of the Lower Ordovician Komstad Limestone and Huk Formations, southern Scandinavia. *Fossils and Strata 38*, 374 pp.

Nikolaisen, F. 1961: The Middle Ordovician of the Oslo Region, Norway. 7. Trilobites of the suborder Cheirurina. *Norsk Geologisk Tidsskrift 41*, 279–309, pls 1–4.

Nikolaisen, F. 1962: En komaspidid fra Oslofeltet. *Fossil-Nytt 1962*, 14.

Nikolaisen, F. 1963: The Middle Ordovician of the Oslo Region, Norway. 14. The trilobite family Telephinidae. *Norsk Geologisk Tidsskrift 43*, 345–399, pls 1–4.

Nikolaisen, F. 1965: The Middle Ordovician of the Oslo Region, Norway. 18. Rare trilobites of the families Olenidae, Harpidae, Ityophoridae and Cheiruridae. *Norsk Geologisk Tidsskrift 45(2)*, 231–248, pls 1–4.

Nikolaisen, F. 1968: Funn av Miracybele i Oslofeltet. *Fossil-Nytt 1967–68*, 2–3.

Nikolaisen, F. 1983: The Middle Ordovician of the Oslo Region, Norway 32. Trilobites of the family Remopleurididae. *Norsk Geologisk Tidsskrift 62* (for 1982), 231–329.

Nikolaisen, F. 1991: The Ordovician trilobite genus *Robergia* Wiman, 1905 and some other species hitherto included. *Norsk Geologisk Tidsskrift 71*, 37–62.

Nilssen, I.R. 1985: *Kartlegging av Langesundshalvøyas kambro-ordoviciske avsetningslagrekke, intrusiver og forkastningstektonikk, samt fullført lithostratigrafisk inndeling av områdets mellom-ordovicium.* Unpublished Cand. scient. thesis, Natural History Museum, University of Oslo, Section for Geology, Norway, *A–I*, 1–176.

Nilsson, R. 1952: Till kännedomen om ordovicium i sydöstra Skåne. *Geologiska Föreningens i Stockholm Förhandlingar 73* (for 1951), 682–694.

Norford, B.S. 1973: Lower Silurian Species of the Trilobite *Scotoharpes* from Canada and Northwestern Greenland. *Bulletin of the Geological Survey of Canada 222*, 9–33.

Nystuen, J.P. 1981: The late Precambrian 'sparagmites' of southern Norway: a major Caledonian allochthon – The Osen-Røa Nappe Complex. *American Journal of Science 281*, 69–94.

Oftedahl, C. 1966: Permian Igneous Rocks. *In* Holtedahl, O. & Dons, J.A. (eds): *Geological Guide to Oslo and Districts 37–44*. 2nd edn. Universitetsforlaget. Oslo.

Owen, A.W. 1981: The Ashgill trilobites of the Oslo Region, Norway. *Palaeontographica A 175*, 1–88, pls 1–17.

Owen, A.W. 1985: Trilobite abnormalities. *Transactions of the Royal Society of Edinburgh: Earth Sciences 76*, 255–272.

Owen, A.W. 1987: The Scandinavian Middle Ordovician trinucleid trilobites. *Palaeontology 30*, 1, 75–103.

Owen, A.W. & Parkes, M.A. 2000: Trilobite faunas of the Duncannon Group: Caradoc stratigraphy, environments and palaeobiogeography of the Leinster Terrane, Ireland. *Palaeontology 43*, 2, 219–269.

Owen, A.W., Bruton, D.L., Bockelie, J.F. & Bockelie, T.G. 1990: The Ordovician successions of the Oslo Region, Norway. *Norges Geologiske Undersøkelse. Special Publication 4*, 54 pp.

Pollitt, J.R., Fortey, R.A. & Wills, M.A. 2005: Systematics of the Trilobite families Lichidae Hawle & Corda, 1847 and Lichakephalidae Tripp, 1957: the application of Bayesian inference to morphological data. *Journal of Systematic Palaeontology 3(3)*, 225–241.

Poulsen, C. 1936: Übersicht über das Ordovizium von Bornholm. *Meddelelser fra Dansk Geologisk Forening 9(1)*, 43–66.

Pärnaste, H. 2004a: Revision of the Ordovician cheirurid trilobite genus *Reraspis* with the description of the earliest representative. *Proceedings of the Estonian Academy of Sciences, Geology 53(2)*, 125–138.

Pärnaste, H. 2004b: Early Ordovician trilobites of suborder Cheirurina in Estonia and NW Russia: Systematics, evolution and distribution. *Dissertationes Geologicae Universitatis Tartuensis 16*. Tartu University Press, Tartu, Estonia. 58 pp. + 5 papers.

Pålsson, C., Månsson, K. & Bergström, S.M. 2002: Biostratigraphy and palaeoecological significance of graptolites, trilobites and conodonts in the Middle–Upper Ordovician Andersö Shale: an unusual 'mixed facies' deposit in Jämtland, central Sweden. *Transactions of the Royal Society of Edinburgh: Earth Sciences 93*, 35–57.

Rasmussen, J.A. & Bruton, D.L. 1994: Stratigraphy of Ordovician limestones, Lower Allochton, Scandinavian Caledonides. *Norsk Geologisk Tidsskrift 74(4)*, 199–212.

Rasmussen, J.A. 2001: Conodont biostratigraphy and taxonomy of the Ordovician shelf margin deposits in the Scandinavian Caledonides. *Fossils and Strata 48*, 179 pp.

Raymond, P.E. 1913a: Some changes in the names of genera of trilobites. *Ottawa Naturalist 26*, 1–6.

Raymond, P.E. 1913b: Subclass Trilobita. *In* Eastman, C.R. (ed.): *Textbook of Palaeontology*. 2nd edn. The MacMillan Company, New York, pp. 692–729.

Raymond, P.E. 1925: Some trilobites of the lower Middle Ordovician of eastern North America. *Bulletin of the Museum of Comparative Zoology at Harvard College 67*, 1–181.

Reed, F.R.C. 1928: Notes on the Family Encrinuridae. *Geological Magazine 65*, 51–77.

Reed, F.R.C. 1930: A Review of the Asaphidæ. Part 1. *Annals and Magazine of Natural History, Serie 10(5)*, 288–320.

Reed, F.R.C. 1931: A review of the British species of the Asaphidae. *Annals and Magazine of Natural History 10(7)*, 441–472.

Reyment, R.A. 1980: Specimens figured in Johan Wilhelm Dalman's 'Om palaeaderna eller de s.k. trilobiterna'. *De Rebus 4*, 1–4. Paleontologiska Museet, Uppsala.

Ribecai, C., Bruton, D.L. & Tongiorgi, M. 2000: Acritarchs from the Ordovician of the Oslo Region, Norway. *Norsk Geologisk Tidsskrift 80*, 251–258.

Ross, R.J. Jr. & Shaw, F.C. 1972: Distribution of the Middle Ordovician Copenhagen Formation and its trilobites in Nevada. *United States Geological Survey Professional Paper 749*, 33 pp., pls 1–8.

Salter, J.W. 1864: *A Monograph of British Trilobites*. The Palæontographical Society, London. 224 pp., 30 pls.

Sars, M. 1835: Ueber einige neue oder unvollständig bekannte Trilobiten. *Isis von Oken 1835*, 4. Leipzig.

Schmidt, F. 1881: Revision der ostbaltischen silurischen Trilobiten nebst geognostischer Übersicht des ostbaltischen Silurgebiets. Abtheilung I. *Mémoires de l'Académie Impériale des Sciences de St-Pétersbourg VII 30(1)*, 1–237.

Schmidt, F. 1885: Revision der ostbaltischen Trilobiten. Abtheilung II: Acidaspiden und Lichiden. *Mémoires de l'Académie Impériale des Sciences de St-Pétersbourg VII 33(1)*, 127 pp., pls 1–6.

Schmidt, F. 1894: Revision der ostbaltischen Trilobiten. Abtheilung IV: Calymeniden, Proetiden, Bronteiden, Harpediden, Trinucleiden, Remopleuriden und Agnostiden. *Mémoires de l'Académie Impériale des Sciences de St-Pétersbourg VII 42(5)*, 93 pp.

Schmidt, F. 1898: Revision der ostbaltischen silurischen Trilobiten. Abtheilung V: Asaphiden. Part 1. *Mémoires de l'Academie Impériale des Sciences de St-Pétersbourg VIII 6(11)*, 45 pp.

Schmidt, F. 1901: Revision der ostbaltischen silurischen Trilobiten. Abtheilung V: Asaphiden. Part 2. *Mémoires de l'Academie Impériale des Sciences de St-Pétersbourg VIII 8*, 113 pp., 12 pls.

Schmidt, F. 1904: Revision der ostbaltischen Trilobiten. Abtheilung V: Asaphiden. Lief 3. *Mémoires de l'Académie Impériale des Sciences de St-Pétersbourg VIII 14(10)*, 68 pp.

Schmidt, F. 1906: Revision der ostbaltischen Trilobiten. Abtheilung V: Asaphiden. Lieferung 4. Enthaltend die Gattung *Megalaspis*. *Mémoires de l'Académie Impériale des Sciences de St-Pétersbourg VIII 19(10)*, 62 pp., pls 1–8.

Schmidt, F. 1907: Revision der ostbaltischen Trilobiten. Abtheilung VI: Allgemeine Übersicht mit Nachträgen und Verbesserungen. *Mémoires de l'Académie Impériale des Sciences de St-Pétersbourg VIII 20(8)*, 104 pp.

Schovsbo, N.H. 2003: The geochemistry of Lower Palaeozoic sediments deposited on the margins of Baltica. *Bulletin of the Geological Society of Denmark 50*, 11–27.

Schrank, E. 1972: *Nileus*-Arten (Trilobita) aus Geschieben des Tremadoc bis tiefern Caradoc. *Berichte der Deutschen Gesellschaft für Geologische Wissenschaften, Reihe A, Geologie und Paläontologie 17(3)*, 351–375.

Seilacher, A. & Meischner, D. 1965: Fazies-Analyse im Paläozoikum des Oslo-Gebietes. *Geologische Rundschau 54(2)*, 596–619.

Sheldon, P.R. 1987: Parallel gradualistic evolution of Ordovician trilobites. *Nature 330*, 561–563.

Siveter, D.J. 1977 (for 1976): The Middle Ordovician of the Oslo Region, Norway. 27. Trilobites of the family Calymenidae. *Norsk Geologisk Tidsskrift 56*, 335–396.

Skaar, F.E. 1972: *Orthocerkalksteinen (Etasje 3c) i Oslofeltet. En undersøkelse av den mineralogiske og kjemiske sammensetningen regionalt og stratigrafisk*. Unpublished Cand. Real. Thesis. 130 pp. University of Oslo, Oslo, Norway.

Skjeseth, S. 1952: On the Lower Didymograptus Zone (3B) at Ringsaker, and contemporaneous deposits in Scandinavia. *Norsk Geologisk Tidsskrift 30*, 138–182, pls 1–5.

Skjeseth, S. 1955: The Ordovician of the Oslo Region, Norway. 5. The Trilobite Family Styginidae. *Norsk Geologisk Tidsskrift 35*, 9–28, pls 1–5.

Skjeseth, S. 1963: Contributions to the geology of the Mjøsa Districts and the classical sparagmite area in southern Norway. *Norges Geologiske Undersøkelse 220*, 1–126.

Spjeldnæs, N. 1957: The Middle Ordovician of the Oslo Region, Norway. 8. Brachiopods of the Suborder Strophomenida. *Norsk Geologisk Tidsskrift 37(1)*, 214 pp., pls 1–14.

Stetson, H.C. 1927: The distribution and relationships of the Trinucleidae. *Bulletin of the Museum of Comparative Zoölogy at Harvard College 68(2)*, 87–104, pl. 1.

Sturesson, U., Popov, L.E., Holmer, L.E., Bassett, M.G., Felitsyn, S. & Belyatsky, B. 2005: Neodymium isotopic composition of Cambrian-Ordovician biogenic apatite in the Baltoscandian Basin: implications for palaeogeographical evolution and patterns of biodiversity. *Geological Magazine 142(4)*, 419–439.

Størmer, L. 1930: Scandinavian Trinucleidae with special references to Norwegian species and varieties. *Skrifter av Det Norske Videnskaps-Akademi i Oslo. I. Mat.-Naturv. Klasse 1930(4)*, 1–111.

Størmer, L. 1940: Early descriptions of Norwegian trilobites; the type specimens of C. Boeck, M. Sars and M. Esmark. *Norsk Geologisk Tidsskrift 20*, 113–151, pls 1–3.

Størmer, L. 1953: The Middle Ordovician of the Oslo Region, Norway 1. Introduction to stratigraphy. *Norsk Geologisk Tidsskrift 31*, 37–141, pls 4.

Størmer, L. 1967: Some aspects of the Caledonian geosyncline and foreland west of the Baltic Shield. *Quarternary Journal of the Geological Society of London 123*, 183–214.

Sundvoll, B. & Larsen, B.T. 1994: Architecture and early evolution of the Oslo Rift. *Tectonophysics 240*, 173–189.

Thomas, A.T. & Holloway, D.J. 1988: Classification and phylogeny of the trilobite order Lichida. *Philosophical Transactions of the Royal Society of London. Series B: Biological Sciences 321*, 179–262, Textfig. 2, pls 1–16.

Thorslund, P. & Asklund, B. 1935: Stratigrafiska och tektoniska studier inom Föllingeområdet i Jämtland. *Sveriges Geologiska Undersökning. Serie C, Avhandlingar och uppsatser 388. Årsbok 29(3)*, 1–61, pls 1–3.

Thorslund, P. 1935: Paleontologisk-stratigrafisk undersökning. *In* Thorslund, P. & Asklund, B.: Stratigrafiska och tektoniska studier inom Föllingeområdet i Jämtland. Chapter 1. *Sveriges Geologiska Undersökning. Serie C. Avhandlingar och uppsatser 388. Årsbok 39(3)*, 5–23, pls 1–2.

Thorslund, P. 1940: On the Chasmops Series of Jemtland and Södermanland (Tvären). *Sveriges Geologiska Undersökning, Serie C, 436*, 1–191.

Thorslund, P. 1960: The Cambro-Silurian of Sweden. Description to accompany the Map of the Pre-Quarternary Rocks of Sweden issued in 1958. *Sveriges Geologiska Undersökning, Serie Ba 16*, 69–110.

Tjernvik, T.E. 1956: On the Early Ordovician of Sweden. Stratigraphy and fauna. *Bulletin of the Geological Institutions of the University of Uppsala 36(2–3)*, 107–284, pls 1–11.

Tomczykowa, E. & Tomczyk, H. 1970: Stratigraphy. *In* Sokolowski, S. (ed.): *Geology of Poland 1*, Part 1, 1–651. Warsaw.

Torsvik, T.H., Smethurst, M.A., Van der Voo, R., Trench, A., Abrahamsen, N. & Halvorsen, E. 1992: Baltica. A synopsis of Vendian-Permian palaeomagnetic data and their palaeotectonic implications. *Earth-Science Reviews 33*, 133–152.

Torsvik, T.H., Smethurst, M.A., Meert, J.G., Van der Voo, R., McKerrow, W.S., Brasier, M.D., Sturt, B.A. & Walderhaug, H.J. 1996: Continental break-up and collision in the Neoproterozoic and Palaeozoic – A tale of Baltica and Laurentia. *Earth-Science Reviews 40*, 229–258.

Tripp, R.P. 1962: Trilobites from the 'confinis' Flags (Ordovician) of the Girvan District, Ayrshire. *Transactions of the Royal Society of Edinburgh 65(1)*, 44 pp., pls 1–4.

Tripp, R.P. 1967: Trilobites from the Upper Stinchar Limestone (Ordovician) of the Girvan District, Ayrshire. *Transactions of the Royal Society of Edinburgh 67(3)*, 93 pp., pls 1–6.

Tripp, R.P. 1976: Trilobites from the basal *superstes* Mudstones (Ordovician) at Aldons Quarry, near Girvan, Ayrshire. *Transactions of the Royal Society of Edinburgh 69*, 369–423.

Tripp, R.P. 1980: Trilobites from the Ordovician Balclatchie and lower Ardwell groups of the Girvan district, Scotland. *Transactions of the Royal Society of Edinburgh: Earth Sciences 71*, 123–145.

Törnquist, S.V. 1884: Undersökningar öfver Siljanområdets Trilobitfauna. *Sveriges Geologiska Undersökning, Serie C 66*, 1–101, pls 1–3.

Viira, V., Löfgren, A., Mägi, S. & Wickström, J. 2001: An Early to Middle Ordovician succession of conodont faunas at Mäekalda, northern Estonia. *Geological Magazine 138(6)*, 699–718.

Vodges, A.W. 1890: A bibliography of Palaeozoic Crustacea from 1698 to 1889, including a list of North American species and a systematic arrangement of genera. *US Geological Survey Bulletin 63*, 177 pp.

Vodges, A.W. 1925: Palaeozoic Crustacea. Part II. A list of the genera and subgenera of the Trilobita. *Transactions of the San Diego Society of Natural History 4*, 87–115.

Wahlenberg, G. [1818] 1821: Petrificata Telluris Svecanae. *Acta Societatis Regiae Scientiarum Upsaliensis 8*, 116 pp. Uppsala.

Waisfeld, B.G. & Vaccari, N.E. 2006: Revisión de la Biozona *Ogygiocaris araiorhachis* (Trilobita, Tremadocian tardío) en la región de Pascha-Incamayo, Cordillera Oriental, Argentina. Parte 2: Sistemática [Revision of tHe *Ogygiocaris araiorrhachis* Zone (Trilobita, Late Tremadocian) in the Pascha-Incamayo Region, Cordillera Oriental, Argentina. Part 2: Systematic]. *Ameghiniana (Asociación Paleontológica Argentina) 43(4)*, 729–744 [In Spanish with English summary].

Wandås, B.T.G. 1982: The area around Vikersundbakken, Modum. *In* Bruton, D.L. & Williams, S.H. (eds): Field Excursion Guide. IV International Symposium on the Ordovician System. *Palaeontological Contributions from the University of Oslo 279*, 132–138.

Wandås, B.T.G. 1984: The Middle Ordovician of the Oslo Region, Norway, 33. Trilobites from the lowermost part of the Ogygiocaris Series. *Norsk Geologisk Tidsskrift 63* [Year 1983], 211–267.

Warburg, E. 1937: Angelin's *Lichas norvegicus* – a Silurian species. *Bulletin of the Geological Institute of Upsala 27*, 212–218.

Warburg, E. 1939: The Swedish Ordovician and Lower Silurian Lichidae. *Kungliga Svenska Vetenskapsakademiens Handlingar. Series 3 17(4)*, 1–162, pls 1–14.

Whittington, H.B. 1950: Sixteen Ordovician genotype trilobites. *Journal of Paleontology 24(5)*, 531–565, pls 68–75.

Whittington, H.B. 1959: Silicified Middle Ordovician trilobites: Remopleurididae, Trinucleidae, Raphiophoridae, Endymioniidae. *Bulletin of the Museum of Comparative Zoology at Harvard College 121(8)*, 371–496, pls 1–36.

Whittington, H.B. 1965a: Trilobites of the Ordovician table head formation, Western Newfoundland. *Bulletin of the Museum of Comparative Zoology, Harvard University 132(4)*, 275–442, pl 68.

Whittington, H.B. 1965b: A monograph of the Ordovician trilobites of the Bala area, Merioneth. Part II. *Palaeontographical Society for 1964, Monographs*, 33–62, pls 9–18.

Yin, G., Tripp, R.P., Zhou, Z-Y., Zhou, Z-Q. & Yuan, W. 2000: Trilobites and biofacies of the Ordovician Pagoda Formation, Donggongsi of Zunyi, Guizhou Province, China. *Transactions of the Royal Society of Edinburgh: Earth Sciences 90*, 203–220.

Zhou, Z., Dean, W.T., Yuan, W. & Zhou, T. 1998: Ordovician trilobites from the Dawangou Formation, Kalpin, Xinjiang, North-West China. *Palaeontology 41(4)*, 693–735.

Plates 1–29

Plate 1

1–8: ***Ogygiocaris dilatata* (Brünnich, 1781)**

1: Cast of lectotype. MGUH 3885, Geological Museum, University of Copenhagen, Denmark. Collected from near Fossum Iron works north of Skien, Skien-Langesund. Brünnich's specimen. Scale bar 1 cm.

2: Partly exfoliated specimen. PMO 206.958. Heggen Member, Elnes Formation. Råen at Fiskum, Eiker Sandsvær. Coll. J. Hurum, 2005. Scale bar 1 cm.

3: Lectotype of *O. dilatata* var. *stroemi* Angelin, 1878. PMO 67032. Heggen Member, Elnes Formation. Eiker in Eiker-Sandsvær. Scale bar 1 cm.

4: External mould with hypostome attached. PMO 206.961. Heggen Member, Elnes Formation. Råen at Fiskum, Eiker Sandsvær. Coll. J. Hurum, 2005. Scale bar 1 cm.

5: Cranidium and free cheek. PMO 60451. Heggen Member, Elnes Formation. Muggerudkleiva at Eiker, Eiker-Sandsvær. Coll. L. Størmer, 1925–27. Scale bar 1 cm.

6: Pygidium and five posterior thoracic segments. PMO 82518. Elnes Formation. Skinviktangen (?) at Rognstranda, Skien-Langesund. Coll. N. Spjeldnæs, 09.09.1960. Scale bar 1 cm.

7: Rubber cast of large hypostome. PMO 82140. Elnes Formation. Probably from Muggerudkleiva at Vestfossen, Eiker-Sandsvær. Scale bar 1 cm.

8: Librigena. PMO 60551. Elnes Formation. Muggerudkleiva at Vestfossen, Eiker-Sandsvær. Coll. L. Størmer, 1925–1927. Scale bar 1 cm.

9–10: ***Ogygiocaris isodilatata* n. sp.**

9: Paratype. Strongly flattened hypostome. PMO 202.871A/1. Middle Sjøstrand Member, Elnes Formation. Level 24.82 m in the Old Quarry at Slemmestad, Oslo-Asker. Coll. T. Hansen, 2003. Scale bar 1 cm.

10: Paratype. Strongly flattened pygidium. PMO 202.864/1. Middle Sjøstrand Member, Elnes Formation. Level 21.11 m in the Old Quarry at Slemmestad, Oslo-Asker. Coll. T. Hansen, 2003. Scale bar 1 cm.

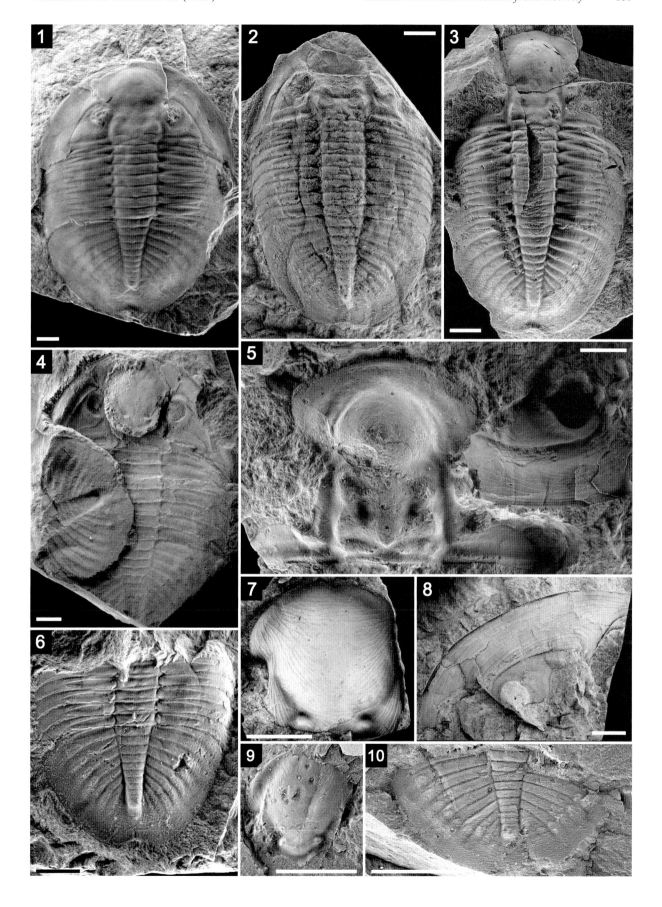

Plate 2

1–6: *Ogygiocaris isodilatata* **n. sp.**

1:   Holotype. Strongly flattened articulated specimen. PMO 202.899C/1. Middle Sjøstrand Member, Elnes Formation. Level 27.24 m in the Old Quarry at Slemmestad, Oslo-Asker. Coll. T. Hansen, 2003. Scale bar 1 cm.

2:   Holotype. Fragmentary cephalon with parts of four thorax segments attached. PMO 202.899A/1. Middle Sjøstrand Member, Elnes Formation. Level 27.24 m in the Old Quarry at Slemmestad, Oslo-Asker. Coll. T. Hansen, 2003. Scale bar 1 cm.

3:   Paratype. Pygidium and five posterior thoracic segments. PMO 202.873C/7. Middle Sjøstrand Member, Elnes Formation. Level 25.02 m in the Old Quarry at Slemmestad, Oslo-Asker. Coll. T. Hansen, 2003. Scale bar 1 cm.

4:   Paratype. Strongly flattened cranidium. PMO 202.886A/1. Middle Sjøstrand Member, Elnes Formation. Level 25.12 m in the Old Quarry at Slemmestad, Oslo-Asker. Coll. T. Hansen, 2003. Scale bar 0.5 cm.

5:   Paratype. Pygidium showing coarse doublural terrace-lines. PMO 202.877A/1. Middle Sjøstrand Member, Elnes Formation. Level 25.07 m in the Old Quarry at Slemmestad, Oslo-Asker. Coll. T. Hansen, 2003. Scale bar 1 cm.

6:   Paratype. Nearly complete specimen. PMO 202.877A/1. Middle Sjøstrand Member, Elnes Formation. Level 25.07 m in the Old Quarry at Slemmestad, Oslo-Asker. Coll. T. Hansen, 2003. Scale bar 1 cm.

7–11: *Ogygiocaris sarsi* **Angelin, 1878**

7:   Partly articulated specimen. PMO 3689. Middle Engervik Member, Elnes Formation. Huk on Bygdøy, Oslo-Asker. Collected 1880. Scale bar 1 cm.

8:   External mould of partly articulated and strongly forelengthened specimen. PMO 56311. Probably lower Engervik Member, Elnes Formation. Hjortnæstangen in Oslo, Oslo-Asker. Scale bar 1 cm.

9:   Nearly complete librigena. PMO 82206. Collected about 13 m above the Huk Formation. Helskjer on Helgøya, Mjøsa. Coll. G. Henningsmoen and K. Egede-Larsen, 26.05.1960. Scale bar 1 cm.

10: Lectotype. Forelengthened hypostome belonging to nearly complete but disarticulated specimen. PMO 20287. Middle Engervik Member, Elnes Formation. Hjortnæstangen in Oslo, Oslo-Asker. Coll. M. Sars. Scale bar 0.2 cm.

11: Lectotype. Plasticine cast of cephalon and two thoracic segments. PMO 20288. Middle Engervik Member, Elnes Formation. Hjortnæstangen in Oslo, Oslo-Asker. Coll. M. Sars. Scale bar 1 cm.

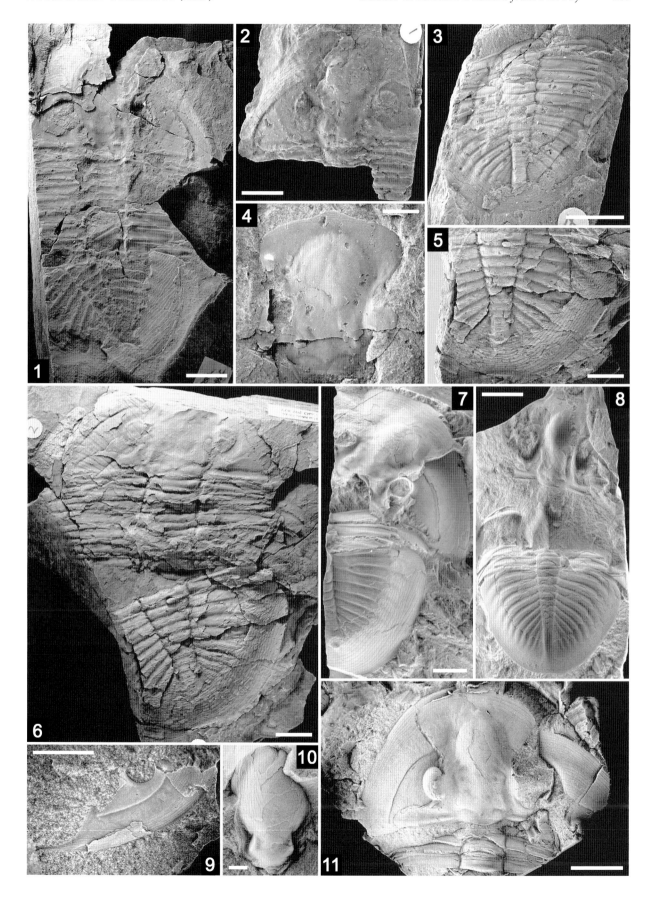

Plate 3

1:   *Ogygiocaris sarsi* **Angelin, 1878**

Lectotype. PMO 20287. Partly disconnected and somewhat forelengthened specimen. Middle Engervik Member, Elnes Formation. Hjortnæstangen in Oslo, Oslo-Asker. Coll. M. Sars. Scale bar 1 cm.

2–8:   *Ogygiocaris henningsmoeni* **n. sp.**

2:   Holotype. Nearly complete small specimen. PMO 206.159/1. Uppermost Sjøstrand Member, Elnes Formation. Level 2.99 m in the road profile along Bøveien just south of Slemmestad, Oslo-Asker. Coll. T. Hansen, 2004. Scale bar 1 cm.

3:   Paratype. Librigena showing doublural terrace-lines. PMO 203.067b/8. Upper Sjøstrand Member, Elnes Formation. Level 22.45 m in the beach profile at Djuptrekkodden, Slemmestad, Oslo-Asker. Coll. T. Hansen, 2003. Scale bar 1 cm.

4:   Paratype. Cranidium of disarticulated specimen. PMO 203.058/4. Upper Sjøstrand Member, Elnes Formation. Level 22.41 m in the beach profile at Djuptrekkodden, Slemmestad, Oslo-Asker. Coll. T. Hansen, 2003. Scale bar 1 cm.

5–6:   Paratype. Dorsal view of cranidium and complete specimen. Note the lack of posterior pygidial truncation. PMO 209.545. Probably from upper part of the Heggen Member, Elnes Formation. Profile 1 (?) at Råen, Eiker-Sandsvær. Coll. J. Hurum. Scale bars 0.5 cm.

7:   Paratype. Fragmentary pygidium. PMO 203.166c/5. Lower half of the Engervik Member, Elnes Formation. Level 31.56 m in the beach profile at Djuptrekkodden, Slemmestad, Oslo-Asker. Coll. T. Hansen, 2003. Scale bar 1 cm.

8:   Paratype. Head and thorax belonging to pygidium shown on 7. PMO 203.166a-c/3. Scale bar 1 cm.

9–11: *Ogygiocaris lata* **Hadding, 1913**

9:   Small hypostome. PMO 72099. Elnes Formation. Hovindsholm on Helgøya, Mjøsa. Coll. G. Henningsmoen, 1957. Scale bar 0.2 cm.

10:   Paratype. Cranidium. PMO 36790. Elnes Formation. Hovindsholm on Helgøya, Mjøsa. Coll. O. Holtedahl, 1906. Scale bar 1 cm.

11:   Paratype. Cephalon. PMO 36722. Elnes Formation. Hovindsholm on Helgøya, Mjøsa. Coll. O. Holtedahl, 1906. Scale bar 1 cm.

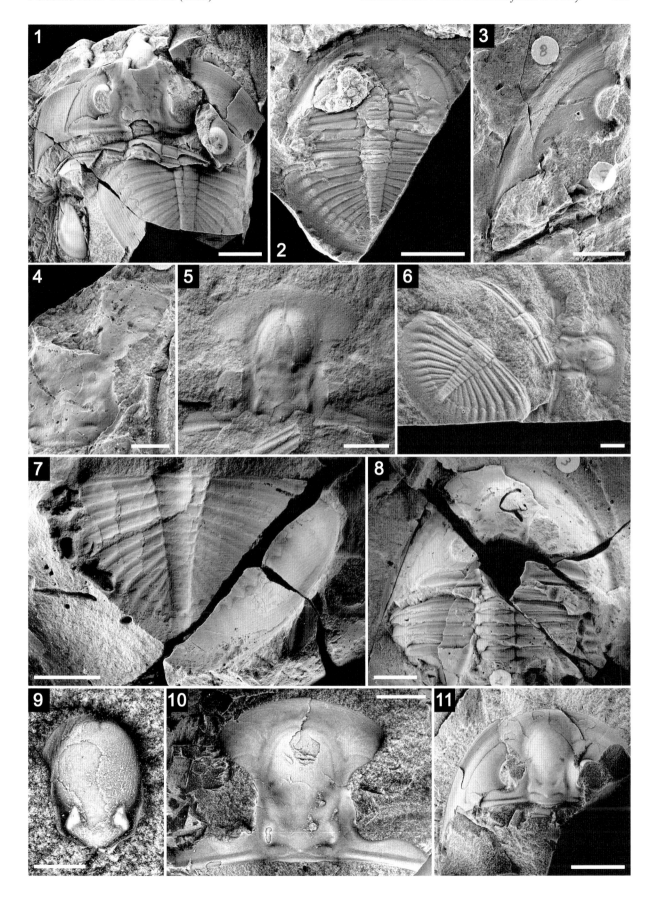

Plate 4

1–5: ***Ogygiocaris lata* Hadding, 1913**

1–2: Holotype. Complete specimen. 1: in rock sample. 2: when surrounding rock has been removed electronically from picture. PMO 72090. Elnes Formation. Toten, Mjøsa. Scale bar 1 cm.

3: Paratype. Librigena showing doublural terrace-lines and parts of border terrace-lines. PMO 36786. Elnes Formation. Hovindsholm on Helgøya, Mjøsa. Coll. O. Holtedahl, 1906. Scale bar 1 cm.

4: Nearly complete small specimen. PMO 33335. Elnes Formation. Hadeland. Scale bar 1 cm.

5: Paratype. Small cranidium. PMO 72099. Elnes Formation. Hovindsholm on Helgøya, Mjøsa. Coll. G. Henningsmoen, 1957. Scale bar 0.5 cm.

6–12: ***Ogygiocaris delicata* Henningsmoen, 1960**

6: Partly articulated small specimen. PMO 72101. Elnes Formation. Hovindsholm Beach on Helgøya, Mjøsa. Coll. G. Henningsmoen, 14.04.1957. Scale bar 1 cm.

7: Fragmentary hypostome. PMO 36879. Elnes Formation. Alfstad (farm?) in Østre Toten, Mjøsa. Gift present by the family of Nils Olsen Alfstad, 1868. Scale bar 0.2 cm.

8: Cranidium showing narrow frontal border. PMO 207.443/11. Elnes Formation. Level 4.53 m on sub-profile 3 at northern road-cut where Fv 84 crosses E6 southwest of Nydal, Mjøsa. Coll. T. Hansen, 18.08.2005. Scale bar 0.5 cm.

9: External mould of cephalon. PMO 82904. Elnes Formation. Huk on Bygdøy, Oslo-Asker. Coll. J. H. L. Vogt, 1907. Scale bar 1 cm.

10: Slightly deformed pygidium. PMO 207.443/2. Elnes Formation. Level 4.52 m on sub profile 3 at northern road-cut were Fv 84 crosses E6 southwest of Nydal, Mjøsa. Coll. T. Hansen, 18.08.2005. Scale bar 1 cm.

11: Pygidium showing outer terrace-lines. PMO 36879. Elnes Formation. Alfstad (farm?) in Østre Toten, Mjøsa. Present from the family of Nils Olsen Alfstad, 1868. Scale bar 1 cm.

12: Librigena showing doublural terrace-lines. PMO 36879. Elnes Formation. Alfstad (farm?) in Østre Toten, Mjøsa. Present from the family of Nils Olsen Alfstad, 1868. Scale bar 1 cm.

13–17: ***Ogygiocaris striolata striolata* Henningsmoen, 1960**

13: Complete dorsal shield. PMO 95829. Heggen Member, Elnes Formation. Loose block collected halfway down the small ski jump at Vikersund Skijump, Modum. Coll. K. Gram, 16.10.1982. Scale bar 1 cm.

14: Cranidium showing the short preglabellar area. PMO 82450. Engervik Member (*Nileus* interval), Elnes Formation. Elnestangen at Bjerkås, Oslo-Asker. Coll. G. Henningsmoen, 29.11.1959. Scale bar 0.5 cm.

15: Paratype. Hypostome showing characteristic dense striation. PMO 72104. Middle Engervik Member, Elnes Formation. Collected from concretionary horizon nearly 7 m above base of member. Elnes Tangen at Bjerkås, Oslo-Asker. Coll. G. Henningsmoen, 1959. Scale bar 0.2 cm.

16: Paratype. Detail of anterior part of cranidium showing fine striations. PMO 72115. Collected from concretionary horizon nearly 7 m above base of member. Elnes Tangen at Bjerkås, Oslo-Asker. Coll. G. Henningsmoen, 1958. Scale bar 0.5 cm.

17: Holotype. Pygidium showing doublural terrace-lines. PMO 72108. Middle Engervik Member, Elnes Formation. Collected from concretionary horizon nearly 7 m above base of member. Elnes Tangen at Bjerkås, Oslo-Asker. Coll. G. Henningsmoen, 1958. Scale bar 1 cm.

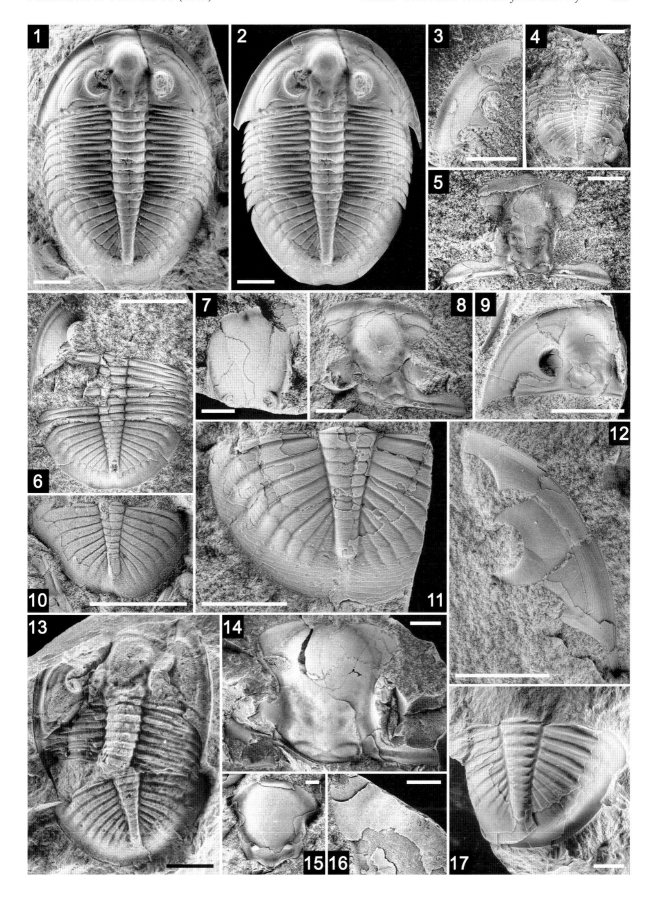

Plate 5

1–4: *Ogygiocaris striolata striolata* **Henningsmoen, 1960**

1:   Paratype. Librigena showing micro-striation on surface. PMO 72112. Elnes Formation. Collected at Djuptrekkodden north of Slemmestad, Oslo-Asker. Coll. G. Henningsmoen, 1958. Scale bar 1 cm.

2:   Fragmentary librigena showing doublural terrace-lines. PMO 102.746. Elnes Formation. Loose block from Vikersundbakken, Modum. Coll. B. Wandås. Scale bar 1 cm.

3:   Paratype. Detail of pygidial surface. PMO 72114. Middle Engervik Member, Elnes Formation. Collected from concretionary horizon nearly 7 m above base of member. Elnes Tangen at Bjerkås, Oslo-Asker. Coll. G. Henningsmoen, 1958. Scale bar 0.5 cm.

4:   Pygidium showing doublural terrace-lines and flattened pleural ribs. PMO 72104. Middle Engervik Member, Elnes Formation. Collected from concretionary horizon nearly 7 m above base of member. Elnes Tangen at Bjerkås, Oslo-Asker. Coll. G. Henningsmoen, 1958. Scale bar 1 cm.

5–10: *Ogygiocaris striolata corrugata* **Henningsmoen, 1960**

5:   Paratype. Fragmentary cranidium showing fine striation around mesial tubercle. PMO 72131. Elnes Formation. Loose limestone lens from Elnestangen at Bjerkås, Oslo-Asker. Coll. G. Henningsmoen, 1959. Scale bar 1 cm.

6:   Pygidium and parts of posteriormost six thoracic segments. PMO 206.253/1. Upper Heggen Member, Elnes Formation. Collected from level 8.79 m in profile 1 at Råen, Eiker-Sandsvær. Coll. T. Hansen, 2004. Scale bar 1 cm.

7:   Detail of pygidium showing dense surface striation. PMO 206.275c/3. Upper Heggen Member, Elnes Formation. Collected from level 8.76 m in profile 1 at Råen, Eiker-Sandsvær. Coll. T. Hansen, 2004. Scale bar 0.5 cm.

8:   Poorly preserved librigena. PMO 208.505/6. Upper Heggen Member, Elnes Formation. Level 9.06 approximatedly, profile 1 at Råen, Eiker-Sandsvær. Coll. T. Hansen, 2004. Scale bar 1 cm.

9:   Same pygidium as Plate 5, fig. 7. Scale bar 0.5 cm.

10:   Fragmentary hypostome showing posterior part of middle body. PMO 203.346/1. Upper Sjøstrand Member, Elnes Formation. Collected from 22.45 m above base of beach profile at Djuptrekkodden near Slemmestad, Oslo-Asker. Coll. T. Hansen, 2003. Scale bar 1 cm.

11–15: *Volchovites perstriatus* **(Bohlin, 1955)**

11, 15: Dorsal and lateral view of articulated specimen. Note the posterolaterally directed ridge on librigena and the pointed pleural terminations on the posterior thoracic segments. PMO 82056. Helskjer Member, 4.7 m above the Huk Formation at Helskjer, Mjøsa. Coll. G. Henningsmoen, 1961. Scale bars 1 and 0.5 cm respectively.

12:   Hypostome in place on compressed specimen. PMO 106.513. Uppermost Holen Limestone. Kinnekulle in Västergötland, Sweden. Coll. J. Johansson, 1980. Scale bar 0.5 cm.

13:   Cephalon and anteriormost thorax. PMO 106.090. Lower Elnes Formation. Collected 7.9 m above base of formation at Hovodden, Randsfjorden, Hadeland. Coll. Ø. Lauritzen, 1967. Scale bar 1 cm.

14:   Detail of fragmentary librigena showing small terrace-lines and pores on anterior part. PMO 33326. Lower Elnes Formation. Hovstangen at Gran, Hadeland. Coll. Th. Münster, 16.09.1893. Scale bar 0.5 cm.

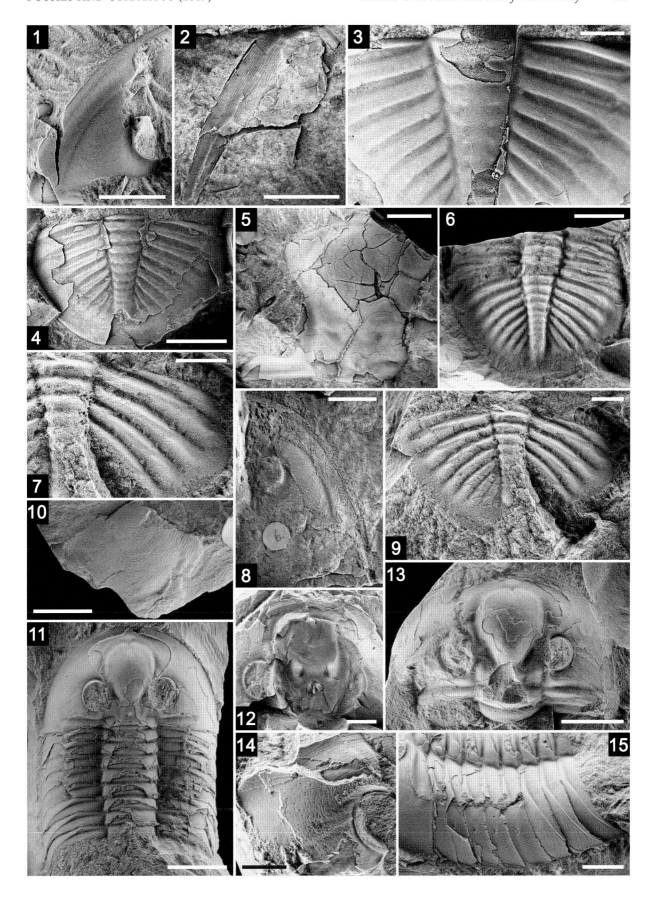

Plate 6

1–8: *Volchovites perstriatus* (Bohlin, 1955)

1: Cephalon in frontal view. PMO 105.899. Lower Elnes Formation. Collected from 7.95 m above base of formation at Hovodden, Randsfjorden, Hadeland. Coll. D. L. Bruton, B. Wandås and D. A. T. Harper, 1980. Scale bar 1 cm.

2: Lateral view of cephalon. PMO 207.380. Helskjer Member. Level 4.92 m in sub-profile 1 along E6 between Furnes and Nydal, Mjøsa. Coll. T. Hansen, 17.08.2005. Scale bar 0.5 cm.

3: Large pygidium showing external and doublural terrace-line patterns. Note tear-like incission on left side. PMO 83240. Uppermost Helskjer Member. Level 7.75 m above the Huk Formation at Furnes Church, Mjøsa. Coll. F. Nikolaisen, 1961. Scale bar 1 cm.

4: Extremely rounded pygidium. Note strong pleural ribs. PMO 83258. Uppermost Helskjer Member. Level 7.75 m above the Huk Formation at Furnes Church, Mjøsa. Coll. G. Henningsmoen, 07.21.1961. Scale bar 0.5 cm.

5: Posterior view of pygidium. Note the convex outer part of the pleural region. PMO 207.519. Uppermost Helskjer Member. Level 10.08 m in sub-profile 1 along E6 between Furnes and Nydal, Mjøsa. Coll. T. Hansen, 31.05.2003. Scale bar 0.5 cm.

6: Oblique posterior view of pygidium. Note the slightly less convex pygidial flanks. PMO 207.381/1. Helskjer Member. Level 5.10 m in sub-profile 1 along E6 between Nydal and Furnes, Mjøsa. Coll. T. Hansen, 17.08.2005. Scale bar 0.5 cm.

7: Dorsal view of small pygidium with very faint pleural ribs. PMO 207.390/4. Helskjer Member. Level 6.81 m in sub-profile 1 along E6 between Furnes and Nydal, Mjøsa. Coll. T. Hansen, 08.17.2005. Scale bar 0.5 cm.

8: Dorsal view of small pygidium with dense terrace-line pattern on doublure. Note slightly concave sides. PMO 67600. Limestone bed from above the Helskjer Member (cit. S. Skjeseth, 1951). Helskjer on Helgøya, Mjøsa. Coll. S. Skjeseth, 1951. Scale bar 0.5 cm.

9–12, 14: *Pseudobasilicus truncatus* n. sp.

9: Holotype. Nearly complete specimen. PMO 202.077A/1. Upper Helskjer Member, Elnes Formation. Level 1.24 m in the southwestern profile in the Old Quarry at Slemmestad, Oslo-Asker. Coll. T. Hansen, 2003. Scale bar 1 cm.

10: Paratype. Articulated specimen showing spine like prolongation of thoracic pleural ends. PMO 103.717. Loose sample from road cut at Vikersund skijump, Modum. Coll. D. L. Bruton and B. Wandås, 22.05.1979. Scale bar 1 cm.

11: Paratype. Anterior part of articulated specimen. PMO 83294. Collected from 0.15 m above the Huk Formation at Molleklev, Rognstranda, Skien-Langesund. Coll. L. Størmer and G. Henningsmoen, 03.08.1950. Scale bar 1 cm.

12: Paratype. Fragmentary pygidium. PMO 202.084A/1. Upper Helskjer Member, Elnes Formation. Collected 1.26 m above the Huk Formation in the southwestern profile in the Old Quarry north of Slemmestad, Oslo-Asker. Coll. T. Hansen, 2003. Scale bar 0.5 cm.

14: Paratype. External mould of pygidium showing surface terrace-lines. PMO 102.652. Lowermost Heggen Member, Elnes Formation. Collected 5.50 m above base of formation in road cut at Vikersund skijump, Modum. Coll. G. Henningsmoen, 20.09.1970. Scale bar 0.5 cm.

13, 16: *P.* (*Pseudobasilicus?*) sp.

13: Nearly complete pygidium. PMO 36588. Probably uppermost Stein Formation or else basal Elnes Formation. Ålset at Snertingdal, Mjøsa. Scale bar 0.5 cm.

16: Dorsal view of pygidium PMO 33327. Elnes Formation. Lynne farm near Gran, Mjøsa. Scale bar 0.5 cm.

15: *Pseudasaphus limatus limatus* Jaanusson, 1953a
    Strongly flattened librigena with genal spine. Uppermost Håkavik Member, Elnes Formation. Collected from level 57.27 m in the beach profile at Djuptrekkodden, Oslo-Asker. Coll. T. Hansen, 2003. PMO 206.137/1. Scale bar 0.4 cm.

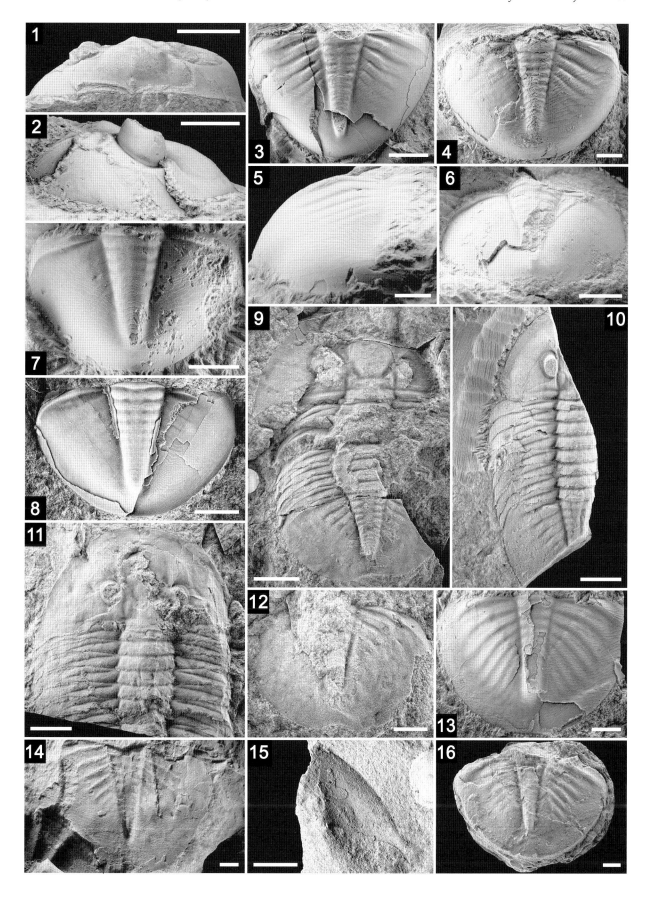

Plate 7

1–2: *Pseudasaphus* **sp. B**

Dorsal view and detail of right anterolateral corner showing abnormal change in terrace-line pattern. Specimen label referring it to being from the Elnes Formation at Huk on Bygdøy, Oslo-Asker. Coll. H. Bäckström, 1886. PMO H2707. Scale bars 1 and 0.5 cm respectively.

3–7: *Pseudobasilicus* **sp. A**

3–4: Dorsal and lateral view of pygidium and posteriormost four thoracic segments. PMO 68334. Lowermost Heggen Member, Elnes Formation. Collected 8.35 m above base of formation on Hovodden at Randsfjorden, Hadeland. Coll. L. Størmer, 1961. Scale bars 0.5 cm.

5–6: Dorsal and oblique posterolateral view of pygidium. Note the fine doublural terrace-line pattern. PMO 105.902. Lowermost Heggen Member, Elnes Formation. Collected from 5.80 m above base of formation on Hovodden at Randsfjorden, Hadeland. Coll. D. L. Bruton and B. T. G. Wandås, 21.09. 1978. Scale bars 0.5 cm.

7: Dorsal view of slightly flattened pygidium. PMO 3392. Elnes Formation. Labelled 'Brocks løkke', district of Oslo region not known. Scale bar 0.5 cm.

8–15: *Pseudasaphus limatus limatus* **Jaanusson, 1953a**

8: Dorsal view of fragmentary cephalon and thorax. PMO 58911. Suggested to have come from the Vollen Formation (Henningsmoen, 1960), though a high content of silt may indicate a stratigraphic closeness to the Håkavik Member of the Elnes Formation. Fornebu old airport, Oslo-Asker. Collected 1935. Scale bar 0.5 cm.

9, 12: Dorsal view of articulated specimen and detail of left side showing panderian openings and posterior part of poorly preserved librigena. PMO 3346. Elnes Formation (Håkavik Mb.?). Professor Dahls gate 48 in Oslo City, Oslo-Asker. Coll. T. Holdt, 1917. Scale bars 1 and 0.5 cm respectively.

10: Partly exfoliated cranidium. PMO 40355. Vollen Formation. Nye Drammensveien at Gyssestad, Oslo-Asker. Coll. A. Heintz, 1933. Scale bar 0.5 cm.

11: Fragmentary hypostome. PMO 40400. Vollen Formation. Nye Drammenveien at Gyssestad, Oslo-Asker. Coll. A. Heintz, 1933. Scale bar 0.5 cm.

13–15: Details of cranidial, thoracic and axial exoskeletal surface respectively. Note the dense pattern of terrace-lines and large pores appearing as granules on the external mould. PMO 58917. Most probably from the Vollen Formation. Fornebu old airport, Oslo-Asker. Collected 1935. Scale bars 0.5, 0.2 and 0.2 cm respectively.

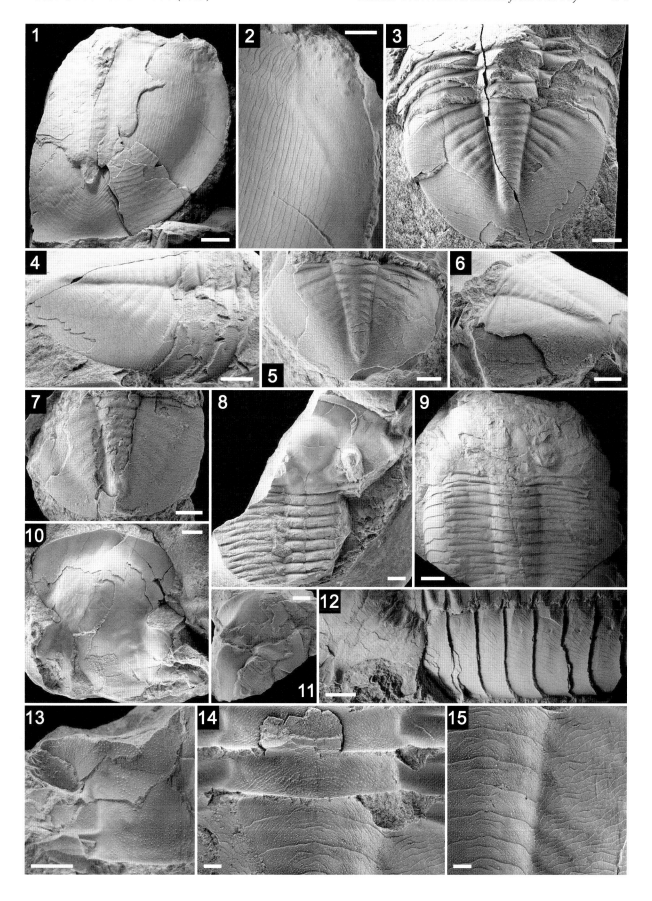

Plate 8

1–3: *Pseudasaphus limatus limatus* Jaanusson, 1953a

1:   Dorsal view of fragmentary pygidium. PMO 3973. Uppermost 4 m of Håkavik Member ('1–2 m above thickest calcarenite bed'). Håkavik, Oslo-Asker. Coll. O. Holtedahl. Scale bar 0.5 cm.

2–3: Dorsal and oblique posterior view of pygidium. PMO 20317. Vollen Formation. Exact locality in Oslo Region unknown. Scale bars 0.5 cm.

4–8: *Pseudasaphus limatus longistriatus* **n. subsp.**

4–6: Dorsal, oblique lateral and posterior view of holotype pygidium. PMO 206.117a/1. Note the long posteriorly directed terrace-lines on inner pleural area. Top Håkavik Member, Elnes Formation. Level 56.01 m in the beach profile at Djuptrekkodden, Oslo-Asker. Coll. T. Hansen, 2003. Scale bars 0.5 cm.

7:   Dorsal view of paratype pygidium. PMO 206.115/2. Top Håkavik Member, Elnes Formation. Level 56.00 m in the beach profile at Djuptrekkodden, Oslo-Asker. Coll. T. Hansen, 2003. Scale bar 0.5 cm.

8:   Dorsal view of strongly flattened pygidium. PMO 206.111. Top Håkavik Member, Elnes Formation. Level 55.83 m in the beach profile at Djuptrekkodden, Oslo-Asker. Coll. T. Hansen, 2003. Scale bar 0.5 cm.

9–11: *Pseudasaphus* **sp. A**

9:   Left side of flattened pygidium. PMO 208.643b/3. Middle Heggen Member, Elnes Formation. Level 1.48 m in profile 6 at Vego, Eiker-Sandsvær. Coll. T. Hansen, 2004. Scale bar 0.5 cm.

10–11: Cast and positive of pygidial mould. PMO 208.643/7. Middle Heggen Member, Elnes Formation. Level 1.48 m in profile 6 at Vego, Eiker-Sandsvær. Coll. T. Hansen, 2004. Scale bars 0.5 cm.

12–17: *Ogmasaphus stoermeri* **Henningsmoen, 1960**

12: Nearly complete exoskeleton preserved in silty mudstone. Holotype PMO 4684. Middle Heggen Member, Elnes Formation. Teigen Slate Pit at Eiker, Eiker-Sandsvær. Coll. M. Langberg. S. Swensen dedit., 1892. Scale bar 0.5 cm.

13: Dorsal view of complete exoskeleton preserved in carbonate concretion. PMO 90560. Heggen Member, Elnes Formation. Point at Brandbu Church, Hadeland. Coll. E. Yochelson and G. Henningsmoen, 25.10.1961. ? Has erroneously been labelled as coming from Skinnvikstangen at Rognstrand, Skien-Langesund, probably by G. Henningsmoen. Scale bar 0.5 cm.

14: Dorsal view of attached hypostome. PMO 4682. Middle Heggen Member, Elnes Formation. Teigen Slate Pit at Eiker, Eiker-Sandsvær. Coll. M. Langberg. S. Swensen dedit., 1892. Scale bar 0.5 cm.

15: Pygidium and part of thorax. Note surface striation. PMO 72069. Middle Heggen Member, Elnes Formation. Probably from Heistad in the Eiker-Sandsvær area. Scale bar 1 cm.

16–17: Dorsal and oblique posterior view of foreshortened pygidium. PMO 82430. Heggen Member, Elnes Formation. Point at Brandbu Church, Hadeland. Coll. L. Størmer, 1925–27. Scale bars 1 cm.

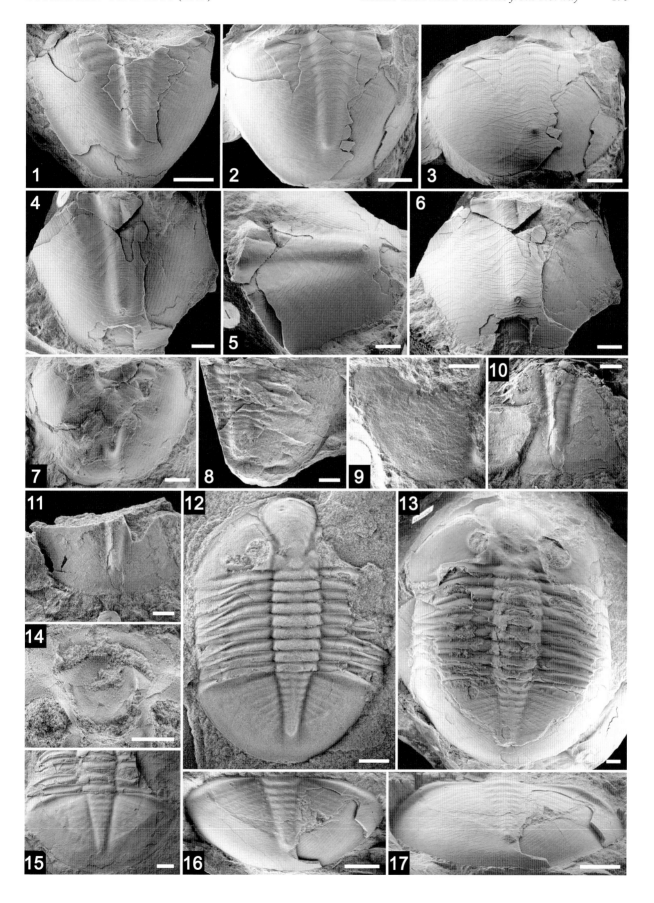

Plate 9

1–9: *Ogmasaphus kiaeri* **Henningsmoen, 1960**

1–2: Anterior dorsal and lateral view of enrolled specimen. Holotype PMO 19112. Vollen Formation. Søndre Kojatangen between Blakstad and Vollen, Oslo-Asker. Coll. J. Kiær, 26.07.1898. Scale bars 0.5 cm.

3: Small pygidium with posterior thoracic segments attached. PMO 3972. Top Håkavik Member or basal Vollen Formation. Håkavik, Oslo-Asker. Coll. O. Holtedahl. Scale bar 0.5 cm.

4–5: Left and right lateral view of librigenal corner and anterior thoracic segments. PMO 58932. Probably Vollen Formation. Sarpsborggata, Oslo. Coll. S. Rönning, 1932. Scale bars 0.5 cm.

6: Dorsal view of small pygidium. PMO 3986. Upper part of the Elnes Formation. Håkavik, Oslo-Asker. Coll. O. Holtedahl, 1906. Scale bar 0.5 cm.

7: Oblique posterior view of pygidium. PMO 3368. Nordraaks gate, Oslo City. Coll. A. V. Jennings, 1903? Scale bar 0.5 cm.

8: Dorsal view of holotype pygidium PMO 19112. Vollen Formation. Søndre Kojatangen between Blakstad and Vollen, Oslo-Asker. Coll. J. Kiær, 26.07.1898. Scale bar 0.5 cm.

9: Pygidium with five posterior thoracic segments attached. PMO 82666. Probably Elnes Formation. Bø farm just north of Skien. Coll. H. H. Horneman, 1905. Scale bar 0.5 cm.

10–15: *Ogmasaphus jaanussoni* **Henningsmoen, 1960**

10: Nearly complete specimen. PMO 20289. Vollen Formation or perhaps upper part of the Elnes Formation. Oslo. Donated by A. G. Ström, 28.11.1891. Scale bar 0.5 cm.

11, 13–15: Dorsal and lateral view and detail of pygidium and cephalon of holotype PMO 61519. Horizon unknown, but may belong to the upper part of the Elnes Formation. Ruseløkkveien, Oslo City. Scale bar 0.5 cm for Fig. 13 and 1.0 cm for the rest.

12: Fragmentary articulated dorsal shield. PMO 82979. Horizon unknown, although the dark concretion and the fossil content of orthid brachiopods would suggest the Vollen Formation as the most likely candidate. Elåsen? on Fornebu, Oslo-Asker. Coll. M. Fossum, 1937. Scale bar 0.5 cm.

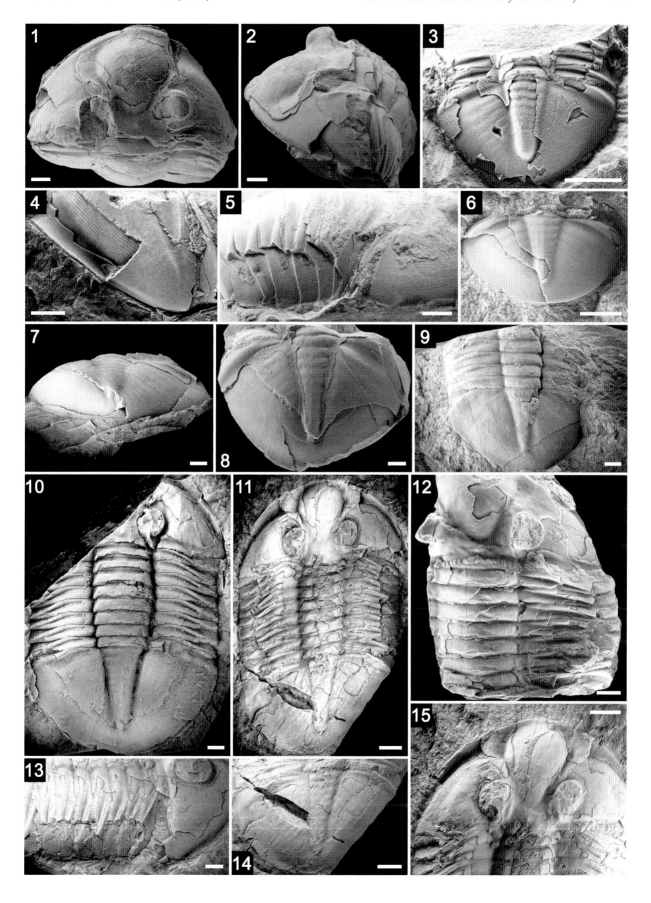

Plate 10

1–7: ***Ogmasaphus multistriatus* Henningsmoen, 1960**
1–2: Dorsal and oblique lateral view of cephalon. PMO 21931. Elnes Formation. Helgøya, Mjøsa. Coll. O. Holtedahl, 1907. Scale bars 0.2 cm.
3: Small cranidium. PMO 36822b. Elnes Formation. Storhamar, Mjøsa. Collected 1883. Scale bar 0.1 cm.
4–5: Dorsal and posterior view of holotype pygidium. PMO 36822a. Elnes Formation. Storhamar, Mjøsa. Collected 1883. Scale bars 0.2 and 0.5 cm respectively.
6: Dorsal view of pygidium. PMO 58928. Possibly Elnes Formation. Fornebu old airport. Collected 1935. Scale bar 0.5 cm.
7: Dorsal view of pygidium. PMO 56580. Elnes or Vollen formations. Oslo City Hall, Oslo-Asker. Coll. A. Heintz, 1933. Scale bar 0.5 cm.

8–13: ***Ogmasaphus* sp. A**
8–9: Dorsal and lateral view of pygidium and posterior thoracic segments. PMO 3466. Elnes Formation. Road cut above Lille Frøen in Oslo, Oslo-Asker. Coll. Holdt, 1918?. Scale bars 1 and 0.5 cm respectively.
10: Dorsal view of exfoliated pygidium. PMO 6007. Most probably Vollen Formation. Kilen on Fornebu, Oslo-Asker. Coll. W. Werenskiold, 1904. Scale bar 0.5 cm.
11–12: Dorsal and posterior view of pygidium. PMO 40345. Vollen Formation. Gyssestad, Oslo-Asker. Coll. A. Heintz, 1933. Scale bars 0.5 cm.
13: Dorsal view of large pygidium. PMO 56581. Elnes Formation or lower part of Vollen Formation. Oslo City Hall, Oslo-Asker. Coll. A. Heintz, 1933. Scale bar 1 cm.

14: ***Ogmasaphus jaanussoni* Henningsmoen, 1960**
Wax cast of nearly complete juvenile specimen. PMO S.1790–91. Håkavik Member, Elnes Formation. Slemmestad and here most probably from the Elnestangen, Oslo-Asker. Coll. L. Størmer. Scale bar 0.5 cm.

15–17: ***Ogmasaphus furnensis* n. sp.**
15: Dorsal view of paratype pygidium. PMO 90498. Top Helskjer Member. Level 8.3 m above the Huk Formation at Furnes Church, Mjøsa. Coll. G. Henningsmoen, 21.07.1961. Scale bar 0.5 cm.
16: Dorsal view of paratype pygidium PMO 90499a. Top Helskjer Member. Level 8.3 m above the Huk Formation in the profile at Furnes Church, Mjøsa. Coll. G. Henningsmoen, 21.07.1961. Scale bar 0.5 cm.
17: Dorsal view of paratype cranidium PMO 90499b. Same sample as PMO 90499a. Scale bar 0.5 cm.

18: ***Pseudobasilicus?* sp.**
Detail of pygidium PMO 33490. Elnes Formation, probably from upper part. Haugslandet at Gran, Hadeland. Coll. T. Münster, 16.07.1891. Scale bar 0.2 cm.

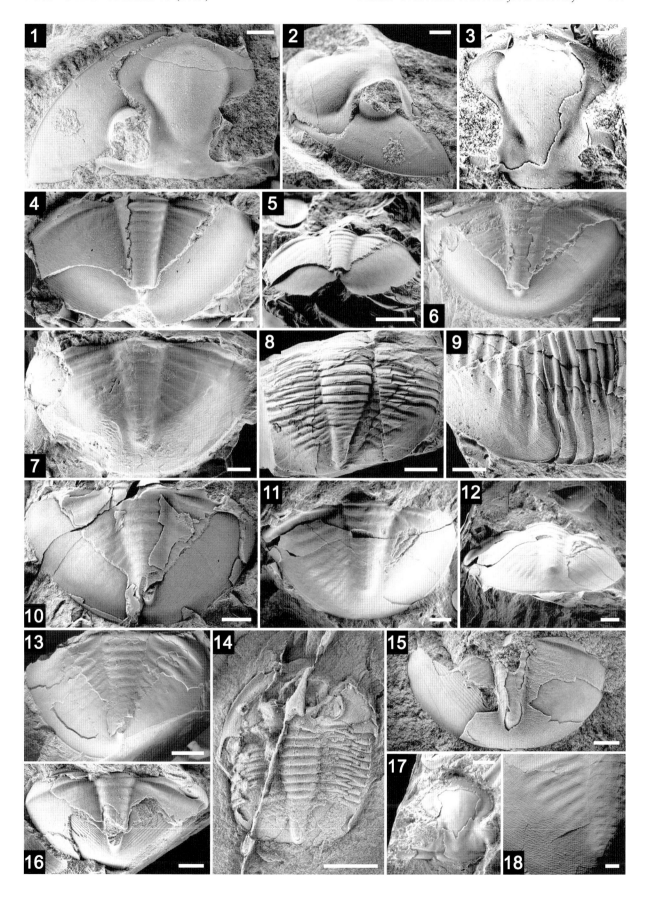

Plate 11

1:   *Pseudobasilicus?* **sp.**
Fragmentary pygidium PMO 33490. Elnes Formation, probably from upper part. Haugslandet at Gran, Hadeland. Coll. T. Münster, 16.07.1891. Scale bar 0.5 cm.

2–3:   *Asaphus ludibundus s.l.* **Törnquist, 1884**
Dorsal view of cranidium and complete specimen. PMO 36941. 'Cephalopod Shale', Elnes Formation. Locality unknown, but the round concretion suggests somewhere in the Mjøsa districts. Scale bars 0.5 and 1 cm respectively.

4–6:   *Subasaphus platyurus* **(Angelin, 1854)**
Dorsal view, detail of same and oblique posterior view of pygidium. PMO 82209. Basal Heggen Member, Elnes Formation. Collected 8.60 m above base of formation at Furnes Church, Mjøsa. Coll. G. Henningsmoen, 21.07.1961. Scale bars 1 cm.

7–14:   *Asaphus sarsi* **Brøgger, 1882**
7:   Cephalic view of enrolled specimen. PMO 56257. Figured in Størmer (1940, pl. 3, figs 12–15). Svartodden Member, Huk Formation. Huk on Bygdøy, Oslo-Asker. Scale bar 0.5 cm.
8:   Cephalic view of enrolled holotype specimen PMO H2618. Svartodden Member, Huk Formation. Oslo City, Oslo-Asker. Scale bar 0.5 cm.
9–10:   Lateral view of enrolled specimen and detail of thoracic pleura. Note the dense pleural terrace-lines in comparison with the librigenal doublure. PMO 1643. Svartodden Member, Huk Formation. Tøyen in Oslo, Oslo-Asker. Coll. Engebretsen. Scale bars 0.5 cm.
11:   Dorsal view of small specimen. PMO 82163. Locality and horizon unknown, but most probably from the uppermost Svartodden Member or lower Helskjer Member in the northern part of the Oslo Region. Scale bar 0.5 cm.
12:   Posterior part of nearly complete specimen. PMO 106.046. Collected 2.8 m above base of Helskjer Member at Helskjer on Helgøya, Mjøsa. Coll. G. Henningsmoen, 27.09.1961. Scale bar 0.5 cm.
13–14:   Posterior and oblique posterolateral view of pygidium belonging to same specimen as figured in Plate 11, fig. 7. PMO 56257. Scale bars 0.5 cm.

15–16:   *Asaphus striatus* **(Sars & Boeck *in* Boeck, 1838)**
15:   Hypostome most probably belonging to this species. PMO 106.551. Top Helskjer Member. Collected 7.75 m above the Huk Formation from road cut at Furnes Church, Mjøsa. Coll. G. Henningsmoen, 1961. Scale bar 0.5 cm.
16:   Exfoliated hypostome. PMO H2634. Helskjer Member, Elnes Formation. Eiker, Eiker-Sandsvær. Coll. M. Sars. Scale bar 0.5 cm.

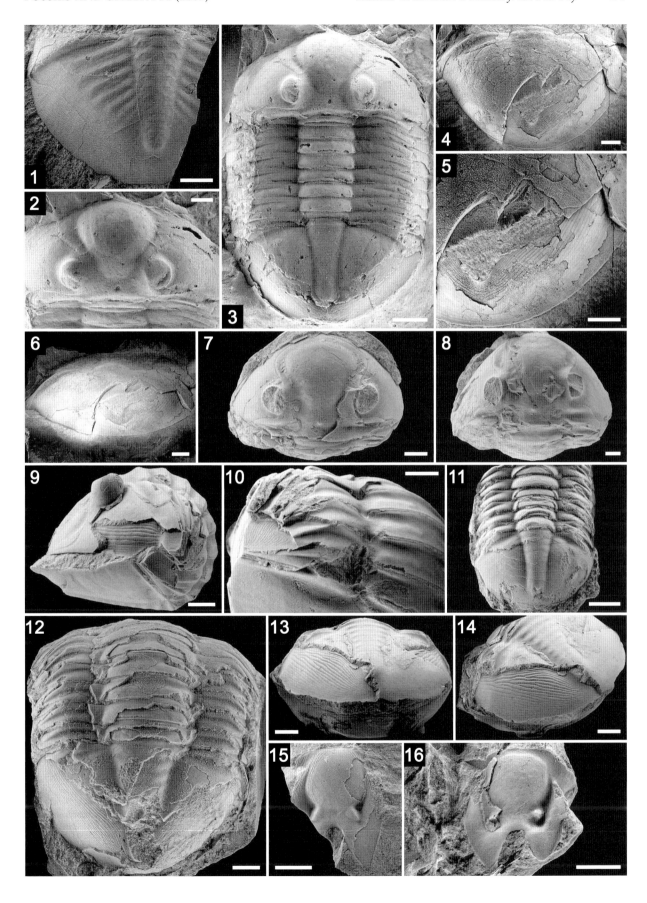

Plate 12

1, 4–8: *Asaphus striatus* (Sars & Boeck *in* Boeck, 1838)

1: Dorsal view of nearly complete specimen PMO 82211. Helskjer Member or basal most 20 cm of the Heggen Member, Elnes Formation. Skinnviktangen (?) at Rognstranda, Skien-Langesund. Coll. G. Henningsmoen, 09.09.1960. Scale bar 0.5 cm.

4–5, 8: Cephalic, lateral and pygidial view of specimen PMO 68413. Collected 6 m above the Huk Formation or approximately 2.5 m above the Helskjer Member. Hovodden at Randsfjorden, Hadeland. Coll. L. Størmer and S. Skjeseth, 21.05.1951. Scale bars 0.5 cm.

6: Lectotype pygidium PMO H2634. Helskjer Member, Elnes Formation. Eiker, Eiker-Sandsvær. Coll. M. Sars. Scale bar 0.5 cm.

7: Posterior view of pygidium. PMO 83010. Upper Helskjer Member. Collected 7.75 m above the Huk Formation from road cut at Furnes Church, Mjøsa. Coll. G. Henningsmoen, 30.08.1961. Scale bar 0.5 cm.

2–3: *Asaphus* sp. A

Dorsal and cranidial view of nearly complete exoskeleton. PMO 72127. Collected 10.6 m above the Huk Formation or 7.0 m above the Helskjer Member. Hovodden at Randsfjorden, Hadeland. Coll. L. Størmer and S. Skjeseth, 21.05.1951. Scale bars 0.5 cm.

9–10, 12–16: *Asaphus raaenensis* n. sp.

9: Hypostome PMO 60390. Heggen Member, Elnes Formation. Muggerudkleiva, Eiker-Sandsvær. Coll. L. Størmer, 1925–27. Scale bar 0.1 cm.

10: Dorsal view of cephalon. PMO 60510. Heggen Member, Elnes Formation. Muggerudkleiva at Hedenstad south of Kongsberg, Eiker-Sandsvær. Coll. L. Størmer, 1925–27. Scale bar 0.5 cm.

12: Exoskeletal view of specimen. PMO 208.524/2. Upper Heggen Member, Elnes Formation. Level 16.82 m in profile 1 at Råen near Fiskum, Eiker-Sandsvær. Coll. T. Hansen, 2004. Scale bar 0.5 cm.

13: Dorsal view of specimen PMO 60375a. Heggen Member, Elnes Formation. Muggerudkleiva at Hedenstad just south of Kongsberg, Eiker-Sandsvær. Coll. L. Størmer, 1925–27. Scale bar 0.5 cm.

14: Somewhat disarticulated specimen preserved in dark mudstone. PMO 60396. Heggen Member, Elnes Formation. Muggerudkleiva at Hedenstad near Kongsberg, Eiker-Sandsvær. Coll. L. Størmer, 1925–27. Scale bar 0.5 cm.

15: External mould of pygidium showing terrace-line pattern. PMO 203.289/1. Engervik Member. Level 36.87 m in the beach profile at Djuptrekkodden north of Slemmestad, Oslo-Asker. Coll. T. Hansen, 2003. Scale bar 0.5 cm.

16: Exfoliated and strongly flattened pygidium showing doublural terrace-lines. PMO 203.455. Engervik Member, Elnes Formation. Collected from 38.29 m in the beach profile at Djuptrekkodden north of Slemmestad, Oslo-Asker. Coll. T. Hansen, 2003. Scale bar 0.5 cm.

11, 17: *Asaphus narinosus* n. sp.

11: Nearly complete exoskeleton preserved in a concretion. Holotype PMO 208.580. Upper Heggen Member, Elnes Formation. Level 2.68 m in profile 4 at Råen near Fiskum, Eiker-Sandsvær. Coll. T. Hansen, 2004. Scale bar 0.5 cm.

17: Dorsal view of somewhat flattened and foreshortened pygidium. Paratype PMO 102.650. Base of Heggen Member, Elnes Formation. Collected 3.35 m above the base of the formation or at the transition between the Helskjer Member and the Heggen Member. Vikersundbakken, Modum. Coll. G. Henningsmoen, 20.09.1970. Scale bar 0.5 cm.

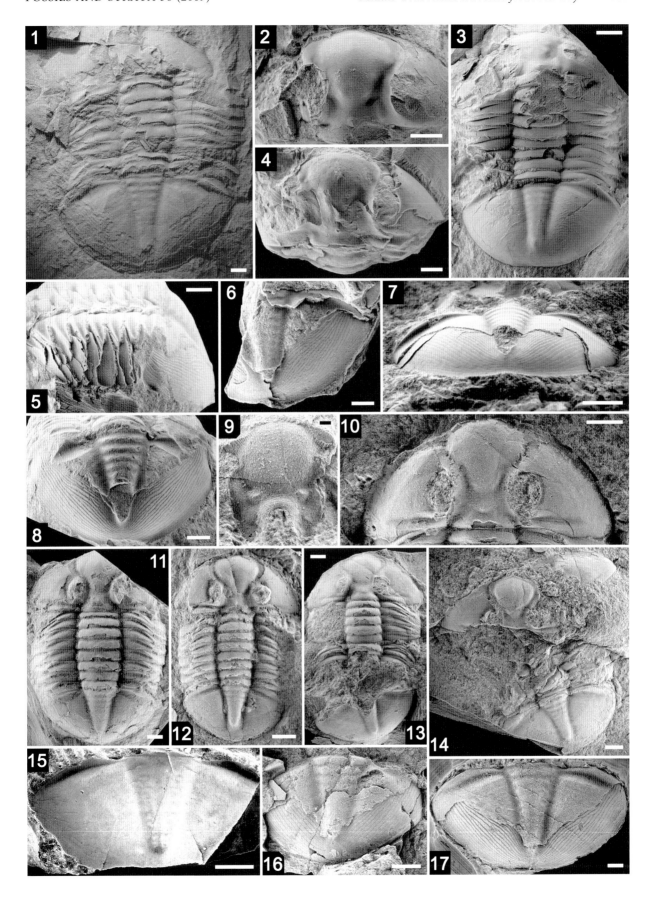

Plate 13

1–12: *Pseudomegalaspis patagiata* (Törnquist, 1884)
1:   Oblique dorsal view of nearly complete librigena. PMO 72079. Engervik Member, Elnes Formation. Elnestangen, Oslo-Asker. Coll. G. Henningsmoen, 1959. Scale bar 0.5 cm.
2, 5, 8: Dorsal shield and closer view of right lateral thoracic termination and cranidium. PMO S.1734. Middle Engervik Member, Elnes Formation. Odden at Bekkebukta on Bygdøy, Oslo-Asker. Coll. L. Størmer. Scale bars 0.5, 0.2 and 0.5 cm respectively.
3, 11: Dorsal view of strongly laterally compressed specimen and close-up of librigena. PMO 56556. Elnes Formation. Vigeland fountain in the Vigeland Park of Oslo, Oslo-Asker. Coll. A. Heintz, 1934. Scale bars 0.5 cm.
4, 6, 12: Dorsal view of cephalon, left lateral thoracic terminations and pygidium. PMO 3695. Upper Elnes Formation. Huk on Bygdøy, Oslo-Asker. Coll. V. M. Goldschmidt, 1906. Scale bars 0.5 cm.
7:   Dorsal view of cranidium PMO 72083. Elnes Formation. Ringerike. Coll. J. Kiær. Scale bar 0.5 cm.
9:   Dorsal view of heavily striated pygidium PMO 206.156c/32. Engervik Member. Level 13.32 m in the Bøveien road profile at Bødalen, Oslo-Asker. Coll. T. Hansen, 2004. Scale bar 0.5 cm.
10: Exfoliated hypostome preserved in mudstone. PMO 203.480b/4. Engervik Member. Level 38.68 m in the beach profile at Djuptrekkodden, Oslo-Asker. Coll. T. Hansen, 2003. Scale bar 0.2 cm.

13–15: *Pseudomegalaspis* cf. *formosa* (Törnquist, 1884)
Dorsal, posterior and oblique lateral view of pygidium PMO 56347. Locality and horizon unknown, but based on the general lithology probably from either the upper part of the Elnes Formation or from the overlying Vollen Formation in the Oslo-Asker district. Scale bars 0.5 cm.

16, 19: *Pseudobasilicus* (*Pseudobasilicus*) sp. B
Dorsal and posterior view of large pygidium. PMO 83251. Helskjer Member.
Level 8.45 m in the profile at Furnes Church, Mjøsa. Coll. G. Henningsmoen, 21.07.1961. Scale bars 0.5 cm.

17:   *Asaphus raaenensis* n. sp.
Dorsal view of large cranidium. PMO 206.963/1. Middle Heggen Member, Elnes Formation. Level 16.81 m in sub-profile 1 at Råen near Fiskum, Eiker-Sandsvær. Coll. T. Hansen, 21.04.2005. Scale bar 0.5 cm.

18, 20–21: *Asaphus narinosus* n. sp.
18: Dorsal view of large cephalon. PMO 208.566a/1. Upper Heggen Member, Elnes Formation. Level 0.93 m in sub-profile 4 at Råen near Fiskum, Eiker-Sandsvær. Coll. T. Hansen, 2004. Scale bar 0.5 cm.
20: Dorsal view of pygidium PMO 208.609/3. Upper Heggen Member, Elnes Formation. Level 2.70 m in sub-profile 4 at Råen near Fiskum, Eiker-Sandsvær. Coll. T. Hansen, 2004. Scale bar 0.5 cm.
21: Dorsal view of cranidium PMO 208.618/3. Upper Heggen Member, Elnes Formation. Level 6.83 m in sub-profile 4 at Råen near Fiskum, Eiker-Sandsvær. Coll. T. Hansen, 2004. Scale bar 0.5 cm.

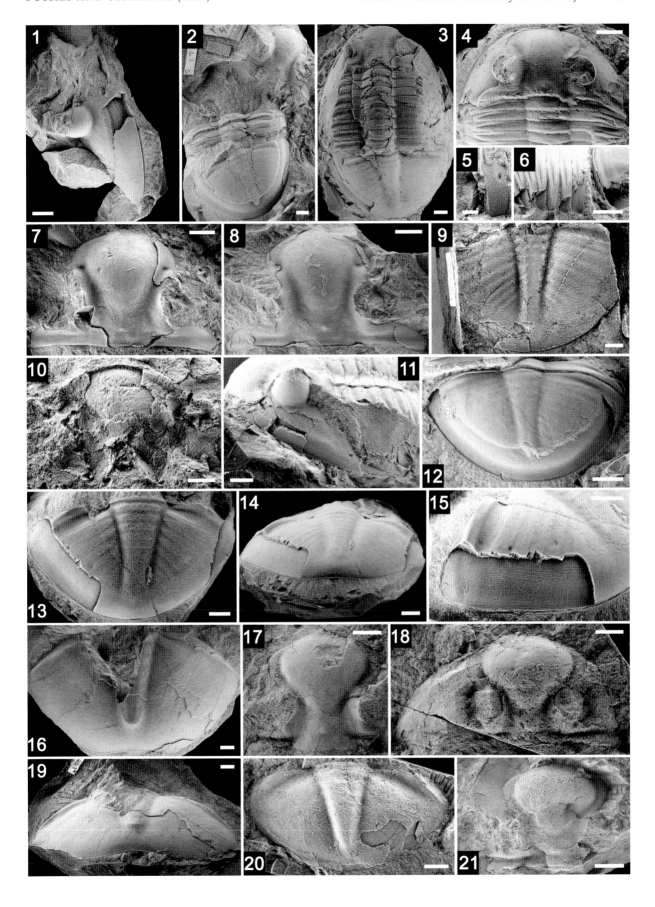

Plate 14
1–3: *Asaphus* cf. *striatus* (Sars & Boeck *in* **Boeck, 1838**)
1:   Dorsal view of fragmentary pygidium. PMO 101.780. Helskjer Member, Elnes Formation. Level 1.16 m in profile at Stavlum near Krekling, Eiker-Sandsvær. Coll. B. Wandås, 11.10.1978. Scale bar 0.5 cm.
2:   Dorsal view of fragmentary small pygidium. PMO 101.737. Helskjer Member, Elnes Formation. Level 0.40 m above the Huk Formation in the profile at Stavlum near Krekling, Eiker-Sandsvær. Coll. B. Wandås 11.10.1978. Scale bar 0.1 cm.
3:   Strongly deformed pygidium. PMO 10007. Top Helskjer or basal Heggen Member, Elnes Formation. Åsen east of Alm, Ringerike. Coll. I. Schetelig, 15.07.1913. Scale bar 0.5 cm.

4–5, 7: *Megistaspis* (*Megistaspidella*) *heroica* **Bohlin, 1960**
4:   Dorsal view of exfoliated cranidium PMO 33199. Helskjer or basal Heggen Member, Elnes Formation. North of Grinaker at Gran, Hadeland. Coll. Th. Münster, 22.08.1893. Scale bar 0.5 cm.
5:   Dorsal view of partly exfoliated cranidium PMO 104.050. Helskjer Member, 3.70 m above the Huk Formation. Helskjer on Helgøya, Mjøsa. Coll. G. Henningsmoen, 22.07.1961. Scale bar 0.5 cm.
7:   Partly exfoliated pygidium PMO 104.048. Helskjer Member, 3.7 m above the Huk Formation. Helskjer on Helgøya, Mjøsa. Coll. G. Henningsmoen, 22.07.1961. Scale bar 1 cm.

6:   *Megistaspis* (*Megistaspidella*) *heroica s.l.* **Bohlin, 1960**
        Nearly complete if somewhat weathered specimen. PMO 104.047. Loose rock sample, probably from the Helskjer Member as it includes small carbonate concretions. Øster Øren farm at Vikersund. Coll. B. Greåker 1974–1975. Scale bar 1 cm.

8–13: *Megistaspis* (*Megistaspidella*) *laticauda* **Wandås, 1984**
8:   External mould of fragmentary librigena showing some of the genal spine. PMO 102.639. Basal Heggen Member, 5.2 m above the Huk Formation. Road-cut at Vikersund skijump, Modum. Coll. B. Wandås, 21.06.1978. Scale bar 0.5 cm.
9, 11: Dorsal and oblique frontal view of holotype cranidium. PMO 102.563. Basal Heggen Member, 5.4 m above the Huk Formation. Road-cut at Vikersund skijump, Modum. Coll. B. Wandås, 20.06.1978. Scale bars 0.5 cm.
10: Strongly compressed cranidium and librigena. PMO 106.539. Basal Heggen Member, approximately 3.5 m above the Huk Formation. Road-cut at Vikersund skijump, Modum. Coll. B. Wandås, 17.07.1980. Scale bar 0.5 cm.
12: External mould of librigena PMO 102.638. Basal Heggen Member, 5.2 m above the Huk Formation. Road-cut at Vikersund skijump, Modum. Coll. B. Wandås, 21.06.1978. Scale bar 1 cm.
13: Large pygidium preserved in marly mudstone. MGUH 28369. Elnes Formation. Probably collected at Slemmestad. Gift to the Geological Museum of Copenhagen, Denmark, from K. R. Pedersen, University of Aarhus. Scale bar 1 cm.

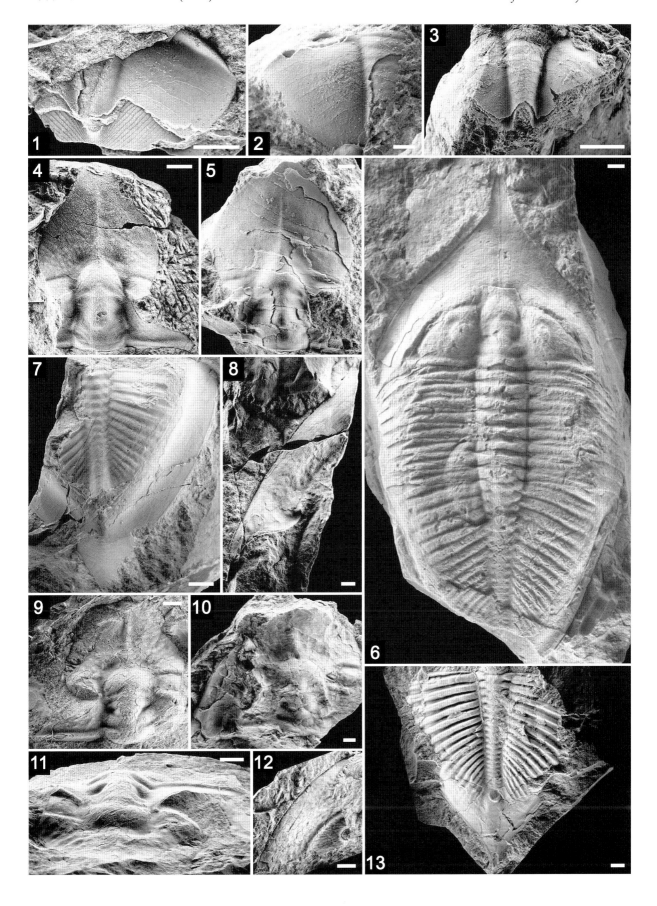

Plate 15
1: *Megistaspis* (*Megistaspidella*) *laticauda* **Wandås, 1984**
Dorsal view of partly exfoliated pygidium. PMO 106.049. Loose rock sample belonging to Helskjer or basal Heggen Member, Elnes Formation. Road-cut at Vikersund skijump, Modum. Coll. D. L. Bruton, G. Henningsmoen, Å. Jensen and L. Koch, 09.05.1974. Scale bar 0.5 cm.

2–10: *Megistaspis* (*Megistaspidella*) *giganteus runcinatus* **n. ssp.**
2: Wax cast of strongly deformed cranidium PMO 102.425. Basal Heggen Member, 5.4 m above the Huk Formation. Road-cut at Vikersund skijump, Modum. Coll. B. Wandås, 20.06.1978. Scale bar 0.5 cm.
3: Compressed cranidium PMO 102.642. Basal Heggen Member, 5.25 m above the Huk Formation. Road-cut at Vikersund skijump, Modum. Coll. G. Henningsmoen, 20.09.1970. Scale bar 0.5 cm.
4: Preglabellar area of cranidium PMO 90584. Helskjer Member, Elnes Formation. Skinnvikstangen at Rognstrand, Skien-Langesund. Coll. E. Yochelsen and G. Henningsmoen, 25.10.1961. Scale bar 0.5 cm.
5: Strongly flattened librigena PMO 102.598. Basal Heggen Member, 5.4 m above the Huk Formation. Road-cut at Vikersund skijump, Modum. Coll. B. Wandås, 20.06.1978. Scale bar 0.5 cm.
6: Fragmentary pygidium preserved in limestone. PMO 201.974/1. Collected 0.09 m below the top of the Huk Formation in the Old Eternite Quarry at Slemmestad, Oslo-Asker. Coll. T. Hansen, 2003. Scale bar 1 cm.
7: Partly exfoliated pygidium preserved in fine-grained marly sandstone. PMO 90569. Helskjer Member, Elnes Formation. Skinnviktangen at Rognstrand, Skien-Langesund. Coll. E. Yochelsen and G. Henningsmoen, 26.10.1961. Scale bar 0.5 cm.
8: Foreshortened and exfoliated pygidium preserved in mudstone. PMO 106.060. Erratic rock sample belonging to the basal Heggen Member. Road-cut at Vikersund skijump, Modum. Coll. D. L. Bruton, G. Henningsmoen, Å. Jensen and L. Koch, 09.05.1974. Scale bar 1 cm.
9: Compressed pygidium PMO 106.435. Basal Heggen Member, 3.70 m above the Huk Formation. Road-cut at Vikersund skijump, Modum. Coll. B. Wandås, 17.07.1980. Scale bar 1 cm.
10: Dorsal view of small pygidium PMO 102.476. Basal Heggen Member, 4.0 to 5.0 m above the Huk Formation. Road-cut at Vikersund skijump, Modum. Coll. B. Wandås, 20.06.1978. Scale bar 0.5 cm.

11–16, 18: *Megistaspis* (*Megistaspidella*) *giganteus giganteus* **Wandås, 1984**
11: Cast of pygidium. PMO 102.585. Lower Heggen Member, 7.1 m above the Huk Formation. Road-cut at Vikersund skijump, Modum. Coll. B. Wandås, 20.06.1978. Scale bar 0.5 cm.
12: Dorsal view of small pygidium PMO 106.016. Helskjer Member, 5.1 m above the Huk Formation. Helskjer on Helgøya, Mjøsa. Coll. B. Wandås, 07.05.1980. Scale bar 0.2 cm.
13: Dorsal view of pygidium PMO 90535. Upper Helskjer Member, 7.75 m above the Huk Formation. Profile at Furnes Church, Mjøsa. Coll. G. Henningsmoen, 1961. Scale bar 0.5 cm.
14: Fragmentary cephalon showing the nearly straight anterolateral margins. PMO 90493. Upper Helskjer Member, 7.75 m above the Huk Formation. Profile at Furnes Church, Mjøsa. Coll. Unknown collector from excursion on 10.06.1962. Scale bar 1 cm.
15: Large cranidium PMO 90509. Note the marked preglabellar inflation and the effaced lateral glabellar lobes. Top Helskjer Member. Profile at Furnes Church, Mjøsa. Coll. H. J. Hässler, 24.07.1964. Scale bar 0.5 cm.
16: Dorsal view of librigena. PMO 90388. Top Helskjer Member, 7.75 m above the Huk Formation. Profile at Furnes Church, Mjøsa. Coll. Unknown, but collected on excursion on 10.06.1962. Scale bar 1 cm.
18: Dorsal view of holotype cephalon PMO 90553. Upper Helskjer Member, 7.75 m above base of Helskjer Member. Profile just north of Furnes Church, Mjøsa. Coll. Unknown, but collected on excursion the 10.06.1962. Scale bar 1 cm.

17, 19–22: *Megistaspis* (*Megistaspidella*) *maximus* **Wandås, 1984**
17: Cephalon and parts of thorax. PMO 106.094. Note the relatively distinct L3 and L4 lobes and the strongly hourglass-shaped glabella. Top Helskjer Member, 8.35 m above the Huk Formation. Profile on Hovodden at the Randsfjord, Hadeland. Coll. L. Størmer, 21.05.1951. Scale bar 0.5 cm.
19: Holotype cephalon showing the strongly developed L3 and L4 lobes. PMO 102.651. Lower Heggen Member, 12 to 14 m above the Huk Formation. Road-cut at Vikersund skijump, Modum. Coll. G. Henningsmoen, 1970. Scale bar 1 cm.
20: Cranidium preserved in mudstone. PMO 102.731. Heggen Member, Elnes Formation. Lane east of Heggen church-yard, Modum. Coll. B. Wandås, 18.07.1978. Scale bar 0.5 cm.
21: Partly exfoliated cranidium. PMO 89790. Lower Heggen Member, 9.15 m above the Huk Formation. Hovodden at the Randsfjord, Hadeland. Coll. F. Nikolaisen, 09.10.1961. Scale bar 0.2 cm.
22: External mould of cranidium. PMO 102.435. Heggen Member (erratic rock). Road-cut at cross-road just south of Vikersund skijump, Modum. Coll. B. Wandås, 30.08.1978. Scale bar 1 cm.

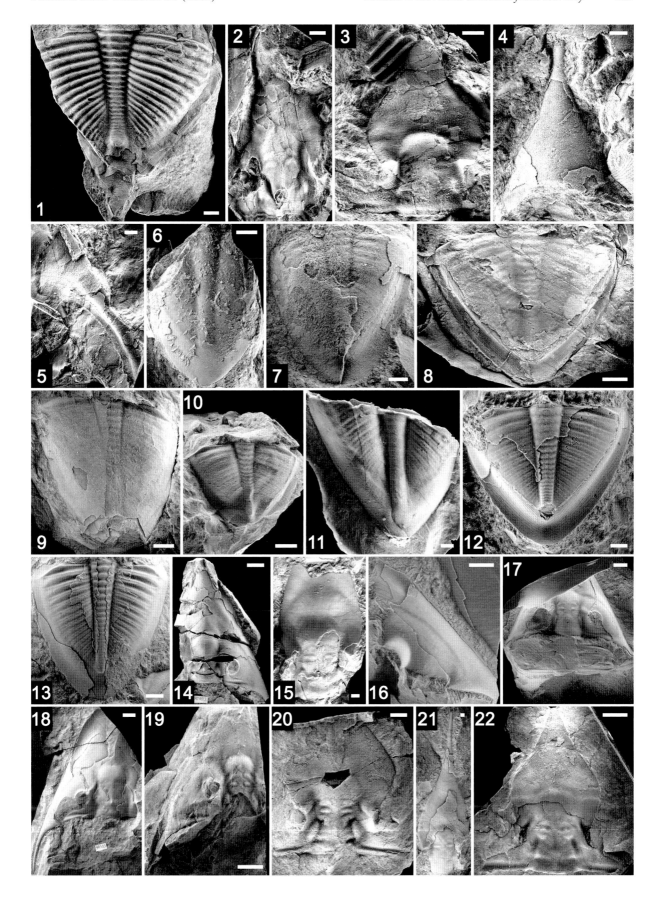

Plate 16

1–2: *Megistaspis* (*Megistaspidella*) *maximus* **Wandås, 1984**

1:   Dorsal view of pygidium. PMO 102.403. Lower Heggen Member, 11.11 m above the Huk Formation. Road-cut at Vikersund skijump, Modum. Coll. B. Wandås, 19.07.1978. Scale bar 0.5 cm.

2:   Dorsal view of pygidium from same rock sample as 1. Scale bar 0.5 cm.

3–13: *Niobe* (*Niobe*) *frontalis* (**Dalman, 1827**)

3:   Large hypostome. PMO 83146. Top Helskjer Member, 7.75 m above the Huk Formation. Road-cut at Furnes Church, Mjøsa. Coll. G. Henningsmoen, 21.07.1961. Scale bar 0.5 cm.

4:   Compressed cranidium preserved in marly mudstone. PMO 106.064. Erratic rock from Helskjer Member or lower Heggen Member. Road-cut at Vikersund skijump, Modum. Coll. D. L. Bruton, G. Henningsmoen, Åge Jensen and L. Koch, 09.05.1974. Scale bar 0.5 cm.

5:   Flattened cranidium. PMO 103.674. Basal Heggen Member, 5 to 6 m above the Huk Formation. Road-cut at Vikersund skijump, Modum. Coll. B. Wandås, 1978. Scale bar 0.5 cm.

6:   Cranidium preserved in limestone. PMO 67593. Top Helskjer Member, 7.75 m above the Huk Formation. Road-cut at Furnes Church, Mjøsa. Coll. S. Skjeseth, 1950. Scale bar 0.5 cm.

7:   Cranidium preserved in limestone. PMO 67184. Top Helskjer Member, probably from limestone bed 7.75 m above the Huk Formation. Road-cut at Furnes Church, Mjøsa. Coll. S. Skjeseth, 1950. Scale bar 0.5 cm.

8:   Anterior fragment of cranidium. PMO 83259. Top Helskjer Member, 7.75 m above the Huk Formation. Road-cut at Furnes Church, Mjøsa. Coll. G. Henningsmoen, 21.07.1961. Scale bar 0.5 cm.

9:   Part of complete specimen preserved in mudstone. PMO 106.053. Erratic rock from basal Elnes Formation. Road-cut at Vikersund skijump. Coll. D.L. Bruton, G. Henningsmoen, Åge Jensen and L. Koch, 09.05.1974. Scale bar 0.5 cm.

10:   Cast of left librigena. PMO 106.378. Elnes Formation, probably limestone bed 7.75 m above the Huk Formation. Road-cut at Furnes Church, Mjøsa. Coll. B. Wandås, 10.07.1979. Scale bar 0.5 cm.

11:   Strongly exfoliated pygidium. PMO 67187. Helskjer Member. Road-cut at Furnes Church, Mjøsa. Coll. S. Skjeseth, 1950. Scale bar 0.5 cm.

12–13: Dorsal and posterior view of pygidium. PMO 90514. Top Helskjer Member, 7.75 m above the Huk Formation. Road-cut at Furnes Church, Mjøsa. Coll. Unknown, collected on excursion on 10.06.1962. Scale bars 0.5 cm.

14–18: *Nileus armadillo* (**Dalman, 1827**)

14, 16–17: Dorsal, frontal cephalic and lateral view of complete specimen PMO 106.514. Upper Elnes Formation, probably Engervik Member. Beach profile at Huk on Bygdøy, Oslo-Asker. Coll. D. L. Bruton, 1982. Scale bar 0.5 cm.

15:   Dorsal view of cephalon. Elnes Formation. Helskjer on Helgøya, Mjøsa. Coll. S. Skjeseth, 10.05.1951. Scale bar 0.5 cm.

18:   Ventral view of small hypostome. Loose rock from Engervik Member, Elnes Formation. Elnestangen just north of Slemmestad, Oslo-Asker. Coll. G. Henningsmoen, 29.11.1959. Scale bar 0.2 cm.

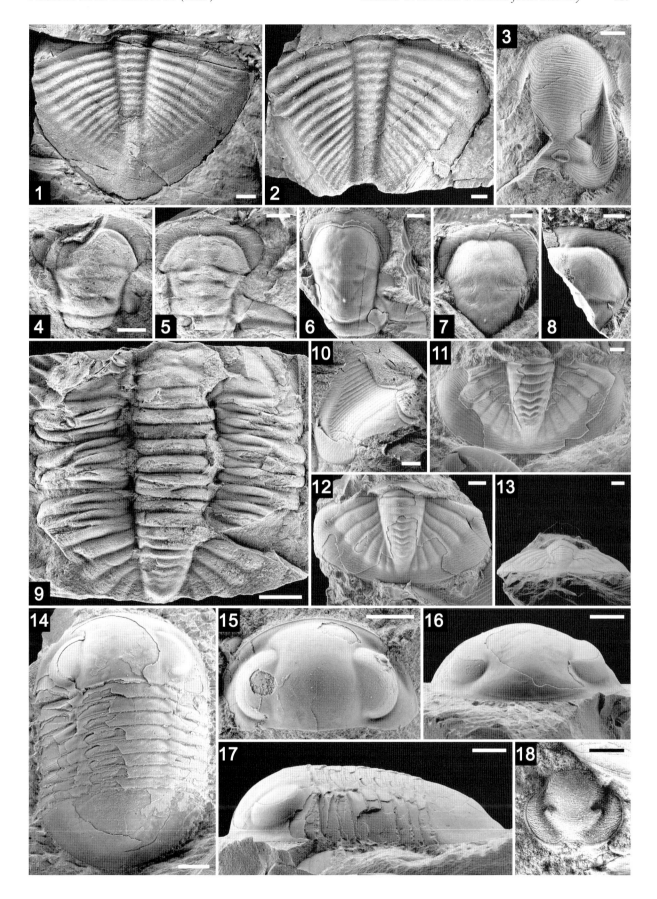

Plate 17

1–11: *Nileus armadillo* (Dalman, 1827)

1: Exfoliated cranidium showing muscle scars and lateral glabellar furrows. PMO 82429. Engervik Member, Elnes Fm. Slemmestad, Oslo-Asker. Coll. Unknown. Scale bar 0.2 cm.

2: Dorsal view of small cranidium. PMO 206.232/3. Top Heggen Member, 3.41 m below datum in profile 1 at Råen near Fiskum, Eiker-Sandsvær. Coll. T. Hansen, 2004. Scale bar 0.2 cm.

3: Ventral view of rostrum. PMO 207.481/9. Basal 'Cephalopod Shale', 18.95 m above datum of sub-profile 3 at northern road-cut where road Fv 84 crosses highway E6 just south-west of Nydal, Mjøsa. Coll. T. Hansen, 19.08.2005. Scale bar 0.2 cm.

4: Dorsal view of moderately large librigena. PMO 207.486/1. Basal 'Cephalopod Shale', 21.45 m above datum of sub-profile 3 at northern road-cut where road Fv 84 crosses highway E6 just south-west of Nydal, Mjøsa. Coll. T. Hansen, 19.08.2005. Scale bar 0.2 cm.

5: Dorsal view of juvenile librigena with genal spine. PMO 207.484/12. Basal 'Cephalopod Shale', 20.25 m above datum in sub-profile 3 at northern road-cut where road Fv 84 crosses highway E6 just south-west of Nydal, Mjøsa. Coll. T. Hansen, 19.08.2005. Scale bar 0.1 cm.

6: Ventral view of hypostome PMO 207.463/6. Top Heggen Member, 10.00 m above datum of sub-profile 3 at northern road-cut where road Fv 84 crosses highway E6 just south-west of Nydal, Mjøsa. Coll. T. Hansen, 19.08.2005. Scale bar 0.1 cm.

7: Detail of pygidial surface just behind axis showing faint radiating pustules. PMO 106.514. Upper Elnes Formation, probably Engervik Member. Beach profile at Huk on Bygdøy, Oslo-Asker. Coll. D. L. Bruton, 1982. Scale bar 0.1 cm.

8: Ventral view of pygidium showing doublure. PMO 207.416/14. Heggen Member, ~1.53 m above datum in sub-profile 2 at northern road-cut where road Fv 84 crosses highway E6 just south-west of Nydal, Mjøsa. Coll. T. Hansen, 18.08.2005. Scale bar 0.2 cm.

9: Dorsal view of pygidium PMO 82329. Note the undeveloped border and sparse terrace-lines. Helskjer Member, road-cut at Furnes Church, Mjøsa. Coll. S. Skjeseth, 1949. Scale bar 0.2 cm.

10: Dorsal view of pygidium PMO 56225 b. Note the wide border covered by terrace-lines. Elnes Formation. Helskjer on Helgøya, Mjøsa. Coll. Unknown. Scale bar 0.2 cm.

11: Dorsal view of partly exfoliated pygidium PMO 207.484/9. Basal 'Cephalopod Shale', 20.25 m above datum in sub-profile 3 at northern road-cut where road Fv 84 crosses highway E6 just south-west of Nydal, Mjøsa. Coll. T. Hansen, 19.08.2005. Scale bar 0.2 cm.

12–13: *Nileus depressus* (Sars & Boeck *in* Boeck, 1838) ssp.

Dorsal and lateral view of small cephalon. PMO 89795. Top Helskjer Member, 7.95 m above the Huk Formation. Road-cut at Furnes Church, Mjøsa. Coll. G. Henningsmoen, 21.07.1961. Scale bars 0.2 cm.

14–19: *Illaenus aduncus* Jaanusson, 1957

14–15: Dorsal and oblique anterior view of cephalon PMO 106.015. Helskjer Member, 5.10 m above the Huk Formation. Helskjer on Helgøya, Mjøsa. Coll. B. Wandås, 07.05.1980. Scale bars 0.5 cm.

16: Lateral view of cephalon and anterior seven thoracic segments. PMO 33440. Elnes Formation. Lynne in Hadeland. Coll. Unknown. Scale bar 0.5 cm.

17: Dorsal view of pygidium PMO 67597 showing external terrace-lines. Helskjer Member. Helskjer on Helgøya, Mjøsa. Coll. S. Skjeseth, 10.05.1961. Scale bar 0.2 cm.

18: Dorsal view of exfoliated pygidium PMO 83107. Helskjer Member, 5.10 m above the Huk Formation. Helskjer on Helgøya, Mjøsa. Coll. F. Nikolaisen, 27.09.1961. Scale bar 0.2 cm.

19: Dorsal view of small pygidium PMO 67598. Helskjer Member. Helskjer on Helgøya, Mjøsa. Coll. S. Skjeseth, 10.05.1961. Scale bar 0.2 cm.

20–23: *Raymondaspis* (*Cyrtocybe*) sp.

20–22: Dorsal, frontal and oblique lateral view of cranidium PMO 83239. Top Helskjer Member, 7.75 m above the Huk Formation. Road-cut at Furnes Church, Mjøsa. Coll. F. Nikolaisen, 27.09.1961. Scale bars 0.1 cm.

23: Dorsal view of incomplete librigena. PMO 67578. Helskjer Member. Road-cut at Furnes Church, Mjøsa. Coll. S. Skjeseth, 1950. Scale bar 0.2 cm. Picture produced by Prof. David L. Bruton.

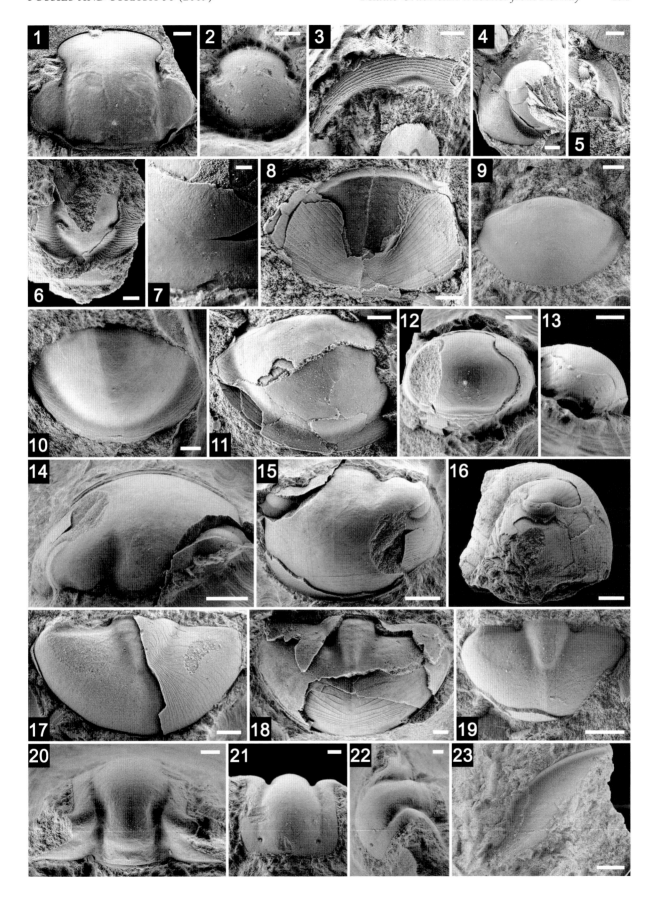

Plate 18

1–4: *Bronteopsis holtedahli* Skjeseth, 1955
1–2: Dorsal and anterior view of holotype cranidium PMO 67011c. Probably uppermost Helskjer Member or perhaps lower Heggen Member. Road-cut at Furnes Church, Mjøsa. Coll. S. Skjeseth, 1953. Scale bars 0.1 cm.
3:   Dorsal view of fragmentary cranidium PMO 67011b. Belonging to same concretion as 1, 2.
4:   Dorsal view of pygidium PMO 67011a from same concretion as 1, 2.

5–9: *Scotoharpes rotundus* (Bohlin, 1955)
5:   Dorsal view of cephalothorax. Plasticine cast of PMO S.1682. Svartodden Member of the Huk Formation. Hukodden on Bygdøy, Oslo-Asker. Coll. L. Størmer, 1937. Scale bar 0.5 cm.
6:   Dorsal view of cephalon PMO 106.152. Top Helskjer Member, probably 7.75 m above the Huk Formation. Road-cut at Furnes Church, Mjøsa. Coll. B. Wandås, 10.07.1979. Scale bar 0.2 cm.
7:   Dorsal view of cephalon PMO 83243. Top Helskjer Member, 7.75 m above the Huk Formation. Road-cut at Furnes Church, Mjøsa. Coll. F. Nikolaisen, 27.09.1961. Scale bar 0.2 cm.
8–9: Dorsal and oblique anterior view of cephalon PMO S.1683. Svartodden Member in the Huk Formation. Hukodden on Bygdøy, Oslo-Asker. Coll. L. Størmer, 1937. Scale bars 0.2 cm.

10–15: *Botrioides simplex* Owen, 1987
10–11: Oblique lateral and dorsal view of holotype cranidium PMO 83021. Top Helskjer Member, probably from 7.75 m above the Huk Formation. Road-section at Furnes Church, Mjøsa. Coll. G. Henningsmoen, 1961. Scale bars 0.1 cm.
12:   Dorsal view of cranidium PMO 90565. Helskjer Member. Road-cut at Furnes Church, Mjøsa. Coll. Unknown, collected on an excursion on 10.06.1962. Scale bar 0.1 cm.
13:   Dorsal view of two complete specimens preserved in mudstone. PMO 104.064. Lower Heggen Member. Level 8.0 m in the road-section at Vikersund skijump, Modum. Coll. B. Wandås, 1979. Scale bar 0.1 cm.
14–15: Casts of two complete specimens from sample PMO 104.055. Lower Heggen Member, 7.9 m above the Huk Formation. Road-cut at Vikersund skijump, Modum. Coll. B. Wandås, 1979. Scale bars 0.1 cm.

16–20: *Botrioides* sp. A
16–17: Dorsal and anterior view of cranidium PMO 82168. Elnes Formation. Road-cut at skijump near Slemmestad, Oslo-Asker. Coll. N. Spjeldnæs, 1951. Scale bars 0.1 cm.
18:   External mould of small pygidium PMO 82165. Elnes Formation. Road-cut at Skijump near Slemmestad, Oslo-Asker. Coll. N. Spjeldnæs, 1951. Scale bar 0.1 cm.
19:   Dorsal view of strongly flattened cranidium PMO 202.292/9. Lower Sjøstrand Member, 5.68 m above the Huk Formation. The Old Eternite Quarry north of Slemmestad, Oslo-Asker. Coll. T. Hansen, 2003. Scale bar 0.1 cm.
20:   Dorsal view of exfoliated cranidium PMO 103.701. Lower Heggen Member, 12.2 m above the Huk Formation. Road-cut at Vikersund skijump, Modum. Coll. B. Wandås, 28.06.1979. Scale bar 0.1 cm.

21–24: *Botrioides impostor* Owen, 1987
21:   Latex cast of nearly complete paratype specimen PMO H0574. Håkavik Member. Elnestangen at Bjerkås, Oslo-Asker. Coll. L. Størmer. Scale bar 0.1 cm.
22:   Cast of nearly complete cranidium PMO 208.582. Heggen Member, 2.70 m below datum in profile 4 at Råen near Fiskum. Coll. T. Hansen, 2004. Scale bar 0.1 cm.
23:   External mould of upper lamella. PMO 104.061. Basal Heggen Member, 8.0 m above the Huk Formation. Road-cut at Vikersund skijump. Coll. B. Wandås, 09.11.1979. Scale bar 0.1 cm.
24:   Holotype cranidium PMO H0566. Elnes Formation. Beach-profile just north of Djuptrekkodden, Oslo-Asker. Coll. Unknown. Scale bar 0.1 cm.

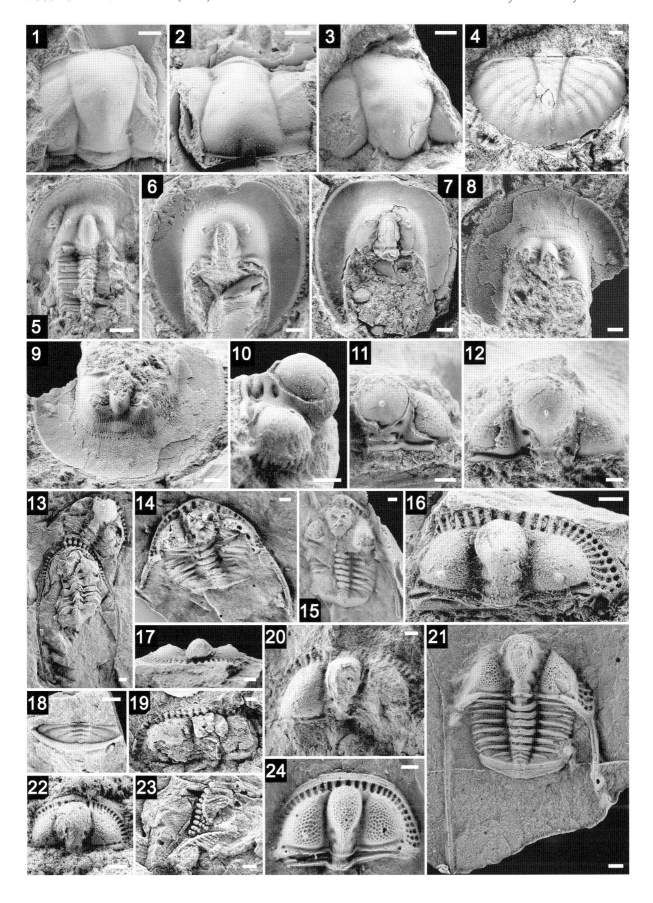

Plate 19

1: *Botrioides impostor* **Owen, 1987**
Paratype lower lamella PMO H401. Upper Elnes Formation. Gomnes in Ringerike. Coll. L. Størmer. Scale bar 0.1 cm.

2–6: *Botrioides bronnii* **(Boeck, 1838)**
2: Dorsal view of lectotype cephalon PMO 61752. Vollen or upper Elnes Formation. Wraatz's Løkke in Oslo. Coll. Sars? Scale bar 0.1 cm.
3: Ventral view of fringe PMO 203.130/25. Upper Sjøstrand Member. Level 23.10 m above datum in the beach profile north of Djuptrekkodden, Oslo-Asker. Coll. T. Hansen, 2003. Scale bar 0.1 cm.
4–5: Frontal and dorsal view of cranidium PMO H481. Upper Elnes Formation. Engervik north of Slemmestad, Oslo-Asker. Coll. J. Kiær. Scale bars 0.1 cm.
6: Posterior view of enrolled specimen PMO H482. Upper Elnes Formation. Engervik north of Slemmestad, Oslo-Asker. Coll. J. Kiær. Scale bar 0.1 cm.

7–11: *Botrioides broeggeri* **(Størmer, 1930)**
7, 11: Dorsal and frontal view of two cephala and lateral view of former in sample PMO 81903. Vollen Formation. Gullerud in Ringerike. Coll. W. C. Brøgger, 1881. Scale bars 0.1 cm.
8: Dorsal view of holotype cephalon PMO H0553. Vollen Formation at Gullerud near Norderhov in Ringerike. Coll. W. C. Brøgger. Scale bar 0.1 cm.
9: Dorsal view of strongly weathered cephalon PMO H0567. Upper Elnes or Fossum Formation. Krekling at Fiskum, Eiker-Sandsvær. Coll. Unknown, collected on excursion in 1977. Scale bar 0.1 cm.
10: Ventral view of lower lamella of fringe. PMO H0563. Vollen Formation. Gullerud at Norderhov in Ringerike. Coll. W. C. Brøgger. Scale bar 0.1 cm.

12–15, 20: *Botrioides efflorescens* **(Hadding, 1913)**
12: Dorsal view of cranidium PMO 203.130/13. Upper Sjøstrand Member, 23.10 m above datum in beach-profile at Djuptrekkodden. Coll. D. L. Bruton, 2003. Scale bar 0.1 cm.
13: Dorsal view of strongly flattened and recrystallised specimen PMO H0570. Eiker, Eiker-Sandsvær. Middle to upper Heggen Member. Coll. Unknown, collected 1878. Scale bar 0.1 cm.
14: Ventral view of lower lamella of fringe PMO 203.161. Lower Engervik Member, 30.66 m above datum in the beach profile at Djuptrekkodden. Coll. T. Hansen, 2003. Scale bar 0.1 cm.
15: Dorsal view of exfoliated pygidium PMO 203.176/14. Lower Engervik Member, 32.01 m above datum of beach profile at Djuptrekkodden. Coll. D. L. Bruton, 2003. Scale bar 0.1 cm.
20: Dorsal view of cranidium PMO 82342. Middle to upper Elnes Formation at Helskjer on Helgøya, Mjøsa. Coll. H. J. Hässler & B. D. Erdtmann, 25.07.1964. Scale bar 0.1 cm.

16–19: *Botrioides foveolatus* **(Angelin, 1854)**
16: Dorsal view of cranidium PMO H391. Upper Elnes Formation. Elnestangen north of Slemmestad. Coll. Unknown, collected on excursion on 24.10.1910. Scale bar 0.1 cm.
17: Dorsal view of strongly flattened cephalon and partly attached thorax. PMO 203.543/3. Engervik Member, 40.33 m above datum in beach profile at Djuptrekkodden. Coll. T. Hansen, 2003. Scale bar 0.1 cm.
18: Dorsal view of fragmentary cranidium showing surface structure. PMO H533. Elnes Formation. Toten at Mjøsa. Coll. Unknown. Scale bar 0.1 cm.
19: Dorsal view of pygidium PMO H0538. Upper Elnes Formation. Huk on Bygdøy, Oslo-Asker. Coll. Unknown. Scale bar 0.1 cm.

21–23: **Trinucleid gen. et sp. indet.**
21–22: Dorsal and oblique anterolateral view of cephalon PMO 87252. Heggen Member, 16 m above Huk Formation at Hovodden, Hadeland. Coll. D. L. Bruton, 1970. Scale bars 0.1 cm.
23: Dorsal view of small and strongly compressed cranidium. PMO 208.642/2. Lower middle Heggen Member. Profile 6 at Krekling, Fiskum, where it was collected 1.25 m below datum. Coll. T. Hansen, 2004. Scale bar 0.1 cm.

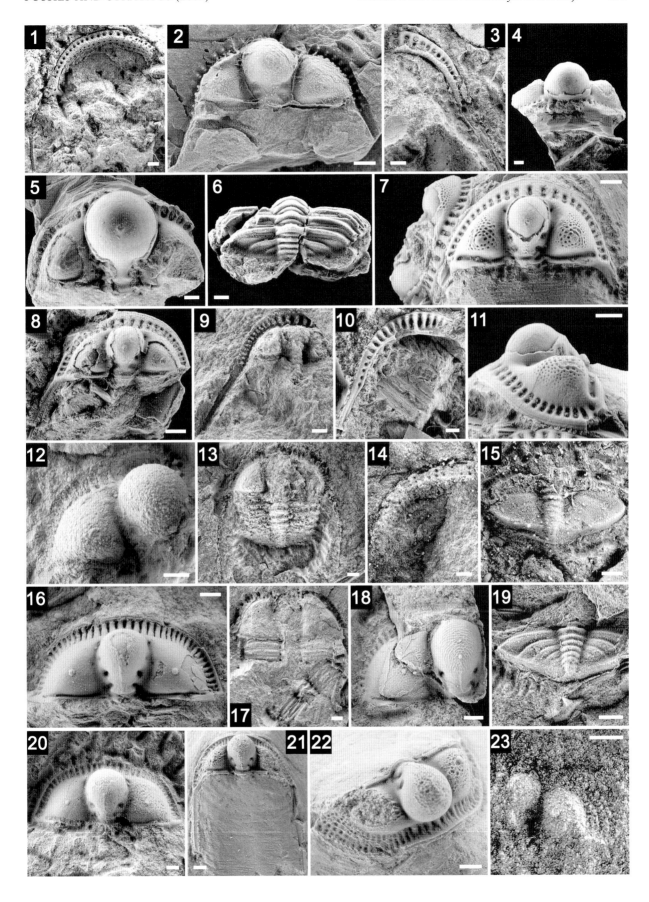

Plate 20

1–5: *Ampyx nasutus* **Dalman, 1827**
1: Dorsal view of cranidium. Cast of PMO 83147. Uppermost Helskjer Member. Collected from 7.75 m above the Huk Formation at Furnes near Hamar, Mjøsa. Coll. G. Henningsmoen, 21.07.1961. Scale bar 0.5 cm.
2: Lateral view of partly exfoliated cranidium. PMO 68447. Collected from the upper part of the Helskjer Member or lower part of the overlying shales at Røykenvik railway station, Hadeland. Coll. L. Størmer, 08.16.1955. Scale bar 0.5 cm.
3: Oblique posterior view of pygidium showing downward curvature of border terrace-lines. PMO 33772. Elnes Formation at Røisumgaard at Gran, Hadeland. Coll. T. Münster, 09.26.1893. Scale bar 0.5 cm.
4: Articulated thorax and pygidium. PMO 206.964. Lower Helskjer Member. Level 1.15 m in sub-profile 1 along E6 at Nydal, north of Hamar, Mjøsa. Coll. T. Hansen, 16.08.2005. Scale bar 0.5 cm.
5: Dorsal view of pygidium showing curving anterior pleural furrow and comparatively narrow pleural field. PMO 67180. Elnes Formation. Nydal-Furnes at Hamar, Mjøsa. Coll. S. Skjeseth, 1950. Scale bar 0.5 cm.

6–12: *Ampyx mammilatus* **Sars, 1835**
6–8: Frontal, dorsal and lateral view of cranidium. PMO 141.903. Sampled from upper Engervik Member or Håkavik Member just below restaurant at Huk on Bygdøy, Oslo-Asker. Coll. R. Eriksen, October 1990. Scale bar 0.5 cm for Figs. 6 and 7 and 0.2 cm for 8.
9: Pygidium with six thoracic segments. PMO 33572. Elnes Formation. Hadeland. Scale bar 0.5 cm.
10: Compressed cranidium with parts of frontal spine intact. PMO 67181. Elnes Formation. Nydal-Furnes, Mjøsa. Coll. S. Skjeseth, 1950. Scale bar 0.2 cm.
11–12: Syntype. Posterolateral and dorsal view of pygidium showing narrow axis and border with terrace-lines semi-parallel to margin. PMO 56393. Engervik Member. Hjortnæstangen in Oslo, Oslo-Asker. Scale bar 0.5 cm.

13–14, 16: *Lonchodomas cuspicaudus* **n. sp.**
13–14: Posterior and dorsal view of pygidium showing the margin-parallel terrace-lines on the border. PMO 3969. Uppermost Håkavik Member at Håkavik, Oslo-Asker. Coll. O. Holtedahl. Scale bars 0.5 cm.
16: Dorsal view of pygidium PMO 206.271/1. Uppermost Heggen Member, Elnes Formation. Level 8.43 m in profile 1 at Råen near Fiskum, Eiker-Sandsvær. Coll. T. Hansen, 2004. Scale bar 0.5 cm.

15, 17: *Ampyx* **sp. A**
15: Dorsal view of compressed pygidium. PMO 208.542/5. Heggen Member, Elnes Formation. Level 19.63 m in profile 1 at Råen near Fiskum, Eiker-Sandsvær. Coll. T. Hansen, 2004. Scale bar 0.2 cm.
17: Fragmentary librigena and corner of pygidium PMO 208.552/2. Upper part of Heggen Member. Level 9.13 m in profile 1 at Råen near Fiskum, Eiker-Sandsvær. Coll. T. Hansen, 2004. Scale bar 0.5 cm.

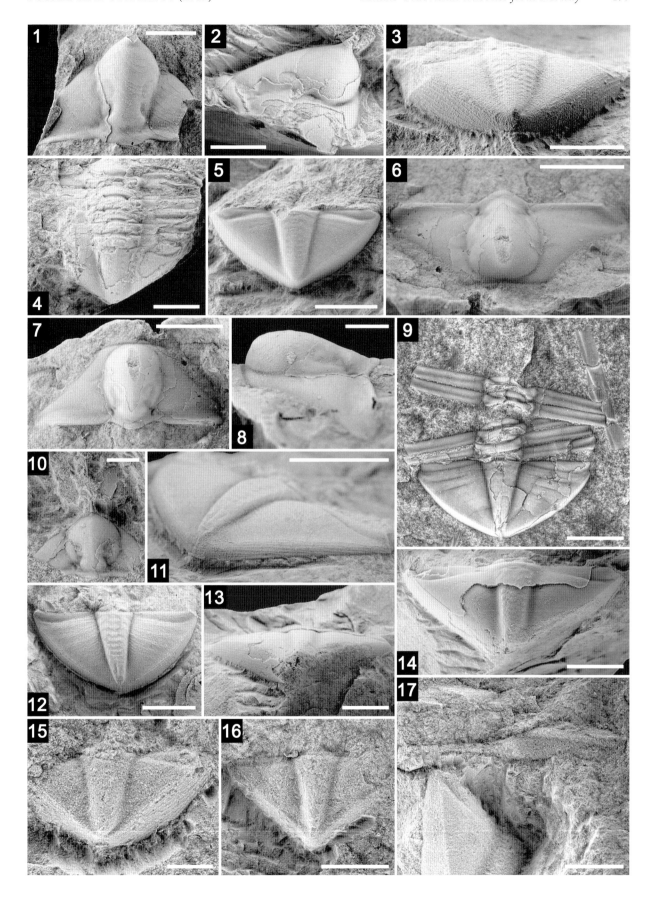

Plate 21

1–7: *Cnemidopyge costata* **(Angelin, 1854)**

1–3: Dorsal, frontal and lateral view of lectotype cranidium PMO 56399. The specimen belongs to the Vollen Formation or perhaps the upper part of the Elnes Formation. Collected on Bygdøy, Oslo-Asker. Scale bars 0.5 cm.

4: Dorsal view of cranidium. PMO 82361. Elnes Formation. Hovindsholm beach on Helgøya, Mjøsa. Coll. G. Henningsmoen, 1957. Scale bar 0.5 cm.

5: Hypotypoid pygidium. PMO 56405. Probably from the Vollen Formation. Hjortnæstangen in Oslo, Oslo-Asker. Scale bar 0.5 cm.

6: Dorsal view of exfoliated pygidium. PMO 82525. Engervik Member. Elnestangen north of Slemmestad, Oslo-Asker. Coll. G. Henningsmoen, 29.11.1959. Scale bar 0.5 cm.

7: Dorsal view of nearly complete specimen. PMO 36868. Elnes Formation (Heggen Member?). Lund at Flakstadelva, Mjøsa. Coll. W. C. Brøgger, 1881. Scale bar 0.5 cm.

8: *Cnemidopyge* **sp. A**

Dorsal view of completely flattened specimen. PMO 202.056/6. Upper half of Helskjer Member. Level 1.15 m above base of formation in the Old Quarry just north of Slemmestad, Oslo-Asker. Coll. T. Hansen, 2003. Scale bar 0.2 cm.

9–18: *Lonchodomas rostratus* **(Sars, 1835)**

9–10: Dorsal and lateral view of cranidium showing faint terrace-lines on anterior part of glabella. PMO 82644. Middle Engervik Member. Elnestangen, Oslo-Asker. Coll. N. Spjeldnæs, 04.10.1951. Scale bars 0.2 cm.

11–12: Oblique lateral and dorsal view showing faint terrace-lines on anterolateral part of fixigena. PMO 207.481/4. Elnes Formation. Level 18.95 m in sub-profile 3 along E6, where road Fv 84 crosses between Nydal and Furnes, Mjøsa. Coll. T. Hansen, 19.08.2005. Scale bars 0.2 cm.

13: Poorly preserved cephalon and parts of pygidium. PMO 205.205a. Top Engervik Member. Level 42.97 m in the beach profile at Djuptrekkodden, Oslo-Asker. Coll. T. Hansen, 2003. Scale bar 0.5 cm.

14, 18: Lectotype. Dorsal and oblique posterior view of articulated specimen. PMO 56407. Vollen Formation. Huk on Bygdøy, Oslo-Asker. Scale bars 0.5 cm.

15: Hypotypoid. Dorsal view of cranidium showing faint muscle scars. PMO 61749a. Vollen Formation. Bygdøy, Oslo-Asker. Scale bar 0.5 cm.

16: Small articulated specimen. PMO 82852. Middle Engervik Member. Level 36.86 m in the beach profile at Djuptrekkodden, Oslo-Asker. Coll. G. Henningsmoen, December 1958. Scale bar 0.2 cm.

17: Hypotypoid. Dorsal view of cranidium showing the distinctly tetragonal frontal spine. PMO 61750. Vollen Formation. Bygdøy, Oslo-Asker. Scale bar 0.5 cm.

19–20: *Lonchodomas cuspicaudus* **n. sp.**

19: Cast of large cranidium. Note the squat outline and tetragonal frontal spine. PMO 206.960. Heggen Member. Collected from talus below road exposure at Råen, Eiker-Sandsvær. Coll. J. H. Hurum, 21.04.2005. Scale bar 0.5 cm.

20: Cast of large cranidium. PMO 206.271/8. Uppermost Heggen Member. Level 8.40 in profile 1 at Råen, Eiker-Sandsvær. Coll. T. Hansen, 2004. Scale bar 0.5 cm.

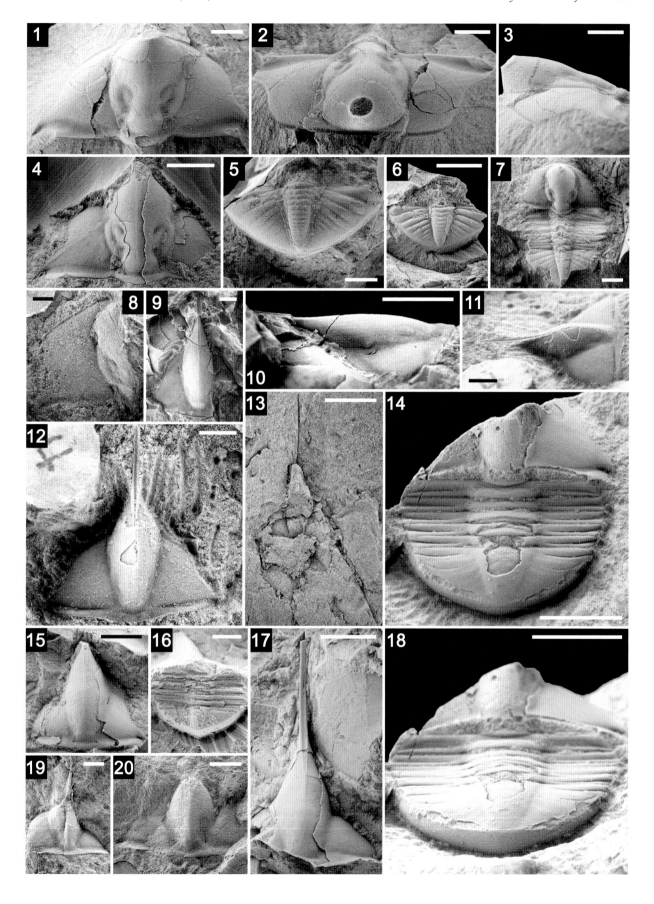

Plate 22

1–8: *Sculptella scripta* Nikolaisen, 1983

1, 3, 4: Dorsal, frontal and lateral view of holotype cranidium PMO 74666. Vollen Formation on the west side of Bygdøy, Oslo-Asker. Coll. F. Nikolaisen, 03.04.1960. Scale bars 0.1 cm.

2: Dorsal view of paratype cranidium PMO 59065. Vollen Formation. Exposure at Saltboden lighthouse at Frierfjorden, Skien-Langesund. Coll. Unknown, collected on excursion on 07.06.1936. Scale bar 0.1 cm.

5: Ventral view of paratype cranidium PMO 74685. Vollen Formation. Exposure on the eastern side of Semsvannet lake, Oslo-Asker. Coll. D. L. Bruton, 08.10.1967. Scale bar 0.1 cm.

6: Dorsal view of nearly complete paratype thoracic segment PMO 88125. Vollen Formation. Huk on Bygdøy, Oslo-Asker. Coll. W. C. Brøgger, 1882. Scale bar 0.1 cm.

7: Dorsal view of fragmentary paratype thorax segment PMO 88126. Vollen Formation. Huk on Bygdøy. Coll. W. C. Brøgger, 1882. Scale bar 0.1 cm.

8: Dorsal view of paratype pygidium PMO 3724. Vollen Formation. Huk on Bygdøy, Oslo-Asker. Coll. W. C. Brøgger, 1882. Scale bar 0.1 cm.

9–15: *Sculptella angustilingua* Nikolaisen, 1983

9–10: Dorsal and frontal view of fragmentary paratype cranidium PMO 74689. Vollen Formation. Gullerud, Ringerike. Coll. W. C. Brøgger, 1881. Scale bars 0.1 cm.

11: Dorsal view of paratype cranidium PMO 20314. Vollen Formation. Gullerud, Ringerike. Coll. W. C. Brøgger, 1881. Scale bar 0.1 cm.

12: Dorsal view of paratype cranidium PMO 74679. Vollen Formation. Gullerud in Ringerike. Coll. W. C. Brøgger, 1881. Scale bar 0.1 cm.

13: Paratype librigena PMO 74681. Vollen Formation. Beach exposure at Gomnes, Ringerike. Coll. J. Kiær, 1914. Scale bar 0.1 cm.

14: External mould of small hypostome. Paratype PMO 4319. Vollen Formation. Beach exposure at Gomnes, Ringerike. Coll. J. Kiær, 27.08.1913. Scale bar 0.1 cm.

15: Thoracic segment and fragmentary pygidium. Paratypes PMO H0551a and b. Vollen Fm. Gullerud in Ringerike. Coll. W. C. Brøgger, 1879–81. Scale bar 0.1 cm.

16: *Sculptella* sp. A

Dorsal view of cranidium PMO 74860. Elnes Formation. Railroad section at Flesberg Station, Eiker-Sandsvær. Coll. G. Henningsmoen, 10.07.1954. Scale bar 0.1 cm.

17–18: *Sculptaspis insculpta* Nikolaisen, 1983

17: Dorsal view of holotype cranidium PMO 3663. Engervik Member. Hukodden on Bygdøy, Oslo-Asker. Coll. Unknown. Scale bar 0.1 cm.

18: Dorsal view of juvenile paratype specimen PMO 74694. Elnes Formation. Quarry at Eternite factory just north of Slemmestad, Oslo-Asker. Coll. G. Henningsmoen, 30.06.1961. Scale bar 0.1 cm.

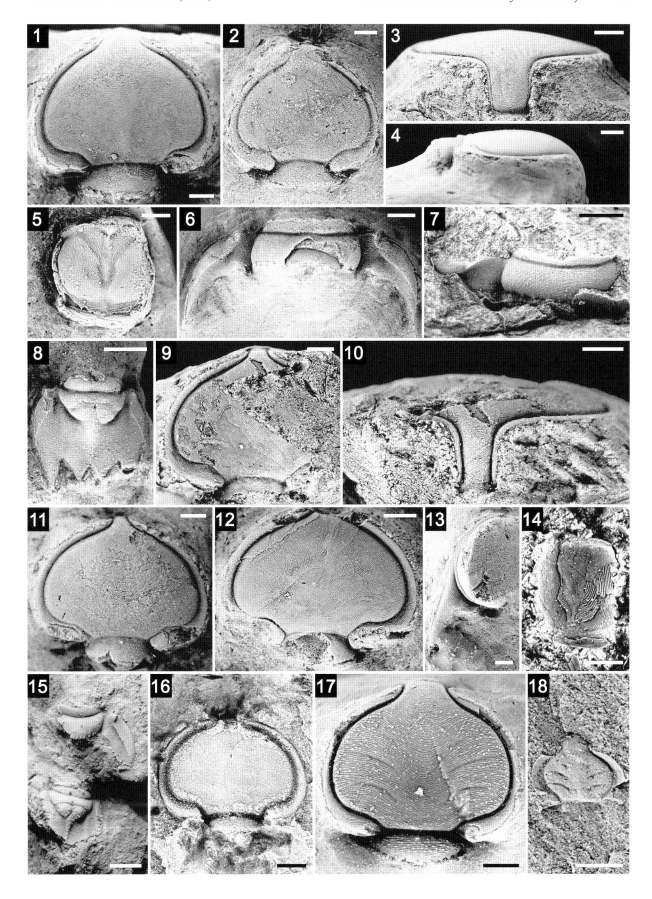

Plate 23

1:   ***Sculptaspis insculpta* Nikolaisen, 1983**
Frontal view of holotype cranidium PMO 3663. Engervik Member. Hukodden on Bygdøy, Oslo-Asker. Coll. Unknown. Scale bar 0.1 cm.

2–8:   ***Sculptaspis erratica* Nikolaisen, 1983**
2–3:   Dorsal and frontal view of cast of paratype cephalon PMO 60393. Heggen Member. Muggerudkleiva at Hedenstad, Eiker-Sandsvær. Coll. L. Størmer, 1925–27. Scale bars 0.1 cm.
4, 6:   Dorsal view and detail of frontal glabellar tongue. Paratype cranidium PMO 67194. Elnes Formation. Sterudstranda, Mjøsa. Coll. S. Skjeseth, 1950. Scale bars 0.1 cm.
5:   Dorsal view of holotype cranidium PMO 74703. Elnes Formation. Road-cut at Furnes Church, Mjøsa. Coll. F. Nikolaisen, 10.05.1969. Scale bar 0.1 cm.
7:   Dorsal view of right librigena PMO 207.448/4. Upper Heggen Member, 6.42 m above datum in sub-profile 3 at northern road-cut where Fv 84 crosses E6 southwest of Nydal, Mjøsa. Coll. T. Hansen, 19.08.2005. Scale bar 0.1 cm.
8:   Dorsal view of poorly preserved articulated specimen PMO 203.341/1. Uppermost Sjøstrand Member, 20.56 m above datum in beach-profile north of Djuptrekkodden, Oslo-Asker. Coll. T. Hansen, 2003. Scale bar 0.1 cm.

9:   ***Sculptaspis* sp. A**
Dorsal view of fragmentary cranidium PMO 74859. Middle Engervik Member, around 37.55 m above datum in the beach profile at Djuptrekkodden, Oslo-Asker. Coll. G. Henningsmoen, 1958. Scale bar 0.1 cm.

10–11:   ***Sculptaspis* sp. B**
10:   Dorsal view of cranidium PMO 74697. Sjøstrand Member. The Old Eternite Quarry at Slemmestad, Oslo-Asker. Coll. G. Henningsmoen, 30.06.1961. Scale bar 0.1 cm.
11:   Dorsal view of librigena PMO 74695. Sjøstrand Member. The Old Eternite Quarry at Slemmestad, Oslo-Asker. Coll. G. Henningsmoen, 30.06.1961.

12–13:   ***Icelorobergia*? sp. A**
Dorsal and oblique frontal view of cranidium PMO 74686. Helskjer Member. Road-cut at Furnes Church, Mjøsa. Coll. G. Henningsmoen and F. Nikolaisen, 27.09.1961. Scale bar 0.1 cm.

14–20:   ***Robergia sparsa* Nikolaisen, 1983**
14:   Dorsal view of paratype cranidium PMO S.1817. Elnes Formation. Slemmestad, Oslo-Asker. Coll. L. Størmer, 1937. Scale bar 0.1 cm.
15–16:   Frontal and dorsal view of holotype cranidium PMO 75055. Lower Elnes Formation. Road-cut at Furnes Church, Mjøsa. Coll. F. Nikolaisen, 06.05.1967. Scale bars 0.1 cm.
17:   Tiny librigena PMO 202.950/2. Sjøstrand Member, 2.19 m above datum in beach-profile at Djuptrekkodden, Oslo-Asker. Coll. T. Hansen, 2003. Scale bar 0.05 cm.
18:   Ventral view of librigena PMO 202.949/5. Sjøstrand Member, 2.17 m above datum in the beach-profile at Djuptrekkodden, Oslo-Asker. Coll. T. Hansen, 2003. Scale bar 0.1 cm.
19:   Dorsal view of paratype pygidium PMO 75056. Lower Elnes Formation. Road-cut at Furnes Church, Mjøsa. Coll. F. Nikolaisen, 1969. Scale bar 0.1 cm.
20:   Dorsal view of paratype pygidium PMO 75058. Lower Elnes Formation. Road-cut at Furnes Church, Mjøsa. Coll. S. Skjeseth, 1950. Scale bar 0.1 cm.

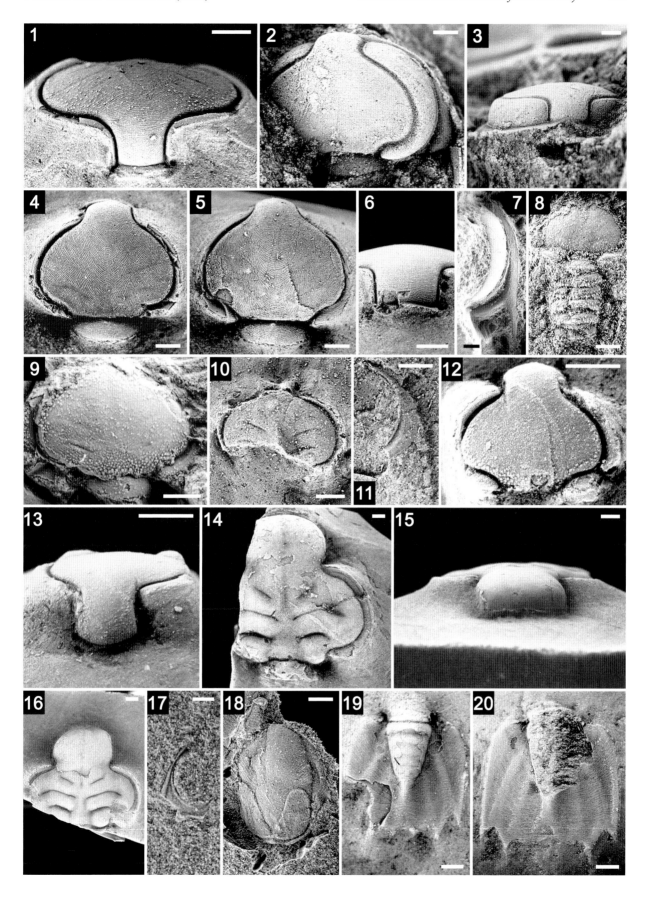

Plate 24

1–2, 15–16: *Telephina bicuspis* (Angelin, 1854)
1:   Dorsal view of cranidium PMO 21920. Heggen Member. Helgøya, Mjøsa. Coll. O. Holtedahl, 1906. Scale bar 0.1 cm.
2:   Dorsal view of cranidium PMO 72696. Heggen Member. Håve near Ringsaker, Mjøsa. Coll. S. Skjeseth, 1948. Scale bar 0.1 cm.
15:   Dorsal view of holotype cranidium of *Telephina vesca* Nikolaisen, 1963. PMO 72693a. Heggen Member. Road-cut at Furnes Church, Mjøsa. Coll. S. Skjeseth, 1950. Scale bar 0.1 cm.
16:   Dorsal view of exfoliated cranidium figured as *T. vesca* by Nikolaisen (1963). PMO 72728a. Håve at Håvesveen in Ringsaker, Mjøsa. Coll. S. Skjeseth, 1948. Scale bar 0.1 cm.

3, 5–7: *Telephina intermedia* (Thorslund, 1935)
3:   Dorsal view of holotype cranidium for *Telephina furnensis* Nikolaisen, 1963. PMO 73013a. Heggen Member. Road-cut at Furnes Church, Mjøsa. Coll. P. Jørgensen, 1962. Scale bar 0.1 cm.
5:   Dorsal view of small cranidium PMO 4247. Engervik Member. Kullerud at Norderhov, Ringerike. Coll. O. Holtedahl, 1915. Scale bar 0.1 cm.
6:   Dorsal view of cranidium PMO 72727a. Heggen Member. Road-cut at Furnes Church, Mjøsa. Coll. S. Skjeseth, 1950. Scale bar 0.1 cm.
7:   Dorsal view of large cranidium. PMO 72727b. Same locality and horizon as 6. Scale bar 0.1 cm.

4, 10: *Telephina norvegica* Nikolaisen, 1963
4:   Dorsal view of cranidium figured by Nikolaisen (1963) as *T. (T.)* aff. *furnensis*. PMO 33550. Heggen Member. South of Skiaker at Gran, Hadeland. Coll. T. Münster, 1893. Scale bar 0.1 cm.
10:   Dorsal view of holotype cranidium PMO 72730a. Heggen Member. Road-cut at Furnes Church, Mjøsa. Coll. S. Skjeseth, 1950. Scale bar 0.1 cm.

8–9: *Telephina mobergi* (Hadding, 1913)
8:   Dorsal view of cranidium PMO 72685. Heggen Member. Road-cut at Furnes Church, Mjøsa. Coll. S. Skjeseth, 1950. Scale bar 0.1 cm.
9:   Dorsal view of exfoliated cranidium. PMO 72686. Heggen Member. Road-cut at Furnes Church, Mjøsa. Coll. S. Skjeseth, 1950. Scale bar 0.1 cm.

11–12: *Telephina skjesethi* Nikolaisen, 1963
11:   Dorsal view of holotype cranidium PMO 72722. Heggen Member. Dyste in Toten, Mjøsa. Coll. S. Skjeseth, 1950. Scale bar 0.1 cm.
12:   Fragment of two thoracic segments. Same sample as 11. Scale bar 0.1 cm.

13–14: *Telephina sulcata* Nikolaisen, 1963
13:   Dorsal view of holotype cranidium PMO 72758a. Helskjer Member. Erratic boulder from road section at Furnes Church, Mjøsa. Coll. G. Henningsmoen, 16.08.1961. Scale bar 0.1 cm.
14:   Dorsal view of exfoliated cranidium PMO 72756. Helskjer Member. Erratic boulder from road section at Furnes Church, Mjøsa. Coll. G. Henningsmoen, 16.08.1961. Scale bar 0.1 cm.

17:   *Telephina viriosa* Nikolaisen, 1963
        Dorsal view of holotype cranidium PMO 72754. Helskjer Member. Road-cut at Furnes Church, Mjøsa. Coll. G. Henningsmoen, 1961. Scale bar 0.1 cm.

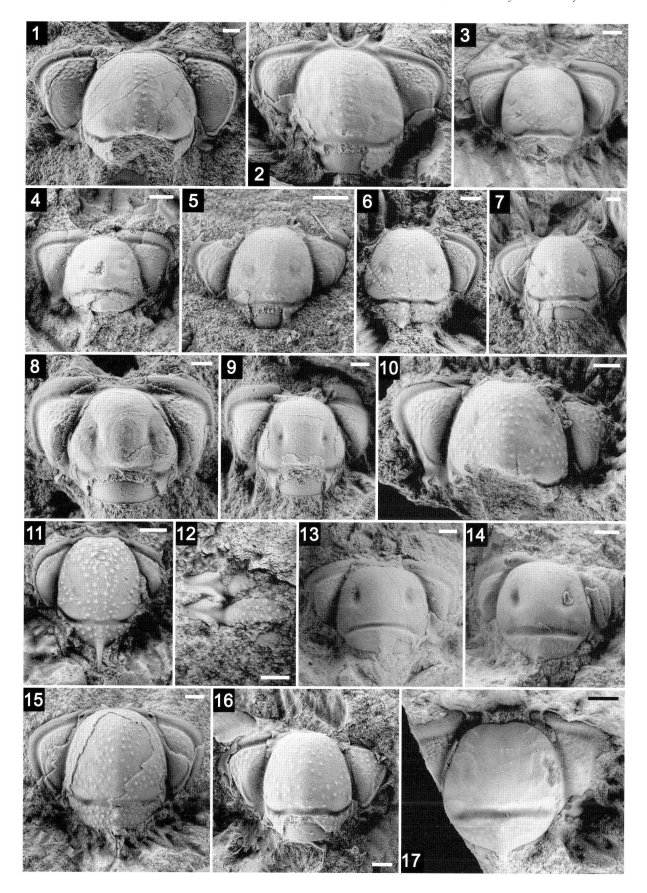

Plate 25

1:   *Telephina bicuspis* **(Angelin, 1854)**
Dorsal view of cranidium figured as *T. (T.)* sp. no. 1 by Nikolaisen (1963). PMO 72687. Heggen Member. Road-cut at Furnes Church, Mjøsa. Coll. S. Skjeseth, 1950. Scale bar 0.1 cm.

2:   *Telephina norvegica* **Nikolaisen, 1963**
Dorsal view of large cranidium figured as *T. (T.)* sp. no. 2 by Nikolaisen (1963). PMO 72721. Engervik Member. Huk on Bygdøy, Oslo-Asker. Coll. G. Henningsmoen & F. Nikolaisen, 1953. Scale bar 0.2 cm.

3–4:   *Telephina* **aff.** *granulata* **(Angelin, 1854)**
3:   Dorsal view of small cranidium PMO 72726b. Heggen Member. Sterudstranda, Mjøsa. Coll. S. Skjeseth, 1950. Scale bar 0.1 cm.
4:   Dorsal view of compressed specimen PMO 33331. Heggen Member. Collected south of Skiaker near Gran, Hadeland. Coll. T. Münster, 1893. Scale bar 0.2 cm.

5–6:   *Telephina invisitata* **Nikolaisen, 1963**
5:   Dorsal view of small specimen PMO 72729. Heggen (?) Member, Mjøsa. Håve at Håvesveen, Mjøsa. Coll. S. Skjeseth, 1948. Scale bar 0.1 cm.
6:   Dorsal view of large cranidium PMO 72723. Heggen Member. Dyste in Toten, Mjøsa. Coll. S. Skjeseth, 1950. Scale bar 0.5 cm.

7–9:   **Elviniid sp.**
Dorsal, lateral and frontal view of cranidium PMO 36358. Helskjer Member, 2 m above the Huk Formation at Hovindsholm on Helgøya, Mjøsa. Coll. Unknown. Scale bars 0.2 cm.

10:   *Porterfieldia humilis* **(Hadding, 1913)**
Dorsal view of cranidium PMO 67208. Heggen Member. Road-cut at Furnes Church, Mjøsa. Coll. S. Skjeseth, 1950. Scale bar 0.1 cm.

11–14: *Primaspis multispinosa* **Bruton, 1965**
11:   Dorsal view of holotype cranidium PMO 72776a. Heggen Member, road-cut at Furnes Church, Mjøsa. Coll. S. Skjeseth, 1950. Scale bar 0.1 cm.
12:   Left librigena. Paratype PMO 72776c. Heggen Member, road-cut at Furnes Church, Mjøsa. Coll. S. Skjeseth, 1950. Scale bar 0.1 cm.
13:   Paratype hypostome PMO 72776c. Same sample as 12. Scale bar 0.1 cm.
14:   Paratype pygidium PMO 72776. Same sample as 12. Scale bar 0.1 cm.

15–17: *Metopolichas celorrhin* **(Angelin, 1854)**
15:   Ventral view of fragmentary hypostome. PMO 106.421. Helskjer Member, 1.75 m above the Huk Formation in the road-cut below Vikersund skijump, Modum. Coll. B. Wandås, 17.07.1980. Scale bar 0.5 cm.
16–17: Lateral and dorsal view of large cranidium PMO 212.657. Helskjer Member, Elnes Formation. Rognstranda, Skien-Langesund. Coll. Bjarne Lie, 03.05.1964. Scale bars 0.5 cm.

18:   **Trochurinae genus et sp. indet.**
External mould of dorsal surface of small cranidium. PMO 211.734. Base of silt bed from Håkavik Member on south-eastern side of Killingen island at Bygdøy, Oslo-Asker. Coll. L. Størmer, 1929. Scale bar 0.1 cm.

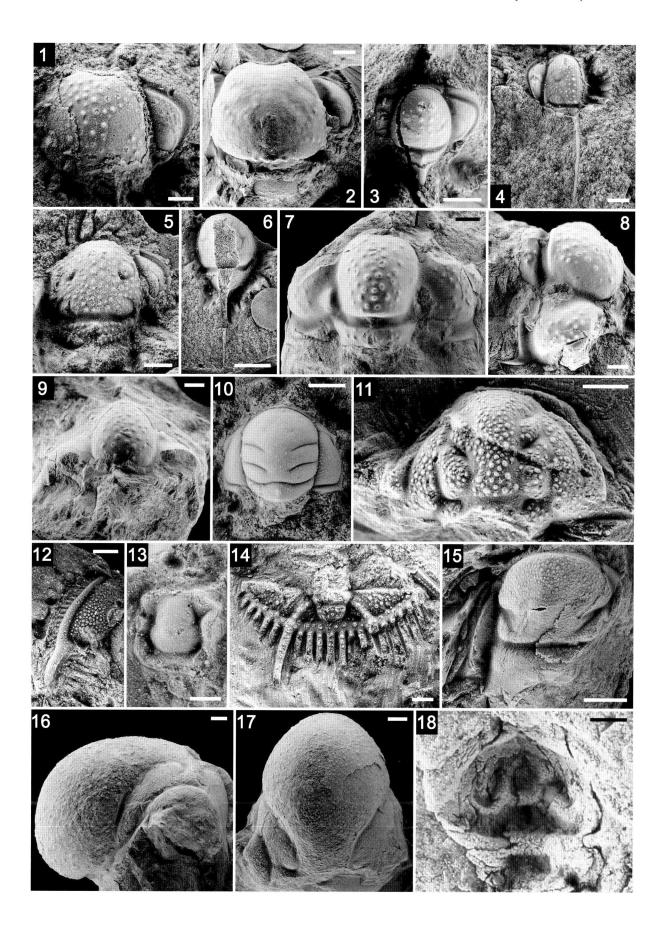

Plate 26

1–2:   Trochurinae genus et sp. indet.
Latex cast and external mould of cranidium PMO 211.733. Base of silt bed from Håkavik Member on south-eastern side of Killingen Island at Bygdøy, Oslo-Asker. Coll. L. Størmer, 1929. Scale bars 0.1 cm.

3:   *Pterygometopus* sp.
Cast of cranidium PMO 207.522. Helskjer Member. Erratic boulder from road-cut at Furnes Church, Mjøsa. Coll. A. Ernst, 30.05.2003. Scale bar 0.1 cm.

4:   *Gravicalymene* sp.
Oblique dorsal view of pygidium PMO 83016. Helskjer Member. Erratic boulder from road-cut at Furnes Church, Mjøsa. Coll. G. Henningsmoen, 1961. Scale bar 0.1 cm.

5–8:   *Gravicalymene capitovata* Siveter, 1977
5:   Dorsal view of internal mould of holotype cranidium PMO 91062. Heggen Member. Erratic boulder from Muggerudkleiva, Eiker-Sandsvær. Coll. D. L. Bruton, 1971. Scale bar 0.5 cm.
6:   Lateral view of right librigena PMO 91036. Heggen Member. Muggerudkleiva, Eiker-Sandsvær. Coll. D. L. Bruton & D. J. Siveter, 1971. Scale bar 0.5 cm.
7:   External mould of hypostome PMO 91051. Heggen Member. Muggerudkleiva, Eiker-Sandsvær. Coll. D. J. Siveter & D. L. Bruton, 1971. Scale bar 0.5 cm.
8:   Oblique dorsal view of pygidium PMO 91061. Heggen Member. Erratic boulder from Muggerudkleiva, Eiker-Sandsvær. Coll. D. L. Bruton, 1971. Scale bar 0.5 cm.

9–12:   *Sthenarocalymene lirella* Siveter, 1977
9, 12: Dorsal view of articulated specimen and close up of pygidium. PMO 91065. Heggen Member. Muggerudkleiva, Eiker-Sandsvær. Coll. N. Spjeldnæs, 1961. Scale bars 0.5 and 0.1 cm respectively.
10:   Dorsal view of internal mould of cranidium PMO 82466. Heggen Member. Muggerudkleiva, Eiker-Sandsvær. Coll. N. Spjeldnæs, 1961. Scale bar 0.1 cm.
11:   Dorsal view of holotype cranidium PMO 81919. Vollen Formation. Western side of Bygdøy, Oslo-Asker. Coll. F. Nikolaisen, 1961. Scale bar 0.1 cm.

13–14: *Geragnostus hadros* Wandås, 1984
13:   Dorsal view of exfoliated cranidium PMO 90496. Helskjer Member, 0.00–0.05 m below the Heggen Member. Road-cut at Furnes Church, Mjøsa. Coll. G. Henningsmoen, 1961. Scale bar 0.1 cm.
14:   Dorsal view of holotype pygidium PMO 102.666. Helskjer Member, 0.2 m above the Huk Formation in the road-cut below Vikersund skijump, Modum. Coll. G. Henningsmoen, 1970. Scale bar 0.1 cm.

15–16: *Geragnostus* sp.
15:   External mould of strongly compressed cranidium shown in inverted colours (Adobe Photoshop). PMO 203.287. Engervik Member, 36.86 m above datum in beach-profile at Djuptrekkodden, Oslo-Asker. Coll. T. Hansen, 2003. Scale bar 0.1 cm.
16:   External mould of pygidium shown in inverted colours (Adobe Photoshop). PMO 206.060/6. Håkavik Member, 51.78 m above datum in the beach-profile at Djuptrekkodden, Oslo-Asker. Coll. T. Hansen, 2003. Scale bar 0.1 cm.

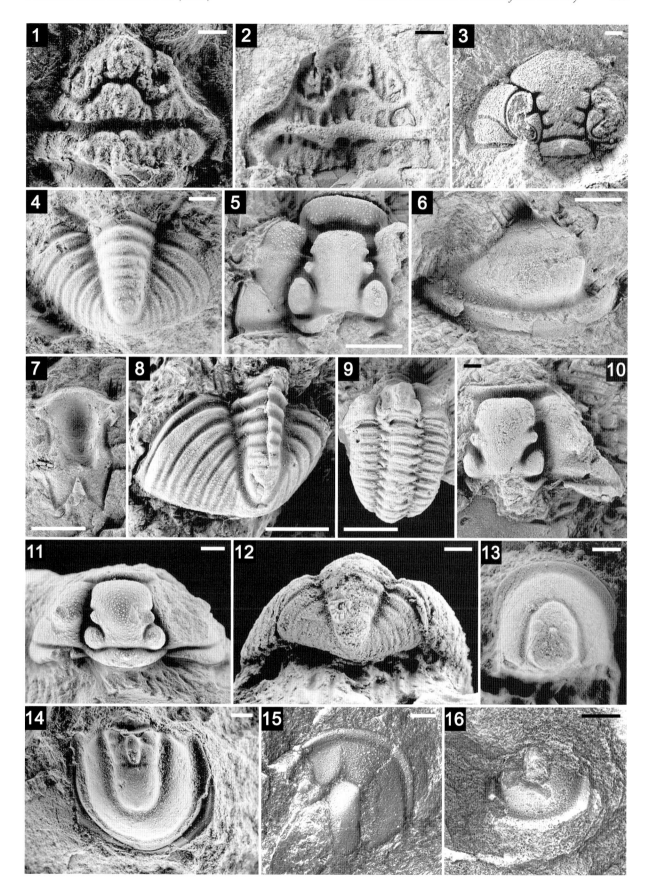

Plate 27

1–4: *Sphaerocoryphe* sp.

1–3: Dorsal, lateral and frontal view of exfoliated cephalon PMO 143.562. Upper Helskjer Member, 7.75 m above the Huk Formation in the road-cut at Furnes Church, Mjøsa. Coll. F. Nikolaisen, 27.09.1961. Scale bars 0.1 cm.

4: Dorsal view of frontal glabellar lobe PMO 143.563. Upper Helskjer Member, 7.75 m above the Huk Formation in the road-cut at Furnes Church, Mjøsa. Coll. F. Nikolaisen, 27.09.1961. Scale bar 0.1 cm.

5–9: *Cyrtometopus clavifrons* (Dalman, 1827)

5: Dorsal view of exfoliated and frontally strongly flattened cephalon PMO 102.648. Basal Heggen Member, 5.5 m above the Huk Formation in road-cut at Vikersund skijump, Modum. Coll. G. Henningsmoen, 20.09.1970. Scale bar 0.5 cm.

6–7: Frontal and lateral view of exfoliated cranidium PMO 1994. Huk Formation at Huk on Bygdøy, Oslo-Asker. Coll. Unknown, collected on excursion on 16.09.1891. Scale bars 0.5 cm.

8–9: Frontal dorsal and posterior dorsal view of enrolled specimen. PMO 1664. Lysaker Member, Huk Formation. Tøyen in Oslo, Oslo-Asker. Coll. T. Münster. Scale bars 0.5 cm.

10–13: *Pliomera fischeri* (Eichwald, 1825)

10: Frontal view of enrolled specimen PMO H2627. Top of Lysaker Member, Huk Formation. Tøyen in Oslo, Oslo-Asker. Coll. Unknown. Scale bar 0.5 cm.

11: Ventral view of exfoliated hypostome PMO 102.814. Heggen Member, 13.90 m above the Huk Formation in road-cut at Vikersund skijump, Modum. Coll. B. Wandås, 23.04.1979. Scale bar 0.1 cm.

12: Dorsal view of exfoliated and somewhat forelengthened specimen PMO 2280. Lysaker Member, Huk Formation. Slemmestad, Oslo-Asker. Coll. H. Fischbein, 1908. Scale bar 0.5 cm.

13: Dorsal view of foreshortened and exfoliated specimen PMO 102.416. Lower Heggen Member, 12.1 m above the Huk Formation in the road-cut at Vikersund skijump, Modum. Coll. B. Wandås, 30.08.1978. Scale bar 0.5 cm.

14–17: *Cybelurus* cf. *mirus* (Billings, 1865)

14–16: Dorsal, oblique lateral and frontal view of cranidium PMO 82991. Upper Helskjer Member, 7.75 m above the Huk Formation in the road-cut at Furnes Church, Mjøsa. Coll. U. Borgen, 01.05.1965. Scale bars 0.5 cm.

17: Latex cast of somewhat compressed cranidium. PMO 36599–36600. Elnes Formation (Heggen Member or 'Cephalopod Shale'). Collected in 1926 by T. Strand just north of Redalen near Biri, Mjøsa. Scale bar 0.5 cm.

18, 21–24: *Atractopyge dentata* (Esmark, 1833)

18, 24: Oblique lateral and dorsal view of exfoliated cranidium PMO 211.428. Heggen Member at Muggerudkleiva in Eiker-Sandsvær. Coll. D. J. Siveter, 1971. Scale bars 0.2 cm.

21: Wax cast of lectotype pygidium PMO 56161a. Elnes Formation or Vollen Formation at the Royal Palace in Oslo. Coll. Unknown. Scale bar 0.1 cm.

22: Wax cast of pygidium PMO 60607. Heggen Member, Muggerudkleiva at Hedenstad, Eiker-Sandsvær. Coll. L. Størmer, 1925–27. Scale bar 0.1 cm.

23: Wax cast of cranidium PMO 211.429, which is the counterpart to PMO 211.428 figured in Plate 27, figs 18, 24. Coll. D. J. Siveter, 1971. Scale bar 0.2 cm.

19–20: *Cybellela* sp.

19: Oblique lateral view of exfoliated librigena. PMO 83178. Upper Helskjer Member, 7.75 m in the road-cut at Furnes Church, Mjøsa. Coll. F. Nikolaisen, 27.09.1961. Scale bar 0.1 cm.

20: External mould of small pygidium. PMO 207.395/3. Helskjer Member. Level 7.32 m in sub-profile 1 along E6, where road Fv 84 crosses between Nydal and Furnes, Mjøsa. Coll. T. Hansen, 17.08.2005. Scale bar 0.05 cm. Photo taken by Y. Candela, National Museums of Scotland.

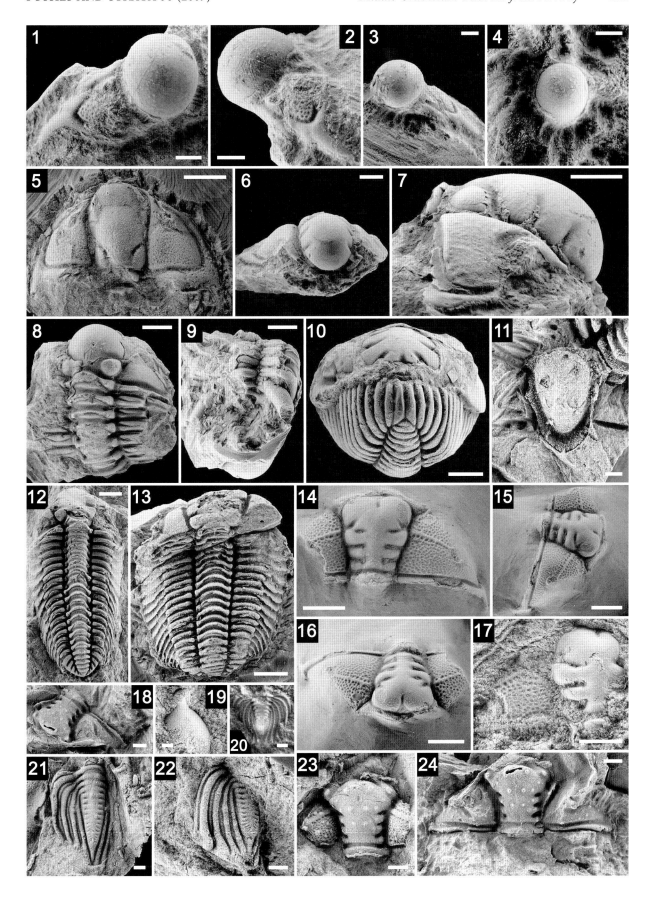

Plate 28
1–7: *Lonchodomas cuspicaudus* **n. sp.**
1–2: Oblique lateral and dorsal view of holotype specimen PMO 4044. Uppermost part of the Elnes Formation (or lower Vollen Formation) on Kojatangen between Vollen and Blakstad, Oslo-Asker. Coll. J. Kiær, 1919. Scale bars 0.5 cm.
3: Lateral view of paratype cranidium PMO 81836a. Vollen Formation. Road-cut about 100 m west of Gyssestad Gård near Sandvika, Oslo-Asker. Coll. J. F. Bockelie, 21.03.1965. Scale bar 0.1 cm.
4–5: Dorsal and anterior view of paratype cranidium PMO 40367a. Vollen Formation. Road-cut along the Drammensveien (Drammen road) at Gyssestad near Sandvika, Oslo-Asker. Coll. A. Heintz, 1933. Scale bars 0.5 cm.
6: Dorsal view of paratype pygidium PMO 40395. The specimen is tectonically deformed. Same horizon, locality and collector as 4–5. Scale bar 0.5 cm.
7: Dorsal view of paratype pygidium PMO 97176. Middle Vollen Formation at Paradisbukta on Bygdøy, Oslo-Asker. Coll. T. Bockelie, 30.03.1972. Scale bar 0.5 cm.

8: *Cybellela* **sp.**
Dorsal view of cast of small cranidium. PMO 83173. Top Helskjer Member, 7.75 m above the Huk Formation. Road-cut at Furnes Church, Mjøsa. Coll. F. Nikolaisen, 26.09.1961. Scale bar 0.1 cm.

9: *Lonchodomas rostratus* **(Sars, 1835)**
Dorsal view of complete specimen PMO 59057 showing the long frontal and genal spines. Fossum Formation. Saltboden lighthouse at Frierfjorden, Skien-Langesund. Collected on excursion 7 June 1936. Scale bar 0.5 cm.

10–11: *Proetus* (*Proetus*) **sp.**
10: Dorsal view of small librigena PMO 36617. Elnes Formation. Tømte Land. Collector unknown. Scale bar 0.1 cm.
11: Dorsal view of small librigena PMO 67604. Uppermost Helskjer Member, probably 7.75 m above the Huk Formation. Helgøya, Mjøsa. Coll. S. Skjeseth, 10.05.1951. Scale bar 0.1 cm.

12: *Atractopyge dentata* **(Esmark, 1833)**
Cast showing dorsal view of cranidium PMO 121.644. Heggen Member, Elnes Formation. Muggerudkleiva near Hedenstad, Eiker-Sandsvær. Coll. D. L. Bruton. Scale bar 0.1 cm.

13: *Metopolichas celorrhin* **(Angelin, 1854)**
Dorsal view of cranidium figured by Brøgger (1882, pl. 5, fig. 12). Uppermost Svartodden Member, Huk Fm., or basal Helskjer Member, Elnes Formation. Eiker, Eiker-Sandsvær. Collector unknown. Scale bar 0.5 cm.

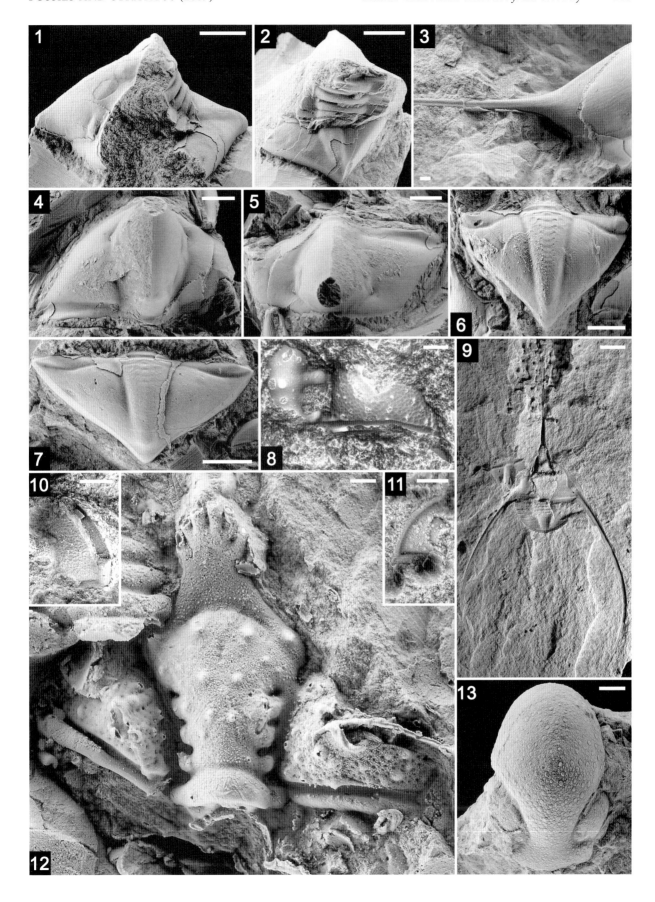

Plate 29

1:   *Pseudasaphus* sp. B
Dorsal view of pygidium PMO 61144. The old airport at Fornebu, Oslo-Asker. Coll. L. Størmer (?), September 1938. Scale bar 0.5 cm.

2:   *Pseudomegalaspis* cf. *formosa* (Törnquist, 1884)
Dorsal view of pygidium attached to posterior four thoracic segments. PMO 112.103. Base of Fossum Formation. Skinnvika East, Skien-Langesund. Coll. Inge Ribland Nilssen, 14.10.1984. Scale bar 0.5 cm.

3–4: *Pseudomegalaspis patagiata* (Törnquist, 1884)
3:   Detail of panderian openings and doublural terrace-lines on right posterior thoracic segments and anterior part of pygidium. PMO S.1734. Middle part of Engervik Member, Elnes Formation. Spit at Bekkebukta on Bygdøy, Oslo-Asker. Coll. L. Størmer. Scale bar 0.5 cm.
4:   Dorsal view of nearly complete specimen. PMO 3695. Upper Elnes Formation. Huk on Bygdøy, Oslo-Asker. Coll. V. M. Goldschmidt, 1906. Scale bar 0.5 cm.

5–7, 9: *Ogmasaphus furnensis* n. sp.
5–7: Dorsal view of complete holotype specimen and detail of cephalon and thoracic pleurae. PMO 67181. Helskjer Member. Road-cut at Furnes Church, Mjøsa. Coll. S. Skjeseth, 1950.
9:   Dorsal view of paratype pygidium PMO 67589a. Helskjer Member. Road-cut at Furnes Church, Mjøsa. Coll. S. Skjeseth, 1950. Scale bars 0.5 cm.

8, 10: *Metopolichas celorrhin* (Angelin, 1854)
8:   Dorsal view of fragmentary pygidium PMO 106.052. Helskjer Member or basal Heggen Member, Elnes Formation. Road-cut at Vikersund skijump, Modum. Collected on an excursion in 1974. Scale bar 0.5 cm.
10:   Lateral view of cranidium figured by Brøgger (1882, pl. 5, fig. 12). Uppermost Svartodden Member, Huk Fm., or basal Helskjer Member, Elnes Formation. Eiker, Eiker-Sandsvær. Collector unknown. Scale bar 0.5 cm.

11:   *Pterygometopus* sp.
Dorsal view of small pygidium PMO 83192. Helskjer Member, 3.70 m above the Huk Formation. Helskjer on Helgøya, Mjøsa (Map position: PN 088 329). Coll. F. Nikolaisen, 27.09.1961. Scale bar 0.1 cm.

12:   Remopleuridid gen. et sp. indet.
Dorsal view of small cranidium PMO 67585. Helskjer Member. Road-cut at Furnes Church, Mjøsa. Coll. S. Skjeseth, 1950. Scale bar 0.1 cm.

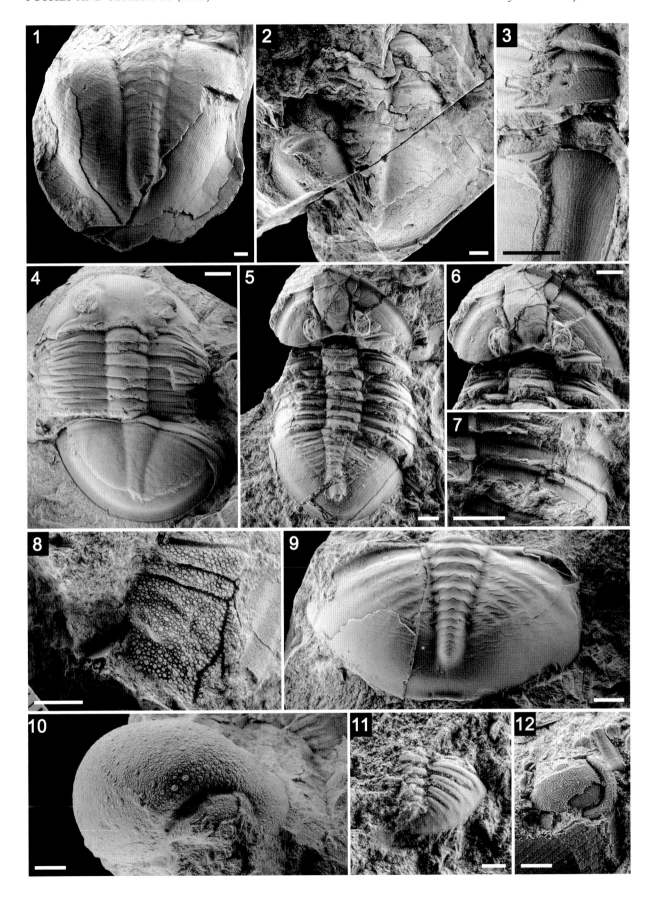